Radioactive Contamination of the Tokyo Metropolitan Area

Hideo Yamazaki

Radioactive Contamination of the Tokyo Metropolitan Area

Five Years after the Fukushima Nuclear Accident

Springer

Hideo Yamazaki
Retired
Kindai University
Naganohara, Japan

ISBN 978-981-15-7370-5 ISBN 978-981-15-7368-2 (eBook)
https://doi.org/10.1007/978-981-15-7368-2

This Springer imprint is published by the registered company Springer Nature Singapore Pte Ltd.
The registered company address is: 152 Beach Road, #21-01/04 Gateway East, Singapore 189721, Singapore

Preface

When the Chernobyl nuclear disaster struck, a famous Japanese professor specialized in nuclear engineering, who also served as a member of a committee in the Japanese government, asserted that such a stupid accident would not occur in Japanese nuclear power plants. Twenty-five years later, the Fukushima nuclear disaster happened, and ANZENSHINWA in Japan collapsed, which is a Japanese word meaning safety myth. This accident was more significant than the Chernobyl accident in terms of emissions of radioactive material. Did Japanese nuclear engineering experts seriously study the experience of the Chernobyl accident? Many nuclear experts in Japan may have assumed that a severe accident leading to nuclear fuel meltdown would not occur. However, in the Fukushima Daiichi Nuclear Power Plant (FDNPP), where six boiling water reactors were operating, three reactors and four reactor buildings collapsed. Moreover, the Tokyo metropolitan area, which is more than 200 km from the FDNPP site, was exposed to a significant radioactive contamination. This event was a severe incident.

This book aims to help readers understand the actual situation of radioactive contamination in the Tokyo metropolitan area due to the Fukushima nuclear disaster. The accident, which occurred between March 11 and 16, 2011, caused the three reactors in operation to meltdown, and the nuclear fuel storage pool of one reactor under periodic inspection nearly collapsed. The FDNPP released more radionuclides to the atmosphere and the ocean than Chernobyl, and the leaking of radionuclides continued for more than 9 years since the crash. The radioactive plume arrived not only in central Tokyo, 220 km from the FDNPP, but also throughout the entire Tokyo metropolitan area and all across Japan. Many published books have analyzed the causes and mechanisms of the FDNPP accident. However, a book summarizing the environmental radioactive contamination in the Tokyo metropolitan area has not yet been published. A slight fluke in the weather protected the Tokyo metropolitan area, where approximately 44 million people live, from devastating radioactive pollution, but it is also a fact that such a situation cannot guarantee security. This book summarizes the results of monitoring the actual state of environmental radioactive contamination, mainly in the Tokyo metropolitan area, during the 5 years following the accident. Unfortunately, some data on environmental radioactive contamination from the FDNPP accident are unreliable, but many fragments are available. Although this book uses reliable data published by public institutions and in academic papers, it is based on data measured by the author.

Fukushima Prefecture and the Tokyo metropolitan area have become experimental fields for investigating the environmental dynamics of radionuclides due to radioactive contamination caused by the FDNPP accident. These regions have an area of 50,680 km^2 and a population of 45.9 million, and approximately 35% of the Japanese people live in these regions. The urban area is approximately 4000 km^2. Fortunately, deadly radioactive contamination was avoided except in a part of Fukushima Prefecture, but the entire Tokyo metropolitan area, which is the center of Japan's politics, economy, and industry, has become a radioactively contaminated area. After the FDNPP accident on April 6, 2011, the Japanese government started an airborne radiation monitoring survey within approximately 80 km of the FDNPP through the MEXT (Ministry of Education, Culture, Sports, Science and Technology, Japan). The MEXT had conducted airborne monitoring in Tokyo from September 14, 2011, but the Japanese government at this time did not recognize the importance of radioactive contamination in the Tokyo metropolitan area due to the Fukushima disaster. Furthermore, this airborne monitoring was only performed once in 2011 and has not been done since then.

On the other hand, the US government issued an evacuation advisory to Americans in the area within 80 km of the FDNPP on the morning of March 17, 2011, in Japan, based on the scientific evidence by the US NRC (United States Nuclear Regulatory Commission). Although it was not based on the actual measurement data except for some data from the US DOE (Department of Energy) and the US DOD (Department of Defense), it seems to have been a symbolic event that shows the difference between the two governments in protecting the security of the people.

The points to be noted in environmental radioactive pollution during a nuclear disaster are as follows:

1. The migration of radioactive materials released from nuclear facilities immediately after the accident should be predicted. Such prediction is needed to clarify the spatiotemporal distribution of the radioactive plume and to elucidate the precipitation behavior of radionuclides to the ground. In Japan, at the time of the FDNPP accident, a System for the Prediction of Environmental Emergency Dose Information (SPEEDI) had been established, but it did not work because of systemic and operational problems. The Japanese government's response to the accident suggests that radiation surveys in the field would have been indispensable in order to promptly control the exposure of the population at the beginning of the accident. However, in reality, neither a survey of residents' radiation

exposure immediately after the accident nor a survey of environmental radioactive contamination was conducted. The author believes that the failure of the Japanese government to issue a child curfew immediately after the FDNPP accident has led to a high incidence of childhood thyroid cancer in Fukushima Prefecture.
2. Understanding the spatial distribution of the large amounts of radionuclides precipitated onto the ground surface from the atmosphere is necessary to evaluate the external exposure dose of the residents. However, in the Tokyo metropolitan area, there has been almost no public investigation by the government regarding environmental radiation contamination other than airborne monitoring. An airborne monitoring survey in Tokyo was conducted only once in September 2011. Additionally, even if some data were collected, they have not been disclosed so that residents can understand them.
3. It is crucial to analyze the dynamics of radionuclides released into the environment. In urban areas, except for gardens and parks, most radionuclides are precipitated on concrete and asphalt, so their behavior depends on rainfall and its wastewater treatment systems. On the other hand, radionuclides precipitated in rural areas and forest areas are selectively adsorbed and fixed to soil, litter, and trees. Their kinetic analysis is necessary to evaluate the cycle of radionuclides in ecosystems.
4. Furthermore, elucidating the radioactive contamination of fish and shellfish in lakes and rivers is also crucial for the recycling of radionuclides in aquatic ecosystems. Therefore, monitoring the spatiotemporal distribution of radioactive contamination due to the redistribution of radionuclides in the environment will ultimately be an essential factor in evaluating internal as well as external exposure of residents.
5. In the initial phase after the accident, radioactive materials released from the reactor were precipitated and distributed in the environment. However, radionuclides move and are redistributed in the environment in the next stage. In this second phase, fine particles contaminated with high concentrations of radionuclides are scattered through the air by wind, which may increase the risk of internal exposure for people who inhale the contaminated particles.

From these points of view, this book describes the results of the analysis and evaluation of the spatiotemporal variations in environmental radioactive pollution in the Tokyo metropolitan area for 5 years after the initial phase of the FDNPP accident. Environmental radioactive contamination in eastern Japan due to the Fukushima disaster has not yet ceased. In addition to the FDNPP, many nuclear power reactors will be decommissioned soon in Japan. The risk of secondary radioactive contamination of residents near the reactor associated with decommissioning operations is necessarily high. Therefore, it is essential to understand the mechanism of environmental radioactive contamination in the Tokyo metropolitan area, where many people live and work in areas with complicated environmental situations. Nine years have passed since the disaster in Fukushima, and various health effects have occurred in the people living in Japan. Multiple incidents of thyroid cancer in Fukushima Prefecture and an increase in heart disease have been confirmed by an epidemiological survey in Japan. The relationship between these health effects and the estimated external exposure dose received by the residents cannot prove a cause-and-effect relationship. A low-dose exposure problem that cannot be assessed with the current knowledge of health physics is occurring in Fukushima Prefecture and the Tokyo metropolitan area. Regarding human health effects caused by radiation exposure, it is necessary to actively recognize not only the external exposure by airborne doses but also the internal exposure. To do so, the kinetics of environmental radioactive contamination in the Tokyo metropolitan area, which is a densely populated area, and in particular the distribution, behavior, and chemical and physical species of radionuclides released from the FDNPP accident must be analyzed.

The Fukushima Daiichi Nuclear Power Plant was devastated by the shaking and tsunami caused by the Great East Japan Earthquake. The TEPCO (Tokyo Electric Power Company reorganized as Tokyo Electric Power Company Holdings in 2016), the Japanese Government, and public institutions analyzed the causes and processes in the early days of the accident, and the results were published in many reports [ ]. Many papers and books on the FDNPP accident were also published based on the data contained in these reports. Therefore, the process of reactor destruction and the release mechanism of radioactive materials in the very early stages of the Fukushima nuclear accident have been evaluated in many papers and books, and it will not be described here. In this book, the impact of environmental radiation contamination in the Tokyo metropolitan area during the 5 years from the Fukushima Daiichi Nuclear Power Plant (FDNPP) accident is discussed based on much monitoring data. More than 9 years had passed since the accident, and the environmental radioactive contamination is considered to have entered the next phase after the initial phase of the accident. In this second phase, radionuclides that were precipitated and held in the environment move and are redistributed. In particular, large amounts of radionuclides held in forest areas will move to urban areas. This redistribution of radionuclides suggests that Japanese residents may be subject to new radiation exposure. The author hopes that our next generation will be able to recover from the radioactive pollution in Fukushima Prefecture and the Tokyo metropolitan area, which will not continue to suffer radiation exposure from the Fukushima disaster.

Radioactivity monitoring related to the Fukushima nuclear disaster has been ongoing since the accident. The data are updated on the websites of the Japanese government and TEPCO. The URLs are listed below for the convenience of the reader [1, 2]. These websites are valid as of June 2020, but the author cannot guarantee the reliability of the data described.

The airborne radioactive pollution monitoring maps quoted in this book can now be viewed at the "Extension site of the distribution map of radiation dose, etc" available from: http://ramap.imc.or.jp/map/eng/. The maps of GSI (Geospatial Information Authority of Japan) are used for these pollution maps.

1. References 2–8 in Chap. 1.
   https://japan.kantei.go.jp/kan/topics/201106/iaea_houkokusho_e.html (English)
   https://radioactivity.nsr.go.jp/en/ (English)
2. https://www4.tepco.co.jp/decommission/data/ (Japanese)
   https://www4.tepco.co.jp/decommission/information (Japanese)
   https://www4.tepco.co.jp/decommission/data/analysis/index-j.html (Japanese)
   https://www7.tepco.co.jp/responsibility/decommissioning/1f_newsroom/data/index-e.html (English)

Naganohara, Japan

Hideo Yamazaki

Acknowledgments

The data described in this book were measured and analyzed by many students who worked in the Laboratory for the Environmental Analytical Sciences and Radioanalytical Chemistry, Kindai University. Dr. Masanobu Ishida cooperated extensively in the author's research on the Fukushima nuclear accident. The author sincerely appreciates their passion for research and had a serious discussion about this research with Dr. Kazuo Jin, former director of the Department of Environmental and Health Science, Hokkaido Institute of Public Health. The collection of fishes from the Kejonuma pond was carried out by Toshio Ikeuchi and the NPO Ecopal Kejonuma. The author thanks them for their kind cooperation. Additionally, many experts and public officials advised this study. An enormous amount of information about the Fukushima nuclear accident has been published by the Tokyo Electric Power Company (TEPCO), the Japanese government, universities, and research institutions. The author proceeded with discussions citing data in these reports and papers.

This study was supported by a Grant-in-Aid for Scientific Research from the Japan Society for the Promotion of Science (JSPA) and a grant from the Japan Association for Chemical Innovation (JACI). The author also thanks Kindai University for their enthusiastic support for this research.

Kita-Karuizawa, Gunma, Japan
June 2020

Hideo Yamazaki

Contents

About the Author

Hideo Yamazaki retired as a Professor at Kindai University and is now a director of an NPO corporation on environmental conservation. He is also a peer reviewer for several academic journals.

After receiving his bachelor's degree in Nuclear Reactor Engineering, he completed his PhD degree at Kindai University with research on the dynamic analysis of heavy metal pollution in the aquatic environment using radiochemical techniques. He was also involved in developing ultra-precision pH measurement methods of geothermal power-plant hot water in Professor Henry Freiser's laboratory at the University of Arizona.

He has received the Encouragement Award from the Japan Association for Chemical Innovation, the Academic Award and Best Paper Award from the Japan Society for Environmental Chemistry, and the Best Paper Award from the Society of Environmental Conservation Engineering.

Chapter 1
Outline of the Fukushima Daiichi Nuclear Power Plant (FDNPP) Accident

Abstract To accurately assess the current status of environmental radioactive contamination in FDNPP accident, it is necessary to understand the dynamics of radionuclides in the environment in the initial phase immediately after the Fukushima nuclear accident. Because the topography and weather in Japan are involved, it is difficult to accurately reproduce the behavior of radionuclides released from the FDNPP reactors. Furthermore, power lines were cut off in many areas of East Japan, including Tokyo, because of the effects of the Great East Japan Earthquake disaster. Therefore, many radiation monitoring systems failed immediately after the earthquake. As a result, available radiation monitoring data in East Japan are very limited. In this chapter, the situation in the early stage of the FDNPP accident is reviewed, and the behavior and amount of radioactive materials released from the broken reactor to the environment are considered. The author was unable to directly measure the status of radioactive contamination and changes in radiation dose on the FDNPP premises during this period. The data used for the discussions are based on those published by TEPCO.

Keywords Fukushima Daiichi Nuclear Power Plant (FDNPP) accident · Accident time series · Radioactive materials · Release amounts

1.1 Introduction

To accurately assess the current status of environmental radioactive contamination, it is necessary to understand the dynamics of radionuclides in the environment in the initial phase immediately after the Fukushima nuclear accident. This chapter discusses the processes of advection and precipitation of radionuclides in the land of Japan, especially in the Tokyo metropolitan area. Because the topography and weather in Japan are involved, it is difficult to accurately reproduce the behavior of radionuclides released from the FDNPP reactors. Furthermore, power lines were cut off in many areas of East Japan, including Tokyo, because of the effects of the Great East Japan Earthquake disaster. Therefore, many radiation monitoring systems

H. Yamazaki, *Radioactive Contamination of the Tokyo Metropolitan Area*,
https://doi.org/10.1007/978-981-15-7368-2_1

failed immediately after the earthquake. As a result, available radiation monitoring data in East Japan are very limited.

In the FDNPP accident, zircaloy on the fuel rods and the cooling water reacted at high temperatures, generating large amounts of hydrogen gas. Then the building of the Unit 1 reactor experienced hydrogen explosions first on the afternoon of March 12, and the buildings of Unit 3 and Unit 4 also underwent hydrogen explosions by the 13th and 14th. There is a suggestion that the Unit 3 reactor was shattered due to a nuclear explosion (prompt criticality) from the nature of the building destruction, but there is no clear, direct evidence. The Unit 3 reactor was loaded with plutonium MOX (mixed oxide) fuel, but the plutonium content was approximately 3–5%. Therefore, the enrichment of ^{235}U was set at approximately 1.2%, which is considerably lower than that of a uranium-only reactor. Therefore, even if such low-enrichment nuclear fuel meltdown, the occurrence of prompt criticality is considered quite a severe condition.

On the other hand, the Unit 2 reactor did not undergo a large-scale explosion, but much radioactive material was released by the dry well vent in the afternoon of the 15th. In the FDNPP accident, the most significant amount of radioactive material is believed to have been released from the Unit 2 reactor, which caused environmental radioactive contamination in the Tokyo metropolitan area and all across Japan. The release of radioactive materials from the FDNPP to the environment continues through the atmosphere and groundwater.

The maximum dose around the FDNPP site during this period was more than 12 mSv/h, according to TEPCO (Tokyo Electric Power Company) measurements [1]. The U.S. DOE (Department of Energy) also measured over 10 mSv/h near the main gate, but the data website is not currently available. Moreover, the dose measured on March 20 by the DOE at a location 13 miles south-southwest of the FDNPP was 3 mR/h (30 μSv/h), which was approximately one-tenth of that on the FDNPP premises on the same day. The beta-emitting nuclide precipitation at the same location was 45.05 μCi/m^2 (1670 kBq/m^2), indicating that the radionuclides released from the FDNPP steadily contaminated the surrounding environment. On the deck of the *USS Ronald Reagan* aircraft carrier operating in the Pacific Ocean 120 miles northeast of the FDNPP, 0.3 mR/h (3 μSv/h) of radiation was detected on March 12 at 14:00. At 16:45, it increased to 0.9 mR/h (9 μSv/h). The precipitation of beta-emitting nuclides at 22:20 on the same day was 0.225 μCi/m^2 (8.33 kBq/m^2) [Reference 12 in Chap. 2]. It is believed that most of the radioactive materials were released from the FDNPP through explosions and venting. These radioactive materials diffused to the Pacific Ocean through the atmosphere and drainage, but some were precipitated onto the land areas of Japan, causing environmental radioactive contamination.

In this chapter, the situation in the early stage of the FDNPP accident is reviewed, and the behavior and amount of radioactive materials released from the broken reactor to the environment are considered. The author was unable to directly measure the status of radioactive contamination and changes in radiation dose on the FDNPP premises during this period. The data used for the discussions are based on those published by TEPCO.

1.2 Time Series of Events Related to Radioactive Contamination Immediately After the FDNPP Accident

The progress of the situation with the FDNPP accident for 2 weeks immediately after the Great East Japan Earthquake is summarized in Table 1.1. The timeline of events in this table is based on a report submitted by the Government of Japan to the IAEA Ministerial Meeting [2].

The Great East Japan Earthquake was a magnitude 9.0 (Mw) earthquake that occurred on the plate boundary at approximately 24 km deep approximately 180 km northeast of the FDNPP at 14:46:18 JST on March 11, 2011. In the FDNPP, the preliminary tremor of this earthquake was detected from 14:46:48 to 14:46:52. The arrival of the first wave of the tsunami occurred at approximately 15:28, but the FDNPP reactors were not affected. The second wave struck at 15:36:10 and reached an elevation of 10 m from the FDNPP site base. The maximum height of this second wave reached 16 m, which submerged the turbine buildings and control buildings of the Unit 1, 2, 3, and 4 reactors. The TEPCO designed the maximum wave height of a tsunami reaching the FDNPP as 6.1 m, and the FDNPP was constructed after review and approval by the Japanese government's NSC (Nuclear Safety Commission). The NSC was reorganized as the NRA (Nuclear Regulation Authority) in 2012. The outline of the FDNPP accident has been published in many reports [3–8]. These reports conclude that the cause of the FDNPP accident was a tsunami caused by the earthquake. However, a report from the National Diet of Japan (NDJ) commission noted that the FDNPP accident was a human-made disaster caused by TEPCO and the Japanese government postponing the safety regulation of nuclear power plants [6]. Nevertheless, the cause of the accident has been discussed based on the data published by TEPCO. Since TEPCO is the party that caused the FDNPP accident, it has the most extensive data on the event. These data are necessary for a systematic analysis of the accident. However, not all data analyses have been completed yet. Some data may have been interpreted and published for their convenience. The TEPCO data quoted in this chapter are as current as possible.

The maximum acceleration at the FDNPP site during this earthquake was approximately 550 gal, 1.3 times the design maximum acceleration for the FDNPP reactors. Due to this earthquake, the external power line was cut off, and the emergency diesel generator and its fuel tank were submerged by the tsunami, so immediately after the earthquake, the Units 1, 3, and 4 reactors entered a complete station blackout (SBO). The Unit 2 reactor was able to use the emergency power supply so that the reactor cooling system could operate for 3 days until March 14. Under these circumstances, the Unit 1, 2, and 3 reactors, which were in steady operation, melted down or melted through immediately after the earthquake and released large amounts of radioactive material into the environment. On the other hand, the Unit 4 reactor building, which was under periodic inspection, was destroyed by the hydrogen gas that flowed from the Unit 3 building into the Unit 4 building. The Unit 2 reactor did not produce a hydrogen explosion, but a large amount of radioactive material was released to the environment by a dry well vent on

Table 1.1 Overview of the FDNPP accident at the beginning

Phase	Date	Time	Mark	Reactor	Event	Remarks	Radioactivity detected
A	March 11, 2011	14:46:18	1		Great East Japan Earthquake	2011 off the Pacific coast of Tohoku Earthquake, Magnitude 9.0	
		14:46:48	2	1, 2, 3, 4	Preliminary tremor	Maximum acceleration 550 gal at the FDNPP	
		15:28	3		Tsunami, No impact on reactors	First wave. Wave height 6 m or less	
		15:36:10	4	1, 2, 3, 4	Tsunami, Reactor building submerged	Second wave. 16 m or more	
B		pm	5	1	Loss of cooling function	Also loss of water injection/ decompression	
		around 20:00	6	1	Meltdown began (Estimated)	Hydrogen generation/leakage into the building	
	12	4:00	7	1	Nuclides began to release from the building	Pressure vessel also damaged	Radioactivity detected in the *USS RR* located 193 km northeast of the FDNPP site. 3.0–9.0 μSv/h (Measured at 1 m above the deck. 16:00–18:00). 5.0–8.3 kBq/m^2 (β-ray survey. 22:20).
		10:10	8	1	Vent	Estimated	
		14:00	9	1	Vent	Estimated	
		15:36	[a]**10**	1	[a]Hydrogen explosion	Damaged Unit 1 Reactor building	
		21:00	11	[b]	[b]	[b]	
	13	2:42	12	3	Loss of cooling function	Also loss of water injection/ decompression	
		am	13	3	Meltdown	Hydrogen gas generation	
		9:05	14	3	Vent (Partially invaded Unit 4 reactor building)	Release from the stack	
		9:25	15	3	Water injection into reactor	Use a fire truck	
		am–pm	16	3	Damaged the pressure vessel		

		14:00	17	3	Released from the building	Estimated	
		am–pm	18	3	Damaged the containment vessel	Hydrogen gas leakage into the building	
		[b]	19	3, 4	Hydrogen gas flows from Unit 3 building to Unit 4 building	Through vent exhaust line	
	14	[b]					Radioactivity detected at the US Yokota Air Base approximately 40 km west from the center of Tokyo. 0.70–4.94 μSv/h.
		2:00	20	3	Released from the building	Estimated	
		7:30	21	3	Released from the building	Estimated	
		11:01	[a]**22**	3	[a]Hydrogen explosion	Damaged Unit 3 Reactor building	
C		13:25	23	2	Loss of cooling function		
		pm	24	2	Meltdown	Depressurize the pressure vessel	
		19:54	25	2	Water injection into reactor	Use a fire truck	
		21:30	26	1 or 2	[b]	Release from the stack	
		pm	27	2	Damaged the pressure vessel		
		pm	28	2	Damaged the containment vessel		
	15	6:14	[a]**29**	4 (3)	[a]Hydrogen explosion (Intrusion from Unit 3 reactor)	Damaged Unit 4 Reactor building	Radioactivity detected at the US Embassy in the center of Tokyo. 0.10–0.27 μSv/h, 0.14–24.0 kBq/m^2 (β-ray survey).
		8:00–14:00	[a]**30**	2	Dry venting	Significant amount radionuclides released	
		am–pm	31		Fuel pool water temperature rises	84 °C. Unit 4 building	
		23:00	[a]**32**	2	[a]Dry venting	Significant amount radionuclides released	

(continued)

Table 1.1 (continued)

Phase	Date	Time	Mark	Reactor	Event	Remarks	Radioactivity detected
D	16	10:00–16:00	[a]**33**	3	Released from the building	Smoke rising from the building	The US Embassy had recommended that Americans residing within 80 km of FDNPP do evacuate as a precautionary measure.
	18						^{131}I detected in Tokyo. 0.051 kBq/m^2.
	19	9:00	34	3	Released from the building		^{131}I is 0.040 kBq/m^2.
	20	8:21	35	4	Water injection into the fuel pool		^{134}Cs and ^{137}Cs detected in Tokyo. 0.55 kBq/m^2 each.
	21						^{134}Cs and ^{137}Cs: 5.3 kBq/m^2 each. ^{131}I: 32 kBq/m^2.
	22	15:00	36	3	Released from the building	Smoke rising from the building	^{134}Cs and ^{137}Cs: 0.33 kBq/m^2 each. ^{131}I: 36 kBq/m^2.
	23	14:00	37	3	Released from the building	Smoke rising from the building	During this period, these nuclides were still detected in the Tokyo areas.
	25	11:00	38	1	Released from the building	Smoke rising from the building	
		20.00	39	1	Released from the building	Smoke rising from the building	
	28	10:00	40	2	Released from the building		
	29	5:00–15:00	41	1	Released from the building		

Mark numbers are the same in Fig. 1.1

[a]Massive radioactive materials released by the hydrogen explosion or dry vent

[b]No information

March 15. Among the collapsed reactors Unit 1, 2, and 3, it was estimated that the Unit 2 reactor released an enormous amount of radioactive material. The spent nuclear fuel storage pool located on the top floor of the broken building of the Unit 4 reactor had 1535 fuel rods that had been moved to a safe storage location by December 22, 2014. Moreover, the MOX fuel was loaded into the Unit 3 reactor for plutonium thermal use. At the time of the accident, 32 of the MOX fuel assemblies were loaded into the Unit 3 reactor. The Japanese government allowed the Unit 3 reactor to use MOX fuel for 240 of the total 548 fuel assemblies.

During the FDNPP accident, two TEPCO employees were killed in the Unit 4 turbine building by the tsunami. On March 24, three people working in the basement of the Unit 3 turbine building received high-dose exposure and were transferred to a specialized hospital for radiation exposure treatment. There are reports that two of them suffered beta-ray burns on their feet, but details of the accident have not been made public. On the premises of the FDNPP, six people were injured during the earthquake, and 15 people were injured during the hydrogen explosion. The exposure doses of thirty people were reported to have exceeded 100 mSv.

Table 1.1 and Fig. 1.1 show the events and the radiation dose changes in chronological order immediately after the accident. These data were prepared mainly based on various measurement reports from TEPCO and the analysis by the Japanese government [1, 2, 9, 10]. Reports on variations in radiation dose and interpretation of accident events are not always consistent. The author expects further analysis within the TEPCO. The purpose of this book is to analyze what role the FDNPP accident played as an environmental pollution source for the Tokyo metropolitan area. Therefore, no nuclear engineering analysis is conducted in this chapter.

The changes in radiation dose measured at the FDNPP site from March 11, 2011, to the end of March, are shown in Fig. 1.1. The numbers in this figure indicate the mark numbers in Table 1.1. The layout of the facilities on the FDNPP premises and the monitoring sites for radiation doses are shown in Fig. 1.2. Many of the monitoring sites failed due to the earthquake, and the power supply was interrupted so that radiation would have been measured by a mobile measuring vehicle at the beginning of the accident. Numbers marked with an asterisk are events when the reactor buildings exploded and the significant dry venting episodes. Most of the air doses were measured at a height of 2 m from the ground by mobile measuring vehicles. The main gate is approximately 1 km west of the reactor buildings that were damaged by the accident. MP-4 is located 1.3 km northwest of the reactor buildings, and the western gate is located northwest of the main gate, 1.3 km from the reactor buildings. The office building is near the reactor buildings, with north station on the back and the south station on the front side of the reactors.

The radiation dose was measured at least 1 km away from the four reactors, except for the office building, so the radiation directly emitted from the damaged reactors had little effect on the air doses. That is, the doses measured at the main gate, western gate, and MP-4 were values from the ground. As described later, the measured values in the office building, which was only approximately 0.2 km from the reactor, do not appear to have been directly affected by the radiation

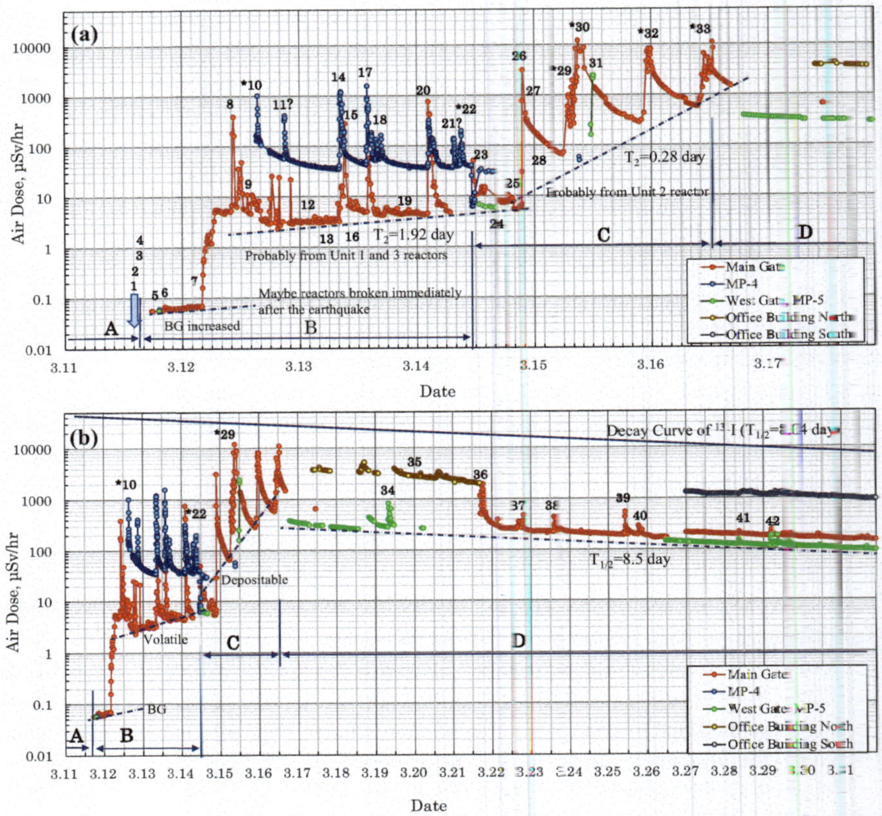

Fig. 1.1 Temporal changes in the air doses at the FDNPP premises immediately after the accident. (**a**) One week after earthquake. (**b**) Three weeks after earthquake. The event numbers in the figure same the mark numbers in Table 1.1. T_2: doubling time. $T_{1/2}$: half-life time

from the reactor. This was probably due to the structures around the broken reactors shielding the radiation. A monitoring survey detector, which is widely used for gamma-ray measurement, detects gamma rays from radionuclides suspended in the atmosphere and from radionuclides precipitated on the ground. Accurate monitoring requires a spatial and temporal detector configuration. Perhaps the TEPCO had many of radiation measurement experts, so the TEPCO data were analyzed on the assumption that they were strictly calibrating the measurement equipment.

The environmental behavior of radioactive materials released from the reactors at the beginning of the FDNPP accident is divided into four stages. In Stage A, the tsunami reached the FDNPP site, and the turbine buildings were submerged. At this point, the FDNPP was mostly SBO (Station Black Out), and the cooling function of the reactors was lost. In Stage B, the Unit 1 and 3 reactors meltdown, and the building exploded with hydrogen. In Stage C, a substantial dry well venting occurred in the Unit 2 reactor, and a significant amount of radioactive material was released

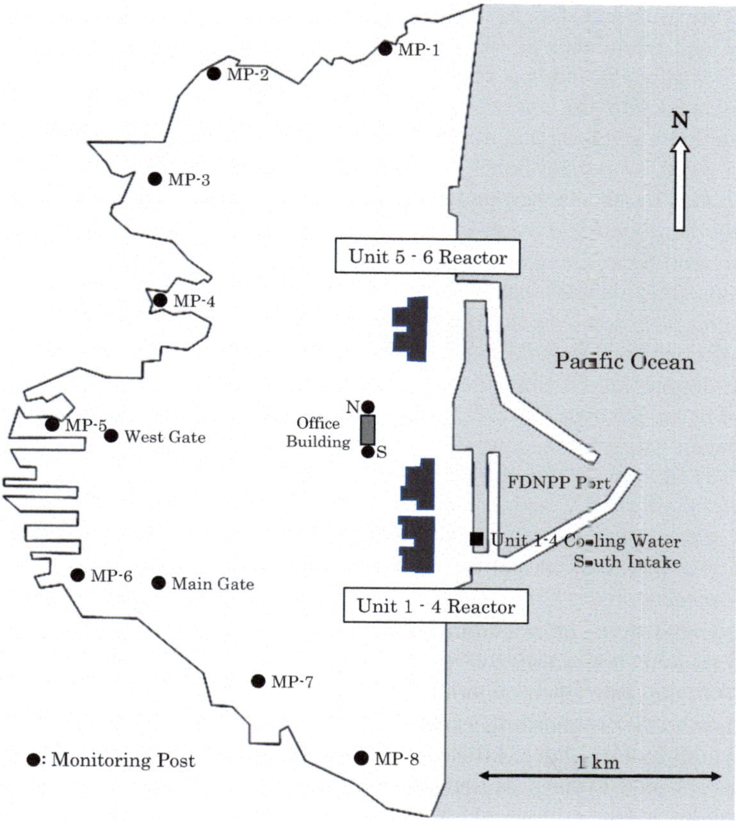

Fig. 1.2 Dose monitoring post on FDNPP premises

into the environment. Until this time, the cooling function of Unit 2 was somehow maintained, but the nuclear fuel meltdown before the venting. Thus, the substantial radioactive materials primarily released with the collapse of the reactors was completed by Stage C. In Stage D, the cores of the broken reactors were still hot enough that volatile radionuclides were continuously released into the atmosphere. Such phenomena indicate that a damaged nuclear reactor could have been the source of environmental radioactive contamination after the accident.

According to the TEPCO reports, approximately 4 h after the tsunami arrived at the FDNPP site, core heating and nuclear fuel meltdown began in the Unit 1 reactor due to the loss of the reactor cooling function. A second tsunami wave with a height of 16 m or more invaded the FDNPP facilities, and the turbine building and control building were submerged at 15:36 on the 11th. It is also thought that the hydrogen gas produced by the reaction between the zircaloy and the reactor cooling water began to accumulate in the pressure vessel. However, at 15:36 on the 12th, just 24 h later, the Unit 1 reactor building was destroyed by a hydrogen explosion. Since the reactor core was shielded by a double pressure and containment vessel, it was designed and manufactured so that even if hydrogen gas were generated in the reactor, it was would not leak easily. However, in the Unit 1 reactor, hydrogen gas easily escaped into the reactor building. The natural hydrogen explosion in the reactor building suggests that the pressure vessel and containment vessel may have been damaged by the earthquake before the meltdown in the Unit 1 reactor. As shown in Fig. 1.1, the background dose at the main gate in Stage A clearly increased before the hydrogen explosion occurred. It increased from 0.041 to 0.080 μSv/h. This increased background dose at a point more than 1 km from the reactors suggests that radioactive nuclides had leaked into the atmosphere. The increase in this background dose supports the possibility that the earthquake damaged the reactor vessel before the meltdown occurred. The volatile radioactive nuclides began to leak from the double vessels immediately after the earthquake. Since the Unit 1 reactor had been in use for over 40 years, the pressure vessel had likely been weakened due to significant radiation damage.

The TEPCO reports that the maximum accelerations in the E-W direction experienced by the Unit 1, 2, and 3 reactors due to the earthquake were 447 gal, 550 gal, and 507 gal, respectively. Compared with the maximum response accelerations of 489 gal, 438 gal, and 441 gal for the reference ground motion, these values were greatly exceeded in the Unit 2 and 3 reactors. This adverse condition easily induced hydrogen explosions. In any case, the continued use of such aging reactors caused not only meltdown and melt-through of nuclear fuel in Stage B but also damage to the reactor pressure vessel and containment vessel. As a result, the fuel rod zircaloy and the reactor coolant reacted to generate large amounts of hydrogen, destroying the reactor buildings. The damage to the pressure and containment vessels meant that the reactor cores were exposed to the atmospheric environment; thus, not only volatile radionuclides such as noble gases, iodine, and cesium but also fine radioactive particles could be released into the air. The release of these radioactive nuclides caused radioactive contamination in the environment.

The variation in the radiation dose at the main gate was 1.92 days for the baseline doubling time in Stage B. It was 0.28 days in Stage C, which was much faster than that in Stage B. The main gate of the FDNPP is approximately 1 km west of the reactor buildings, but this baseline value may be considered to reflect the amounts of radioactive nuclides precipitated on the ground at the main gate dosimetry point. On the other hand, the change in the peak is considered to have detected the radioactive nuclide activity in the radioactive plume moving in the atmosphere. In other words, the radioactive material detected in Stage B was released from the Unit 1 and 3 reactors but migrated to the monitoring site through the atmosphere without being deposited near the reactors. The radioactive material released from the Unit 2 reactor in Stage C was not transported far from the reactor compared to that in Stage B but was prone to precipitate near the FDNPP site. However, this interpretation does not mean that the radioactive materials released from the Unit 2 reactor were not transported to the Tokyo metropolitan area. The amounts of radioactivity released by dry venting at Unit 2 on March 14 and 15 were 190 PBq for radioactive noble gases, 140 PBq for ^{131}I, 2.7 PBq for ^{134}Cs, and 2.1 PBq for ^{137}Cs. These activities are equivalent to 38%, 35%, 27%, and 26% of the total emissions estimated by TEPCO, respectively [10]. The data published by TEPCO are incomplete and not perfect, but they provide much information about the behavior of radioactive materials from the broken reactor in the early days of the FDNPP accident.

The radiation dose in Stage D decreased at every monitoring point. Spike-like peaks suggest that radioactive material was intermittently released into the atmosphere from the broken reactors in this stage. However, the background dose decreased with a half-life of 8.5 days, which is very close to the ^{131}I half-life of 8.04 days. This coincidence of half-life means that ^{131}I mainly controlled the air dose in Stage D, and of course, there are also nuclides with longer half-lives. The estimated releases of radionuclides for each stage shown in Table 1.2 indicate that ^{131}I contributed significantly to the air dose in all stages.

1.3 Dynamics of Release and Diffusion of Radioactive Materials in the Early Stage of the FDNPP Accident

Radioactive materials are also released into the environment from nuclear power plants and nuclear reprocessing plants during regular operations. Each country regulates the concentrations and amounts of radionuclides released from a nuclear facility. For example, the concentration limits of ^{131}I and ^{137}Cs in Japan are 2 kBq/m^3 and 3 kBq/m^3 for air and 40 Bq/L and 90 Bq/L for wastewater, respectively. There is no legal limit for ^{134}Cs. If an accident occurs at a nuclear facility and the accident becomes uncontrollable in the facility isolated from the outside, radioactive materials leak outside the facility and pollute the surrounding environment and affect the general population. When the impact of an accident extends beyond the nuclear facility, the highest priority should be given to controlling the radiation exposure of residents. To this end, it is vital to quickly grasp the amounts of radioactive materials released from the site immediately after the accident and their behaviors. Of course, if radioactive material that exceeds the legal regulation value leaks from the accident site, a monitoring survey is also essential.

Large-scale disasters at nuclear facilities occurred continuously in the late 1950s. Examples include the reactor fire at Windscale in England (1957), the nuclear facility accident in Chelyabinsk in the former Soviet Union (1957), details of which are still unknown, and the critical accident in Los Alamos in the United States (1958). These facilities are related to the production of nuclear weapons. The accidents at commercial nuclear power plants related to environmental radioactive pollution and radiation exposure to residents must include the TMI (Three Mile Island) nuclear accident (1979) [11–14] and the Chernobyl nuclear accident (1986) [15–17]. The outlines of these nuclear disasters and the amounts of radioactive materials released from them can quickly be found using websites such as Wikipedia, so this book rarely mentions them.

Table 1.2 Estimation of the amounts of radionuclides released at each stage and their activity ratio

Period[a]	TEPCO's estimated amounts[b], PBq				Fraction for each stage				Ratio to ^{137}Cs		
	Noble Gas	^{131}I	^{134}Cs	^{137}Cs	Noble Gas	^{131}I	^{134}Cs	^{137}Cs	Noble Gas	^{131}I	^{134}Cs
March 12–31 (20 days)	497	473	12.5	10.1	1.0	1.0	1.0	1.0	49	47	1.2
Phase B (3 days)	137	61.5	1.27	0.89	0.28	0.13	0.10	0.10	155	69	1.4
Phase C (2 days)	253	217	4.19	3.78	0.51	0.46	0.33	0.33	67	57	1.1
Phase D (15 days)	107	194	7.07	5.42	0.22	0.41	0.56	0.56	20	36	1.3

[a]The period of each phase is shown in Table 1.1 and Fig. 1.1
[b]Cumulative amounts at each phase released from the FDNPP reactors. These values are estimated by TEPCO

The cause of most nuclear accidents is that the critical state becomes uncontrollable and the nuclear reaction in the reactor goes out of control. Then, in a short period, catastrophic reactor destruction and radioactive material emission occurs. In the TMI accident, the Chernobyl accident, and the FDNPP accident, massive amounts of radioactive materials leaked into the environment in very short times. In the case of the FDNPP accident, the cause of the reactor collapse was a great earthquake, so the power supply not only on the reactor premises but also in the surrounding area was interrupted for a long time. In addition to the external power supply, the internal power supply collapsed immediately after the earthquake due to the tsunami, so the FDNPP effectively experienced station black out (SBO), making it impossible to control the reactor. Briefly surviving storage batteries and powered vehicles became the only sources of power. It was no longer possible to replenish fuel for the powered vehicles from outside. Therefore, most of the existing monitoring posts and stack monitors that monitored FDNPP radiation were not available. Under such circumstances, the status of radioactive contamination on the premises was surveyed by data from several monitoring posts and mobile measuring vehicles that were operating. Figure 1.1 was created using data measured under conditions.

To analyze the behavior of radionuclides emitted by nuclear accidents, it is necessary to evaluate the release processes and transport routes from the reactor. In a typical reactor collapse accident, the radionuclides in the reactor core are released into the atmosphere. According to a brief from the MOE (Ministry of the Environment, Japan) [18], the radionuclides released from the reactor in the Chernobyl accident included approximately 100% for ^{133}Xe, 50% for ^{131}I, and 30% for ^{137}Cs. In the FDNPP accident, the release rates were estimated to be approximately 60% for ^{133}Xe, 2–8% for ^{131}I, and 1–3% for ^{137}Cs. However, if nuclear reactor meltdown or melt-through occurs and the reactor pressure vessel and containment vessel collapse, the coolant leaks out of the core as highly polluted radioactive water. This contaminated water mixes with rainwater and groundwater and migrates into the environment. In this case, the chemical properties and the physical and chemical species of the radionuclides are essential to understanding the transport processes through the aquatic environment. The release rates of radionuclides from the cores in the FDNPP reactors were considerably lower than those in the Chernobyl accident, so it seems that most of these radionuclides remained in the collapsed cores as radioactive debris. In other words, in the FDNPP accident, not only the estimated amounts of radionuclides released into the atmosphere but also the release process of radionuclides via groundwater played vital roles in environmental radioactive contamination.

For example, if radioactive cesium is present as monovalent cations in groundwater, it adsorbs strongly to 2:1-type clay minerals such as chlorite, biotite, and vermiculite in the soil during the process of underground leaching [19–30]. Moreover, the radioactive cesium is trapped in the soil unless the soil moves and the radionuclides adsorbed on the soil are fixed. However, when radioactive cesium forms water-soluble species with organic substances such as litter, humic acid and humus, it can quickly move through the soil [31–33]. Furthermore, a radionuclide that has formed such water-soluble organic species has an extremely high risk of entering the ecosystem via the food chain. It is necessary to monitor radioactive materials leaked into the geosphere and hydrosphere in addition to the atmosphere because the FDNPP was installed on land where much rainfall occurs and groundwater is present. As discussed in Chap. 8, the radioactively contaminated groundwater diffusing under the FDNPP reactors is still springing from the seabed offshore the FDNPP. Radioactively contaminated groundwater is responsible for radioactive contamination of marine products.

Various institutions and researchers have estimated the amount of radionuclide materials released to the environment by the FDNPP accident, but the values vary greatly depending on the estimation method. However, in most cases, the estimated data are based on TEPCO measurements. Additionally, many of the published estimates of the amount of released radionuclides do not indicate the basis for the estimation. The author proceeds with the discussion based on the report titled "Estimation of the amount of radioactive material released into the atmosphere at the Fukushima Daiichi Nuclear Power Plant accident (Japanese)" published by TEPCO in May 2012 [10]. This report is used because it explains the estimation method most logically. The results of the estimation for release into the atmosphere are shown in Table 1.3.

TEPCO evaluated the released amounts of radioactive noble gases, ^{131}I, ^{134}Cs, and ^{137}Cs. The report also discusses the time series of the releases and the effects of dry venting from the containment vessel. Table 1.2 was obtained using the published time series data of released radionuclides. Radionuclide emissions were estimated based on the radiation dose rates on the FDNPP premises and the meteorological data by AMEDAS (Automated Meteorological Data Acquisition System) of the Japan Meteorological Agency around the FDNPP. TEPCO's DIANA (Dose Information Analysis for Nuclear Accident) was used for the calculation. This system could analyze the behaviors of radioactive noble gases, radioactive iodine, and radioactively contaminated particles released around the FDNPP with a three-dimensional advection-diffusion model in the atmosphere, and estimate the air dose at any point around the FDNPP. The precipitation of each radioactive material onto the ground surface was evaluated for dry deposition and wet deposition. Table 1.3 shows the estimated values of radionuclides released from the FDNPP evaluated by TEPCO and NISA (Japan Nuclear and Industrial Safety Agency). NISA's estimated values are those reported to the IAEA by the Japanese government in June 2011, and they represent the official view of the Japanese government. The values estimated by TEPCO are approximately three times those reported by the NISA for ^{131}I, 0.56 times for ^{134}Cs, and 0.57 times for ^{137}Cs. It is difficult to evaluate which estimate is correct. However, the measurement for radioactive contamination in various samples from the Tokyo metropolitan area by the author and the isotope ratios between ^{131}I, ^{134}Cs, and ^{137}Cs are consistent with the values from TEPCO, so the estimated values of TEPCO are better. They are taken as highly reliable compared with the NISA values.

The precipitated amounts of nuclides at each air dose measurement point in Stage D, when the accident was considered to have shifted to a steady state, were estimated from the time series of air doses on the FDNPP premises shown in Fig. 1.1.

Table 1.3 Estimation of the amounts of radionuclides released from the FDNPP and the precipitated amount of the nuclides on the monitoring sites

Released amounts from the FDNPP, Bq				March 21, 2011 Office building north		March 27, 2011 Office building south	Main gate	West gate	
Nuclide	Estimated by TEPCO	Estimated by NISA[a]	Activity ratio[b]	Estimated precipitation, kBq/m^2	Decay corrected activity ratio[c]	Estimated precipitation, kBq/m^2			Decay corrected activity ratio[c]
^{131}I	5.0×10^{17}	1.6×10^{17}	50	1.5×10^{6} [d]	30	7.8×10^{5}	1.3×10^{5}	8.0×10^{4}	20
^{134}Cs	1.0×10^{16}	1.8×10^{16}	1	5.0×10^{4}	0.99	3.9×10^{4}	6.5×10^{3}	4.0×10^{3}	0.99
^{137}Cs	1.0×10^{16}	1.5×10^{16}	1	5.0×10^{4}	1	3.9×10^{4}	6.5×10^{3}	4.0×10^{3}	1
	Author's estimated dose, mSv/h[e]			2.33		1.31	0.218	0.134	
	TEPCO's measured dose, mSv/h			2.4		1.3	0.21	0.14	

[a]Estimated values were officially reported to the IAEA by the Japanese government in June 2011 [2]. These values were estimated by the NISA (Japan Nuclear and Industrial Safety Agency)
[b]Calculated with the TEPCO estimate
[c]Decay-corrected value to March 16, 2011
[d]Decay-corrected value to March 27 is 8.0×10^{5} kBq/m^2
[e]Calculated based on the IAEA-TECDOC-1162 using the estimated precipitation of each nuclide

Since these time series were derived at the time of the accident, it seems that various factors governed the air dose, so a straightforward estimation was made to ignore their effects. Because the air dose was measured very close to the broken reactor, the physical and chemical behaviors of radioactive iodine and radioactive cesium on the FDNPP premises were the same. Therefore, after release from the reactors, the changes in the isotope ratios were calculated on the simple assumptions that they were fixed even when precipitated on the premises, that the half-life of ^{131}I was short, and that the effects of its radioactive decay were taken into consideration. As a result, the precipitation amounts of ^{131}I, ^{134}Cs, and ^{137}Cs on March 21 and 27, shown in Table 1.3, were obtained.

Moreover, the estimated air doses calculated from these precipitation amounts using the method of IAEA-TECDOC-1162 were in good agreement with the values measured by TEPCO. Looking at the values on March 27, the precipitated amounts of radioactive cesium were approximately one-tenth times lower at the main gate and the western gate, which were approximately 1 km from the reactor, than at the measurement site in the office building near the reactor. The precipitation rate of radionuclides released from the reactor into the atmosphere by hydrogen explosions and dry venting appears to have decreased at a slight distance from the reactor. This phenomenon means that of the radionuclides released into the atmosphere, the majority were released from the Unit 2 reactor, which had the highest emissions among the FDNPP reactors, and were likely to be transported long distances and radioactively contaminate the environment far from the FDNPP. The Tokyo metropolitan area became a highly radioactively contaminated area due to this mechanism.

The amounts of radionuclides released at each stage shown in Fig. 1.1 were calculated using the values estimated by TEPCO related to the time variation in the released amounts of these nuclides. Approximately half of ^{131}I and one-third of ^{134}Cs and ^{137}Cs were released in the 2 days of Stage C when dry venting occurred in the Unit 2 reactor. On the other hand, more than half of the radioactive cesium, which has a longer half-life than ^{131}I, was released in Stage D. In addition to the explosions and venting, continuous leakage of radionuclides from the collapsed is suggested by these numbers.

At present, nuclear power plants operating in Japan use seawater to cool turbines. Therefore, all nuclear power plants in Japan are located on the coast. The FDNPP reactors took seawater from the coast in front of the turbine building for turbine cooling. The concentrations of radioactive cesium in seawater collected at the south intake are shown in Fig. 1.3, where the Unit 1-4 turbine cooling water intake is shown in Fig. 1.2. These time series data were published on the TEPCO website [34]. The values are plotted in Fig. 1.3. The transition in radioactive cesium activity in seawater offshore the FDNPP clearly shows that radionuclides continue to leak from the FDNPP reactors through routes other than the atmosphere. The hill on which the FDNPP was built a right catchment area, and abundant groundwater flows from the plateau to the coast. Groundwater emerges from the coastal cliffs or springs out of the seabed and flows into the Pacific Ocean.

In the FDNPP accident, it was believed that the pressure vessel and containment vessel of the Unit 1, 2, and 3 reactors were damaged, and nuclear fuel that meltdown

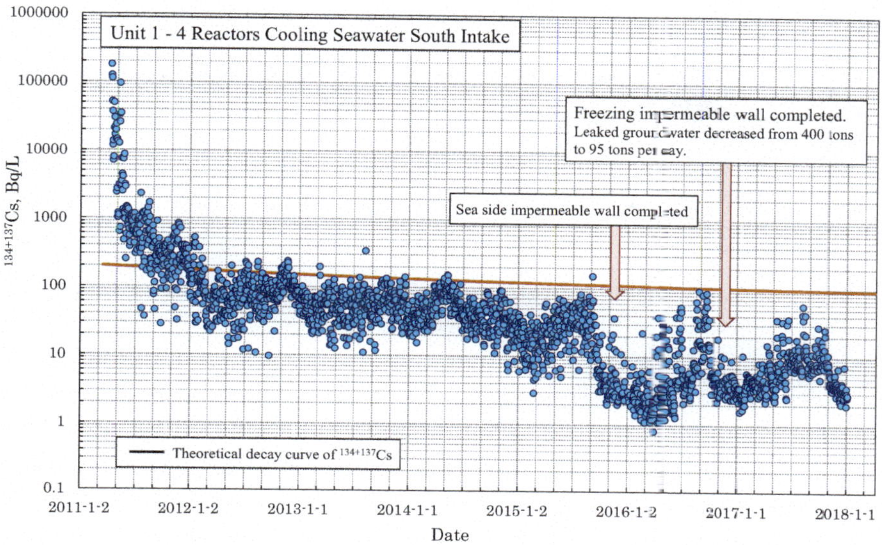

Fig. 1.3 Changes in the concentrations of radioactive cesium in seawater at the reactor cooling water south intake. Amounts of contaminated groundwater leakage from the FDNPP sites in 2019 was estimated to be 170 ton/day by TEPCO

or melt-through was exposed underground. Nine years have passed since the accident, but cooling water continues to be injected into the broken core to remove decay heat. Although some of the core cooling water is recovered and recycled after removing some of the radionuclides, TEPCO announced that 170 ton of cooling water is still leaking underground a day. This contaminated groundwater moves toward the sea, springs from the bottom of the sea, and is a source of radioactive contamination of seawater and local organisms including seafood (see Chap. 8). TEPCO installed a water barrier on the quay in front of the FDNPP port in 2015 to prevent leakage of radioactively contaminated groundwater. Furthermore, in 2016, frozen impermeable walls were installed around the reactor building to prevent inflow and leakage of groundwater. However, the radioactive cesium concentration in seawater in front of the reactors [34] shown in Fig. 1.3 has been changing in the range of 1–10 Bq/L, and the measures to prevent leakage of radioactively contaminated groundwater into the ocean have not been effective at all. Even now, a considerable amount of electricity continues to be consumed to freeze the underground soil.

References

1. TEPCO (2019) Monitoring post measurement status at Fukushima Daiichi Nuclear Power Plant site boundary [in Japanese]. http://www.tepco.co.jp/decommission/data/monitoring_post/index-j.html
2. Japanese Government (2011) Report of Japanese Government to the IAEA Ministerial Conference on Nuclear Safety - The Accident at TEPCO's Fukushima Nuclear Power Stations. https://japan.kantei.go.jp/kan/topics/201106/iaea_houkokusho_e.html
3. IAEA (International Atomic Energy Agency) (2015) The Fukushima Daiichi Accident. Report by the Director General
4. NA Independent (The Independent Investigation on the Fukushima Nuclear Accident) (2014) The Fukushima Daiichi Nuclear Power Station Disaster: investigating the myth and reality. Routledge, London
5. TEPCO (Tokyo Electric Power Company) (2012) The investigation reports of the Fukushima Nuclear accident [in Japanese]. http://www.tepco.co.jp/cc/press/betu12_j/images/120620j0303.pdf
6. NDJ (The National Diet of Japan) (2012) The official report of the Fukushima Nuclear Accident Independent Investigation Commission. https://www.nirs.org/wp-content/uploads/fukushima/naiic_report.pdf
7. Investigation Committee on the accident at the Fukushima Nuclear Power Stations of Tokyo Electric Power Company (2012) Final report. http://www.cas.go.jp/jp/seisaku/icanps/eng/final-report.html
8. NISA (Nuclear and Industrial Safety Agency) (2011) Evaluation on the state of core in units 1, 2 and 3 related to the accident of TEPCO's Fukushima Daiichi Nuclear Power Station [in Japanese]. http://warp.da.ndl.go.jp/info:ndljp/pid/3491887/www.meti.go.jp/earthquake/nuclear/pdf/20110606-1nisa.pdf
9. TEPCO (2019) Past measurement results at Fukushima Daiichi Nuclear Power Station [in Japanese]. http://tepco.co.jp/nu/fukushima-np/f1-rt/html-j/f1-mp-20190701-j.html
10. TEPCO (2012) Estimation of the amount of radioactive material released into the atmosphere at the Fukushima Daiichi Nuclear Power Plant accident [in Japanese]. http://www.tepco.co.jp/cc/press/betu12_j/images/120524j0105.pdf
11. NSAC (Nuclear Science Advisory Committee) (1980) Analysis of Three Mile Island-Unit 2 Accident. NSAC-80-1
12. NRC (U.S. Nuclear Regulatory Commission) (1989) Programmatic Environmental Impact Statement related to decontamination and disposal of radioactive wastes resulting from March 28, 1979 accident. Three Mile Island Nuclear Station, Unit 2
13. Talbott EO, Youk AO, McHugh-Pemu KP, Zborowski JV (2003) Long-term follow up of the residents of the Three Mile Island Accident Area: 1979-1998. Environ Health Presp 111:341–348
14. Mongano J (2004) Three Mile Island: health study meltdown. Bull At Sci 60:30–35. https://doi.org/10.2968/060005010
15. IAEA (International Atomic Energy Agency) (1992) The Chernobyl accident: updating of INSAG-1, Safety Series, INSAG-7
16. IAEA (International Atomic Energy Agency) (2006) Environmental consequences of the chernobyl accident and their remediation: twenty years of experience. IAEA Radiological Assessment Reports Series
17. IAEA (International Atomic Energy Agency) (2006) Chernobyl's legacy: health, environmental and socio-economic impacts and recommendations to the governments of Belarus, the Russian Federation and Ukraine. The Chernobyl Forum 2003–2005, Second Revised Version
18. MOE (Ministry of the Environment, Japan) (2017) Unified basic data on health effects of radiation [in Japanese]. https://www.env.go.jp/chemi/rhm/h29kisoshiryo/h29kisoshiryohtml.html

19. Sawhney BL (1972) Selective sorption and fixation of cations by clay minerals: a review. Clay Clay Miner 20:93–100
20. Evans DW, Alberts JJ, Clark RA III (1983) Reversible ion-exchange fixation of cesium-137 leading to mobilization from reservoir sediments. Geochim Cosmochim Acta 47:1041–1049. https://doi.org/10.1016/0016-7037(83)90234-X
21. Maes A, Verheyden D, Cremers A (1985) Formation of highly selective cesium-exchange sites in montmorillonites. Clay Clay Miner 33:251–257
22. Comans RNJ, Hilton J, Voitsekhovitch O, Laptev G, Popov V, Madruga MJ et al (1998) A comparative study of radiocesium mobility measurements in soils and sediments from the catchment of a small upland oligotrophic lake (Devoke Water, UK). Water Res 32:2846–2855. https://doi.org/10.1016/S0043-1354(98)00038-4
23. Bostick BC, Vairavamurthy MA, Karthikeyan KG, Chorover J (2002) Cesium adsorption on clay minerals: an EXAFS spectroscopic investigation. Environ Sci Technol 36:2670–2676
24. Zachara JM, Smith SC, Liu C, McKinley JP, Serne RJ, Gassman PL (2002) Sorption of Cs^{+} to micaceous subsurface sediments from the Hanford site, USA. Geochim Cosmochim Acta 66:193–211
25. Qin H, Yokoyama Y, Fan Q, Iwatani H, Tanaka K, Sakaguchi A et al (2012) Investigation of cesium adsorption on soil and sediment samples from Fukushima Prefecture by sequential extraction and EXAFS technique. Geochem J 46:297–302. https://doi.org/10.2343/geochemj.2.0214
26. Kogure T, Morimoto K, Tamura K, Sato H, Yamagishi A (2012) XRD and HRTEM evidence for fixation of cesium ions in vermiculite clay. Chem Lett 41:380–382. https://doi.org/10.1246/cl.2012.380
27. Motokawa R, Endo H, Yokoyama S, Nishitsuji S, Kobayashi T, Suzuki S et al (2014) Collective structural changes in vermiculite clay suspensions induced by cesium ions. Sci Rep 4:6585. https://doi.org/10.1038/srep06585. PMID: 25300233
28. Aoi Y, Fukushi K, Itono T, Kitadai N, Kashiwaya K, Yamada H et al (2014) Distribution and mineralogy of radioactive Cs in reservoir sediment contaminated by the Fukushima nuclear accident. J Mineral Petrol Sci 109:23–27. https://doi.org/10.2465/jmps.130620c
29. Mukai H, Hirose A, Motai S, Kikuchi R, Tanoi K, Nakanishi TM et al (2016) Cesium adsorption/desorption behavior of clay minerals considering actual contamination conditions in Fukushima. Sci Rep 6:21543. https://doi.org/10.1038/srep21543
30. Kogure T, Mukai H, Kikuchi R (2019) Weathered biotite: a key material of radioactive contamination in Fukushima. In: Nakanishi TM, O'Brien M, Tanoi K (eds) Agricultural implications of the Fukushima Nuclear Accident (III). Springer, New York, NY, pp 59–75. https://doi.org/10.1007/978-981-13-3218-0_7
31. Ritchie JC (1962) Distribution of fallout cesium-137 in litter, humus, and surface soil layer under natural vegetation in the Great Smoky Mountains. Dissertation, University of Tennessee, Knoxville. http://trace.tennessee.edu/utk_gradthes/1427
32. Valcke E, Cremers A (1994) Sorption-desorption dynamics of radiocaesium in organic matter soils. Sci Total Environ 157:275–283
33. Helal AA, Arida HA, Rizk HE, Khalifa SM (2007) Interaction of cesium with humic materials: a comparative study of radioactivity and ISE measurements. Radiochemistry 49:458–463. https://doi.org/10.1134/S1066362207050141
34. TEPCO (2012) Analytical results of radioactive materials around Fukushima Daiichi Nuclear Power Station [in Japanese] https://www.tepco.co.jp/decommission/data/analysis/index-j.html, http://www.tepco.co.jp/decommission/data/analysis/pdf_csv/2020/1q/intake_canal-newest01-j.csv

Chapter 2
Advection and Diffusion of Radioactive Materials Released in the FDNPP Accident into the Central Area of Japan

Abstract Various studies have focused on the routes of advection and diffusion of radioactive materials released in the FDNPP accident via the atmosphere. Many of them have solved a three-dimensional advection-diffusion equation that combines a weather forecasting model and a mass transport model. The materials are usually assumed to be particles or puffs, but it is also possible to simulate the movement of the materials by mixing them and tracking them in a Lagrangian manner. Of course, it is indispensable to verify the results of the simulation by comparing them with the actual values measured in the field. This chapter suggested that the advection and diffusion through the atmosphere of radioactive materials released from the accident site had significant influences on environmental radioactive contamination and contamination processes by radioactive nuclides in the Tokyo metropolitan area at the beginning of the FDNPP accident. Additionally, the actual situation of environmental radioactive contamination in the Tokyo metropolitan area at the beginning of the accident is shown. Furthermore, the actual state of the radioactive plume diffusing over the land of Japan was verified by comparing the results of the simulation using the advection-diffusion model with the measurement results for radioactive nuclide concentrations in the atmospheric dust. In the process of measuring atmospheric dust radioactivity, high boiling-point fission nuclides were detected at an observation point more than 500 km from the FDNPP. This fact provided valuable information about the process of reactor collapse and the transport of radionuclides by plumes.

Keywords FDNPP accident · Radioactive plume · Diffusion · Tokyo metropolitan area · Meltdown

---

Electronic Supplementary Material The online version of this chapter (https://doi.org/10.1007/978-981-15-7368-2_2) contains supplementary material, which is available to authorized users.

H. Yamazaki, *Radioactive Contamination of the Tokyo Metropolitan Area*,
https://doi.org/10.1007/978-981-15-7368-2_2

2.1 Introduction

Various studies have focused on the routes of advection and diffusion of radioactive materials released in the FDNPP and Chernobyl accidents via the atmosphere [1–10]. Many of them have solved a three-dimensional advection-diffusion equation that combines a weather forecasting model and a mass transport model. The materials are usually assumed to be particles or puffs, but it is also possible to simulate the movement of the materials by mixing them and tracking them in a Lagrangian manner. Of course, it is indispensable to verify the results of the simulation by comparing them with the actual values measured in the field.

However, many published simulation results only track the movement of radionuclides in the atmosphere. For example, Fig. 2.1 shows the path of a radioactive plume estimated by Dr. Masanobu Ishida, taking into account the distributions of radioactive iodine and radioactive cesium deposited on the ground surface and the weather conditions (private communication). However, this route was estimated long after

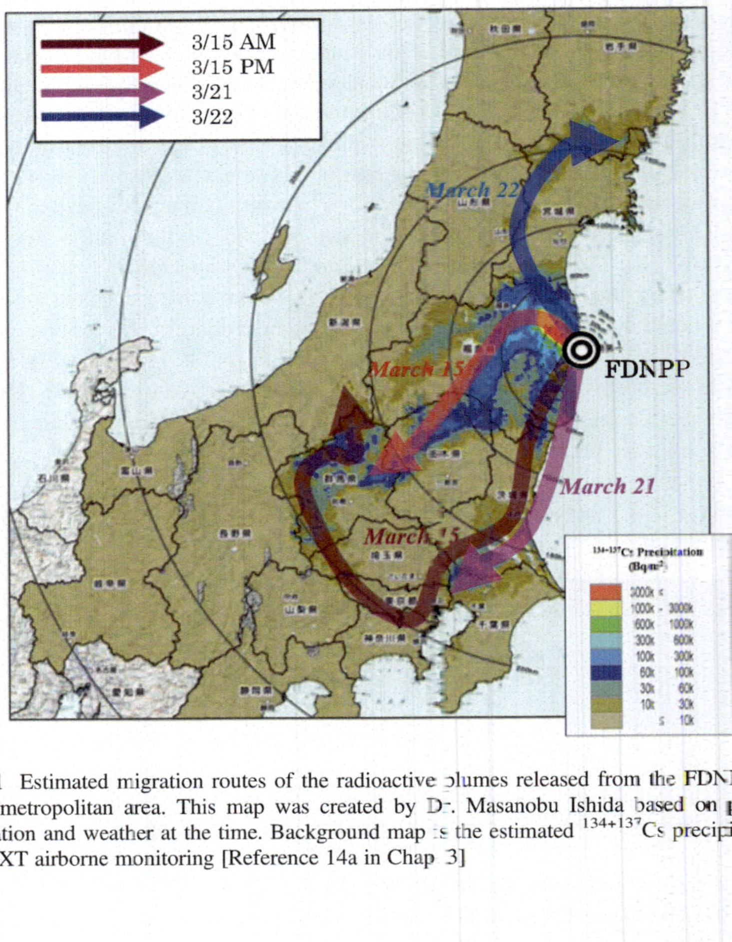

Fig. 2.1 Estimated migration routes of the radioactive plumes released from the FDNPP to the Tokyo metropolitan area. This map was created by Dr. Masanobu Ishida based on published information and weather at the time. Background map is the estimated $^{134+137}Cs$ precipitation by the MEXT airborne monitoring [Reference 14a in Chap. 3]

the accident. These simulations were inadequate for the residents living around the reactor that caused the accident. These people could not know the behavior of the radioactive materials that had fallen on their bodies. If a system that cannot cope with emergency evacuation immediately after an accident is constructed, such a simulation is useless from the viewpoint of the radiation exposure of residents and nuclear workers. Of course, from a scientific point of view, it is essential to evaluate the global spatiotemporal dynamics of radionuclides from nuclear sources.

In the Japanese government, a simulation system called SPEEDI (System for Prediction of Environmental Emergency Dose Information), which was operated by the Nuclear Safety Technology Center of the MEXT (Ministry of Education, Culture, Sports, Science, and Technology), was used to predict changes in radiation doses in the event of an emergency such as a nuclear disaster. However, the SPEEDI system could not be put into operation when the FDNPP accident occurred, which showed that this system would not be useful in an emergency. Six months after the FDNPP accident, on September 6, 2011, the JAEA (Japan Atomic Energy Agency) published the results of a simulation of the atmospheric fall of ^{137}Cs by SPEEDI [11]. According to the results, from 09:00 on March 15 to 09:00 on March 16, little deposition of ^{137}Cs was observed in central Tokyo. However, at this time, ^{137}Cs producing 10–1000 kBq/m^2 precipitated in the northern and western areas of Tokyo (Tochigi, Gunma, Nagano, western Saitama, Yamanashi, western Kanagawa, and eastern Shizuoka Prefectures). On the other hand, between 09:00 on March 20 and 09:00 on March 23, ^{137}Cs producing 100–1000 kBq/m^2 precipitated in central Tokyo. However, publishing the results of such a simulation 6 months after the accident would not help at all to protect the population from radiation. It is inevitable to point out the negligence of the Japanese government in this emergency.

Of course, radioactive nuclides are precipitated successively from a radioactive plume moving in the atmosphere, which becomes the source of radioactive contamination on the ground surface. Therefore, for radioactive contamination of the environment, it is not sufficient to evaluate the advection and diffusion processes of radioactive materials, and it is crucial to understand the depositional processes from the atmosphere to the surface. As mentioned in Sect. 1.3, TEPCO analyzed the dynamics of radionuclides released in the FDNPP accident into the atmosphere using the DIANA system. DIANA uses the atmospheric stability, the precipitation mechanism (dry or wet deposition), and the physical and chemical properties of nuclides as parameters in combination with the AMEDAS system of the Japan Meteorological Agency. However, even if the results are obtained using a high-performance computer, the reliability of the simulation system cannot be guaranteed unless the results can be verified by comparing them with the actual values measured in the field. Fortunately, the typical radionuclides ^{131}I, ^{134}Cs, and ^{137}Cs released into the atmosphere in the FDNPP accident are strongly adsorbed onto clay minerals and litter in the soil and retained in the place where they were deposited. This process means that if the changes in the geographic and temporal distributions of these nuclides are known, the analytical results can be used as dynamic markers for radionuclides precipitated on the surface from the radioactive plume.

This chapter suggested that the advection and diffusion through the atmosphere of radioactive materials released from the accident site had significant influences on environmental radioactive contamination and contamination processes by radioactive nuclides in the Tokyo metropolitan area at the beginning of the FDNPP accident. Additionally, the actual situation of environmental radioactive contamination in the Tokyo metropolitan area at the beginning of the accident is shown. Furthermore, the actual state of the radioactive plume diffusing over the land of Japan was verified by comparing the results of the simulation using the advection-diffusion model with the measurement results for radioactive nuclide concentrations in the atmospheric dust. In the process of measuring atmospheric dust radioactivity, high boiling-point fission nuclides were detected at an observation point more than 500 km from the FDNPP. This fact provided valuable information about the process of reactor collapse and the transport of radionuclides by plumes.

2.2 Migration Route of Contaminated Radioactive Plume to the Tokyo Metropolitan Area

The metropolitan area began to be exposed to radioactive contamination immediately after the FDNPP accident. Because this exposure stared immediately after the massive earthquake, information on radioactive contamination was not available either in Fukushima Prefecture, where the FDNPP was located, or in the Tokyo metropolitan area. The scene of the explosion at the Unit 3 reactor was broadcast live on television, but it was much later when the public knew that the Japanese land had begun to be contaminated by radioactive materials released from the FDNPP. On March 17, Japan time, US Ambassador to Japan Roos issued through the embassy a precautionary evacuation recommendation for Americans within an 80 km radius from the FDNPP. Many Japanese people who heard this statement noted that the FDNPP had a radioactive leakage accident similar to the Chernobyl accident. At this time, however, all significant events in the early days of the accident shown in Table 1.1 had been completed. The reason for issuing the evacuation advisory to Americans in Japan was thought to be the increases in radiation doses at the U.S. Embassy in Japan and at Yokota Air Base, as shown in Fig. 2.2.

In Japan, MEXT started airborne monitoring outside the 30 km range from the FDNPP on March 25, 2011. In April, the United States DOE (Department of Energy) participated, and the creation of a radioactive contamination map around the FDNPP began with the cooperation of the governments of Japan and the United States. This work continued until the end of 2011, and a map of radioactive contamination in East Japan to Shizuoka, Gifu and Toyama Prefectures, was completed. The map showed the geographic distributions of air doses 1 m from the surface and the deposition of ^{134}Cs and ^{137}Cs. In parallel with this airborne monitoring, various institutions and researchers measured, estimated, and analyzed the situations of radioactive contamination in East Japan.

The pollution map by airborne monitoring used in the background of Fig. 2.1 is shown again in Figs. 3.1 and 3.2. The radioactive plume released from the FDNPP moved in the northwest direction from the FDNPP while precipitating radionuclides and then changed course to the southwest with widening of the contamination range.

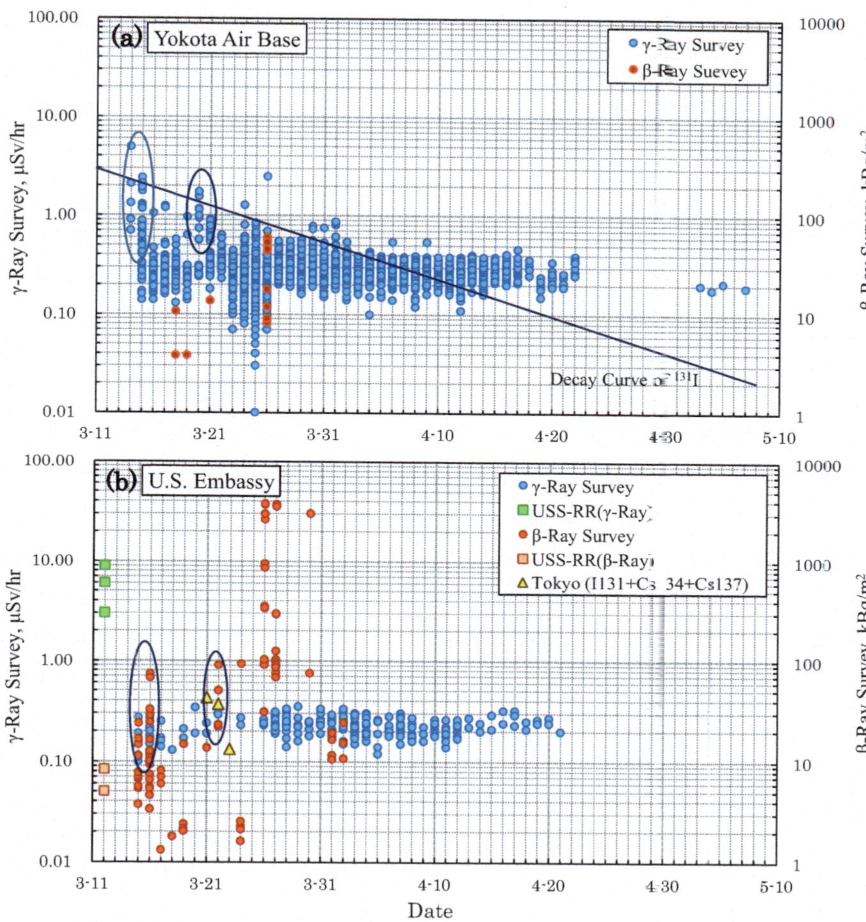

Fig. 2.2 Radioactive contamination measured in Tokyo immediately after the FDNPP accident. (**a**) The Yokota Air Base is located 40 km west from central Tokyo. (**b**) The U.S. Embassy is located in central Tokyo. The aircraft carrier USS RR was operating offshore the FDNPP (see Table 1.1). These figures were prepared by citing data published by the U.S. DOE [12]. The values of Tokyo are the precipitations of ^{131}I, ^{134}Cs, and ^{137}Cs

There was also a direct route to Tokyo. Figure 2.1, summarized by Dr. Ishida, shows the advection route of the radioactive plume to the Tokyo metropolitan area. This advection route is similar to those published by other researchers [4, 9]. These estimated maps show that the radioactive plume released from the FDNPP invaded Tokyo twice, on March 15 and 21. Because the weather was rainy during these days, the radionuclides in the radioactive plume are considered to have moved to the ground via a wet deposition process. These estimates are consistent with the geographic distributions of radioactive cesium precipitation and air doses measured by airborne monitoring. However, a radioactive plume that did not deposit radioactive material might have passed over the Tokyo metropolitan area because this airborne monitoring analysis shows the distribution of radioactive material precipitated on the ground surface. However, from the time variation of the ^{131}I radioactivity in the atmospheric dust shown in Fig. 2.3c, the highly polluted

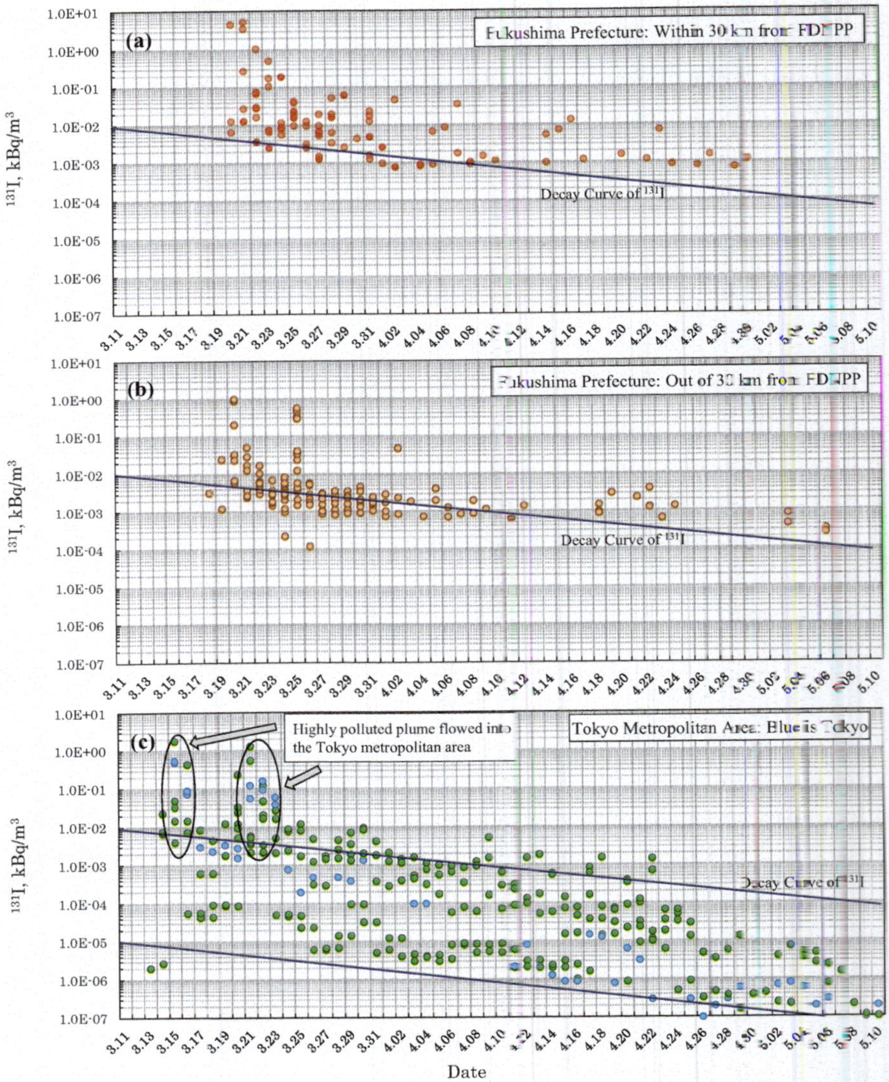

Fig. 2.3 Temporal changes in the ^{131}I activity in atmospheric dust at the early stage of the FDNPP accident. (**a**) Within 30 km from the FDNPP. (**b**) Outer than 30 km from the FDNPP in Fukushima Prefecture. (**c**) Tokyo metropolitan area. Blue filled circles are Tokyo. These figures were created by citing many published data (ex. [13–17]). However, some data published by local governments and research institutes are now deleted from the website

radioactive plume invaded the Tokyo metropolitan area on March 15–16, 19–22, and 29–30. The arrival of the radioactive plume in the Tokyo metropolitan area is consistent with the estimation in Fig. 2.1. In other words, the radioactive plume that was meaningful for radiation exposure and health effects of the residents was transported over the Tokyo metropolitan three times in the early days of the FDNPP accident.

2.3 Radioactive Contamination in Tokyo at the Beginning of the FDNPP Accident

2.3.1 Radioactive Contamination of Tokyo Immediately After the Accident

When the FDNPP accident contaminated the Tokyo metropolitan area is not well understood. There is no direct evidence of a change in the radioactive nuclide concentrations measured in the field. There may exist data measured by public authorities, but they have not been released. Infrastructure was disrupted due to the earthquake, and immediately after the earthquake, the Japanese government did not announce the damage to the FDNPP reactors, so the public could not obtain information on radioactive contamination. Numerous radiation monitoring posts are operated by various institutions in Japan, but many of them were temporarily unavailable in eastern Japan. Therefore, the data that can be used to evaluate environmental radioactive contamination immediately after the FDNPP accident are minimal.

According to Fig. 1.1, on the morning of March 12, 2011, when meltdown began in the Unit 1 reactor, a sufficient amount of radioactive material was released from the reactor to contaminate the surrounding environment. The author believes that the data published on the U.S. DOE (U.S. Department of Energy) website was the first public information suggesting environmental radioactive contamination from the FDNPP accident. The information published at the beginning of the FDNPP accident has been modified and uploaded to the current U.S. DOE website [12]. The top row of the spreadsheet contains data measured by the U.S. Navy aircraft carrier *Ronald Reagan* (RR). The RR was operating at 16:00 on March 12, 2011, at 38.465° north latitude, 142.79° east longitude, and approximately 120 miles northwest of the FDNPP. The gamma-ray dose 1 m above the deck was 0.3 mR/h (3 μSv/h). The value increased to 0.9 mR/h (9 μSv/h) at 16:45. Regarding the beta-ray intensity, the value was recorded 208 miles NW of the FDNPP at 39.63° north latitude and 143.65°east longitude at 22:20 on the same day, and the value was 0.225 μCi/m^2 (8.33 kBq/m^2). These detections of radioactivity mean that the radioactive plume released from Unit 1 at the earliest stage of the FDNPP accident reached the U.S. aircraft carrier RR, taking into account the passage of time. The radioactive intensities recorded on the RR are plotted in Fig. 2.2b.

Figure 2.2 shows the time-dependent changes in radiation intensity at the initial stage of the FDNPP accident, which were measured at the U.S. Embassy in Japan and the U.S. Yokota Air Base (Fussa City, Tokyo) and published on the U.S. DOE website [12]. Looking at the data, it seems that measurements were not necessarily made at the same location and under the same conditions. Therefore, the reliability of these measured values is not high. Nevertheless, these data represent values measured continuously in Tokyo in the early days of the FDNPP accident, so they are essential data for assessing environmental radioactive contamination in the Tokyo metropolitan area. The Yokota Air Base is located in Fussa City, approximately 40 km from the central part of Tokyo, and approximately 8800 Americans including family members are stationed there. Except for the Yokota Airfield, it is a suburban residential area at a slight distance from the city center of Tokyo. At the Yokota Air Base, the values from 21:00 on March 14, 2011, measured by the DOD (U.S. Department of Defense), and those from the U.S. Embassy at 15:06, measured by the DOE, were recorded on the DOE website. Measurements continued for approximately 40 days at these locations after the FDNPP accident. The data in Fig. 2.2 were measured at various locations within each facility. On the U.S. DOE website, the units of radiation intensity were roentgen (R) for gamma rays and curie (Ci) for beta rays. The values converted into sievert (Sv) and becquerel (Bq) by the author are plotted in Fig. 2.2.

It is difficult to evaluate the intensity of the average background radiation across Japan. The primary sources of 1-m gamma rays measured routinely are the radioactive nuclides in the disintegration series of uranium and thorium that exist underground and the cosmic rays emitted from the sun. The background radiation is significantly changed by rainfall because the natural radionuclides in the atmosphere are is washed away. The background radiation varies depending on geographic conditions such as the concentrations of uranium and thorium in the crust, altitude and topography. In urban areas, the shielding effects of asphalt and concrete laid on the ground and the scattering effects of buildings cannot be ignored. In basements and underground malls, the background is affected by the noble gas radon leaking from the crust. Some granite products have high concentrations of uranium and thorium, so a building that uses them in large quantities may have a high background. The author has empirically determined that the typical background radiation intensity in Japan is less than 0.05 μSv/h. The charged particle beams of beta rays and alpha rays interact strongly with substance, so it is difficult to estimate the average background value in the natural environment.

The fluctuations in radiation intensity at U.S. facilities shown in Fig. 2.2 indicate that the gamma-ray intensity at all facilities is higher than the average background radioactivity in Japan. These data mean that a radioactive plume entered into the Tokyo metropolitan area immediately after the FDNPP accident, and radioactive contamination then occurred. The gamma-ray doses at the Yokota Air Base and the amounts of beta-nuclide deposits at the U.S. Embassy were high on March 14–15 and 19–22, consistent with the estimations shown in Fig. 2.1. These values are also consistent with the time series fluctuations of ^{131}I in atmospheric dust in the Tokyo metropolitan area shown in Fig. 2.3. The changes in gamma-ray intensity at the Yokota Air Base until the end of March appear to be in harmony with the attenuation caused by the ^{131}I radioactive decay. This consistency means that ^{131}I decay was observed at the Yokota Air Base in the early days of the accident, and the gamma-ray intensity after April was due to the deposited radioactive cesium, which is understandable. However, such a trend was not observed in the U.S. Embassy data. The gamma-ray dose was several times the background value, and the radionuclide precipitation at this location may have been small. As shown in Fig. 4.2, the precipitation of ^{131}I at the U.S. Embassy may have been less than that in the surroundings areas due to the local heterogeneity of the radioactive plume. Additionally, the U.S. DOE reported extremely high radioactivities from the beta surveys on March 25, 26, and 30 at the U.S. Embassy. The highest value was almost 4000 kBq/m^2. If the results of this survey are correct, the gamma-ray dose measured on the same day did not change, and beta-emitting nuclides such as ^{90}Sr that do not emit gamma rays had precipitated at the embassy. The results of such surveys are valuable because there is little information on the environmental radioactive contamination by ^{90}Sr in Tokyo caused by the FDNPP accident. Nevertheless, the ^{90}Sr-^{90}Y radioactive equilibrium nuclides have long half-lives (28.8 year and 64 h, respectively), so it is difficult to explain the results of the U.S. Embassy beta-ray survey from the deposition of ^{90}Sr.

2.3.2 Radioactive Iodine That Entered the Tokyo Metropolitan Area Immediately After the Accident

To verify the arrival of radioactive plumes in the Tokyo metropolitan area in the early days of the FDNPP accident, the concentration of ^{131}I in atmospheric dust measured in various parts of eastern Japan [4, 13–17] is summarized. Figure 2.3 shows the temporal changes in ^{131}I concentration in atmospheric dust in Fukushima Prefecture and the Tokyo metropolitan area. Many of these measurement results have already been deleted from the website.

As shown in Fig. 2.3a, the measurement of atmospheric dust finally resumed in the area within 30 km of the FDNPP 10 days after the earthquake. Therefore, the activity of the radioactive plume was not measured when a massive amount of radioactive material was released from the Unit 2 reactor immediately after the accident. Hence, the dynamics of the radioactive plume at the beginning of the accident can only be evaluated by computer simulation using an advection-diffusion model. Although it is possible to estimate the movement of the radioactive plume from the analysis of the radionuclide concentrations that precipitated on the ground surface, it should be noted that the values from the accident to sampling are cumulative.

The ^{131}I activity in atmospheric dust in the Tokyo metropolitan area was measured from March 13. Figure 2.3c confirms that a highly polluting radioactive plume arrived in the Tokyo metropolitan area on March 15–16 and 20–23. These dynamics of ^{131}I are consistent with the prediction in Fig. 2.1. The ^{131}I activity in atmospheric dust at this peak in the Tokyo metropolitan area exceeded 1000 Bq/m^3. The base activity of ^{131}I excluding the peak time was approximately 0.1–10 Bq/m^3 and decreased according to the half-life of ^{131}I (8.04 days). The sampling points for atmospheric dust in the Tokyo metropolitan area are approximately 100–250 km from the FDNPP, so the radioactive plume released from the FDNPP reactors that reached these sampling points within approximately a day can be considered. Therefore, the highly radioactive plume on March 15–16 can imply that ^{131}I released in Stage C shown in Fig. 1.1 had reached the metropolitan area. However, the peak corresponding to March 20–23 cannot be found in Fig. 1.1. Perhaps this lack is the result of a highly polluting radioactive plume released from the reactors in the early part of Stage D arriving in the Tokyo metropolitan area as the atmosphere advected. Figure 2.3b shows the changes in ^{131}I activity in atmospheric dust in Fukushima Prefecture outside the 30 km range from the FDNPP. Since this area is adjacent to the Tokyo metropolitan area, radioactive plumes that reach the Tokyo metropolitan area are likely to pass through this area. The narrow peak on March 20–21 in Fig. 2.3b corresponds to the peak on March 20–23 in the Tokyo metropolitan area. The peak spread out in the Tokyo metropolitan area. The reason is that the plume diffused during the process of advection with the atmosphere. The peak on March 25 did not reach the Tokyo metropolitan area. The wind direction seems to have changed during the plume advancement in Fukushima Prefecture.

In Fukushima Prefecture, which is within 30 km of the FDNPP, ^{131}I activity in atmospheric dust exceeded 5000 Bq/m^3 even on March 20 a week after the accident. It can be estimated that the concentration was even higher immediately after the accident. Furthermore, the rate of decrease in ^{131}I radioactivity in the atmosphere in Fukushima Prefecture was much slower than that in the Tokyo metropolitan area. This difference in the rate of decrease means that ^{131}I was selectively precipitated faster during the advection of the radioactive plume and deposited on the ground surface. In any case, the residents of Fukushima Prefecture had inhaled air containing highly radioactive ^{131}I over a long period. Additionally, residents in the Tokyo metropolitan area were breathing air containing ^{131}I at almost the same concentration as those in Fukushima Prefecture during the early period of the FDNPP accident. Since iodine and cesium in the atmosphere are considered to behave almost the same, these residents breathed not only radioactive iodine but also radioactive cesium.

2.4 The Spread of Contaminated Radioactive Plumes from the FDNPP

2.4.1 Evidence for the Precipitation of Radioactive Plumes Diffused in the Atmosphere

The radioactive plume released in the FDNPP accident spread not only in Fukushima Prefecture and the Tokyo metropolitan area but also throughout Japan. The dynamics of the radioactive plume can be evaluated by monitoring the radioactivity of atmospheric dust. However, it is difficult to monitor air dust anywhere in Japan, especially in forests and mountainous areas. Therefore, the behavior of the radioactive plume is often estimated by analyzing the geographic distribution of radionuclides precipitate on the ground surface. When various radionuclides precipitated on the ground surface, they are strongly adsorbed by clay minerals in the soil and retained in the soil. However, since soil particles are often physically disturbed by rainfall and wind, care must be taken when collecting soil samples. Radionuclides that precipitate on the surface of a lake settle statically into its sediment and are fixed to the lake bottom sediments. If the sediment core is not bioturbated, the temporal variations in the nuclides that precipitated from the plume can be evaluated by their vertical distributions. The monitoring of the ^{131}I and ^{137}Cs concentrations in the atmosphere using an aircraft also performed partially [18]. However, since airborne monitoring is not as accurate as ground-based monitoring, it is not very useful for the dynamic analysis of radioactive plumes. Figure 6.9 shows the vertical distribution of radioactive cesium from the FDNPP accident recorded in sediments in the central part of Tokyo Bay. However, Tokyo Bay is not suitable for analyzing the local geographic distribution of radioactive plumes due to its large catchment area. Therefore, traces of radioactive cesium deposited from the radioactive plume of the FDNPP accident were sought in the sediment of an isolated small lake with a small catchment.

The first sediment core was collected at Lake Numazawa in Fukushima Prefecture, located at an elevation of 475 m, 120 km west of the FDNPP on March 31, 2014. Lake Numazawa is a caldera lake with a depth of approximately 100 m, a surface area of approximately 3 km^2, and a catchment area of approximately 4 km^2. According to the MEXT airborne monitoring, the precipitation of $^{134+137}$Cs in the catchment area immediately after the accident was 30–60 kBq/m^2. The sediment core was collected from a depth of 86 m in the center of the lake using an undisturbed acrylic pipe core sampler with an inner diameter of 8 cm. The core was 27 cm long and sliced into layers with a thickness of 1 cm while being pushed upward from the core sampler at the sampling site. The sedimentary age was identified using the ^{210}Pb method by measuring 46.5 keV gamma rays from ^{210}Pb with a Ge detector for low-energy detection. The ^{210}Pb sedimentary age was calculated by determining the porosity of the sediment using the water content of each slice layer and assuming that the density of the deposited material was 2.5 g/cm^3. The error in the sedimentary age estimated from the counting error of ^{210}Pb was approximately ± 2 years.

The analytical results for Lake Numazawa core are shown in Fig. 2.4. Traces of the FDNPP accident and atmospheric nuclear tests from the 1950s to the 1960s were found in the sediment core. However, no trace of ^{137}Cs that came to Japan due to the Chernobyl accident was detected. The average sedimentation years of the core surface layers were 2013.9 at 0–1 cm, 2011.4 at 1–2 cm, 2008.3 at 2–3 cm, and 2004.5 at 3–4 cm. The presence of ^{134}Cs was recognized from the surface to the layer at 3–4 cm. The ^{134}Cs derived from the accident was detected in the layers deposited before the FDNPP accident because the surface sediment layers were slightly mixed due to physical disturbance. Figure 2.4 shows the changes in the ^{137}Cs global fallout measured by the NIRS (National Institute of Radiological Sciences) in Chiba City on

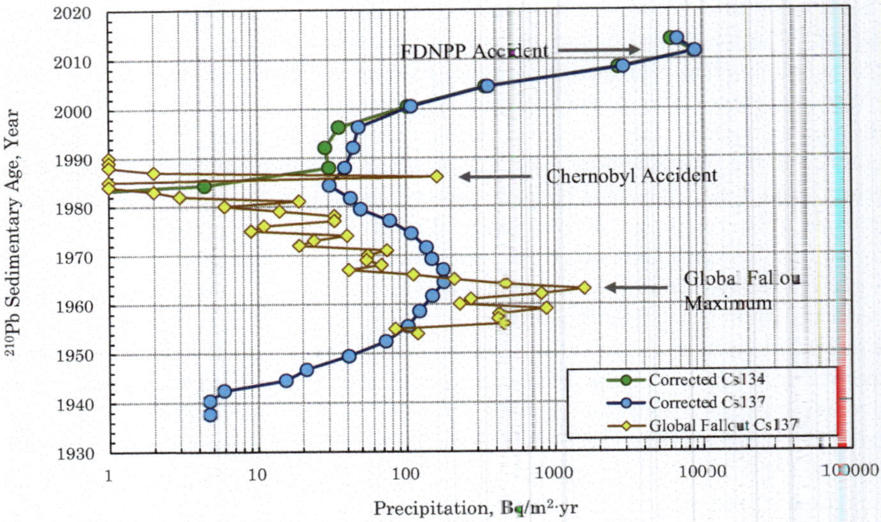

Fig. 2.4 Historical trends of the radioactive cesium distribution recorded in the sediment of Lake Numazawa, Fukushima Prefecture. Core sediment was collected from a water depth of 36 m on May 31, 2014, by co-researcher Mr. Ryoichi Hinokio, Ryukoku University. The ^{137}Cs global fallout was measured by the NIRS (National Institute of Radiological Sciences of Japan) in Chiba City [19]. After 1988, the global fallout was 1 Bq/m^2 year or less

the east coast of Tokyo Bay [19]. The maximum fallout from atmospheric nuclear tests was observed in 1963 in Japan. Traces of global fallout were found in Lake Numazawa sediments in the 1964.4 layer (7–8 cm). The maximum age of global fallout in both Lake Numazawa core and Chiba City agreed well within the errors of the sedimentary ages. The radioactive cesium concentrations were the highest in the 1–2 cm layer, with ^{134}Cs at 5230 Bq/kg and ^{137}Cs at 14,600 Bq/kg. The amount of $^{134+137}$Cs deposited in Lake Numazawa sediment at the time of the accident calculated from these radioactivity values was 72.8 kBq/m^2, which was almost the same as the $^{134+137}$Cs precipitation by MEXT airborne monitoring. This inventory of radioactive cesium derived from the FDNPP accident accumulated in Lake Numazawa sediment is reasonable considering the inflow from the catchment area. In this way, if the amount of radionuclide precipitation recorded in aquatic sediments can be analyzed quantitatively in a time series, the dynamics of the radioactive plume can be evaluated.

On the other hand, the presence of radioactive cesium released from the FDNPP accident was also confirmed in the sediment of a pond in a mountainous area farther from the FDNPP than Lake Numazawa. As already mentioned in Chap. 1, during the FDNPP accident, there were three massive explosions and several primary dry venting. In the explosions of the Unit 1 and 4 reactors, both hydrogen explosions destroyed the reactor building, and white smoke rose several hundred meters. In the explosion of the Unit 3 reactor, black smoke also rose to heights of several hundred meters, and a large amount of what appeared to be building debris was scattered. In the dry venting of the Unit 2 reactor, highly contaminated radioactive pollutant gas in the reactor pressure vessel was released from the exhaust tower. This tower was 120 m high and shared with the Unit 1 reactor. Such events formed a radioactive plume that spread to Japan and around the world. Empirical verification of the altitude reached by these radioactive plumes is vital for understanding the behavior of radionuclides in the atmosphere. The radioactive cesium contamination in plateau soil is discussed in Chap. 3. Here, the reach of the radioactive plume is examined using pond sediments on a plateau in Nagano Prefecture.

Ichinuma is a small pond located on the plateau called Shiga-Kogen, 230 km west-southwest of the FDNPP, at an elevation 1420 m. Surrounding the pond is a forest, and a variety of aquatic plants grow in the pond. The pond Shirakomanoike is located 290 km southwest of the FDNPP and has an elevation of 2115 m on the Yatsugatake-Kogen highland (see Figs. S2.2 and S2.3). The circumference is approximately 1 km, and the pond is surrounded by virgin forest. Sediment cores were collected from these ponds in August 2015. The Ichinuma core was 40 cm long, and the Shirakomanoike core was 73 cm long. The cores were sliced in to layers with a thickness of 2 cm and used for analysis. Due to the high elevations of the ponds, the excess ^{210}Pb activity in the sediments was deficient, and the ages of the sediments could not be measured. Because more than 4 years had passed since the accident, ^{131}I was not detected in the sediments, but ^{134}Cs was detected with sufficient radioactivity, which proves that it was released in the FDNPP accident.

The vertical distributions of ^{134}Cs and ^{137}Cs in the sediment cores are shown in Fig. 2.5. Because the sedimentary ages could not be estimated in these ponds, the effect of sediment mixing by bioturbation on the vertical distributions of radioactive cesium were unknown. Nonetheless, radioactive cesium released by the FDNPP accident was also deposited in a pond on the plateau 230–290 km from the FDNPP, contaminating its natural environment. In the decay correction value to March 16, 2011, the highest activity of $^{134+137}$Cs was 144 Bq/kg in the surface layer (0–2 cm) from the Shirakomanoike pond. This value decreased with depth, and at depths greater than 30 cm, ^{134}Cs was not detected. In the Ichinuma pond, the 12–14 cm layer had the highest radioactivity, which was 328 Bq/kg. Traces of ^{134}Cs were detected down to 60–62 cm. In this core, ^{134}Cs and ^{137}Cs not detected in the surface layer (0–2 cm).

Despite the high elevations and great distances from the FDNPP, radioactive contamination was found on the Yatsugatake-Kogen and Shiga-Kogen plateaus of Nagano Prefecture in the summer of 2015. Considering the surrounding environments of these ponds, radioactive cesium was directly precipitated from the atmosphere to the pond surface immediately after the FDNPP accident. The influxes of radioactive cesium from the catchments of these ponds seemed to be negligible. The measurements were 5.3 kBq/m^2 for Ichinuma and 1.7 kBq/m^2 for Shirakomanoike in terms of the values immediately after the accident. These values, as shown in Table 3.4 in Sect. 3.6, provide evidence that these plateaus received approximately the same level of radioactive contamination as the Tokyo metropolitan area. The fact that radioactive contamination due to the FDNPP accident was confirmed even in high elevation aquatic environments far from the FDNPP is valuable information for

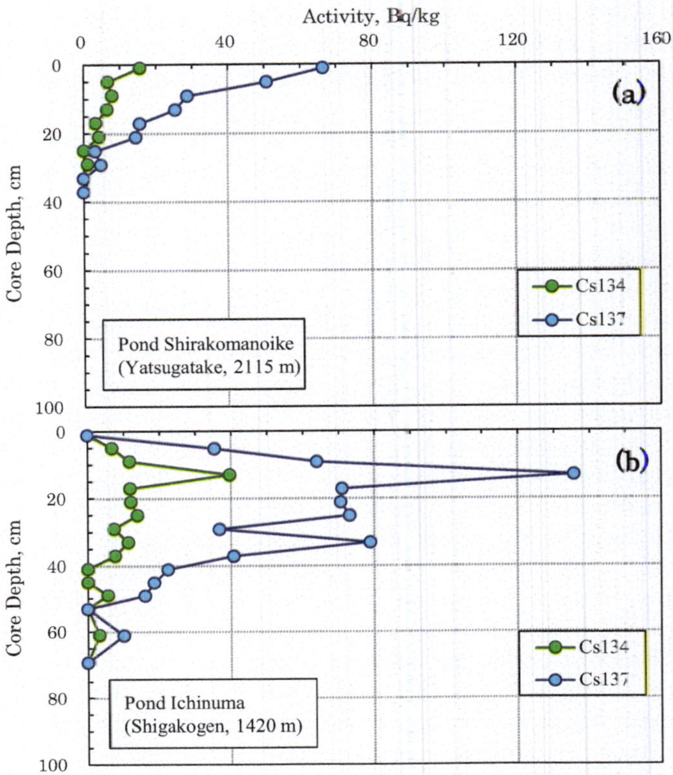

Fig. 2.5 Vertical distribution of radioactive cesium in pond sediment on the plateau. (**a**) Shirakomanoike is a circumference of approximately 1 km, located 290 km southwest from the FDNPP, and an elevation of 2115 m. (**b**) Ichinuma is a small pond located 230 km west–southwest from the FDNPP and an elevation of 1420 m. These core sediments were collected on August 27 and 28, 2015, by Professor Kazuo Kamura, Waseda University

evaluating the dynamics of contaminated radioactive plumes. Additionally, it is surprising that the radioactive plume had spread to an elevation of 2000 m and approximately 300 km from the FDNPP.

2.4.2 A Radioactive Plume from the FDNPP Accident Detected in Western Japan

The author measured atmospheric dust radioactivity in Higashi-Osaka City, Osaka Prefecture, located 580 km southwest of the FDNPP starting on March 15, 2011, immediately after the FDNPP accident. This radioactivity was monitored to confirm the spread of the radioactive plume to western Japan far from the FDNPP. A high-volume air sampler was installed on the roof of a building 15 m above the ground. To collect atmospheric dust, a glass fiber filter with a 99.99% collection efficiency of particles having a diameter of 0.3 μm was used. Dust collection started at 10:00 am, and the air was sucked in at a flow rate of 20–30 m^3/h until the collection volume reached 390 m^3. Depending on the atmospheric conditions, the suction time required approximately 15–20 h. The atmospheric dust collected by the filter was immediately measured with a Ge detector. The measurement time was 70,000–100,000 s. In Fig. 2.6, the typical gamma-ray spectra are shown. A tiny 364 keV gamma-ray peak

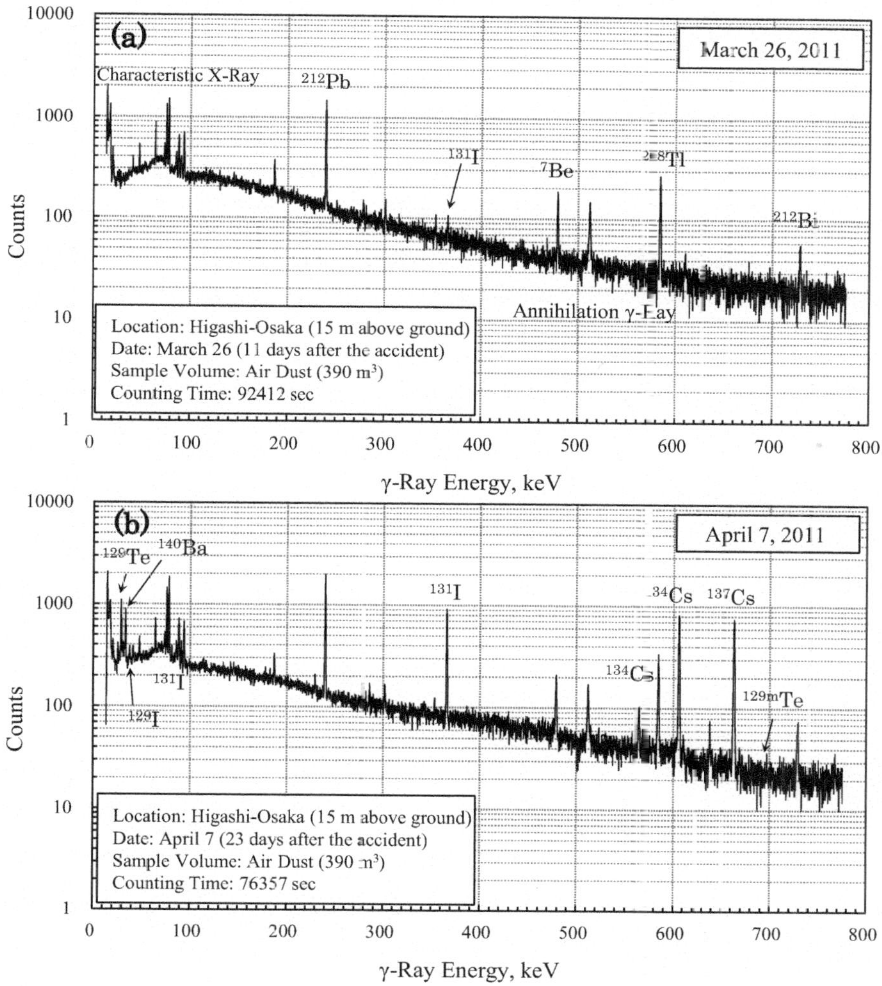

Fig. 2.6 Gamma-ray spectra of atmospheric dust in Higashi-Osaka City, Osaka Prefecture, located 580 km southwest from the FDNPP. (**a**) Background spectrum immediately before the radioactive plume arrives. (**b**) Spectrum of the day when the highest concentration of radioactive plume arrived

of ^{131}I was detected in the spectrum of the filter collected from March 23–25 in Fig. 2.6a. This spectrum was the first detection of a radionuclide released from the FDNPP in Higashi-Osaka City. The ^{131}I peak intensity was very small compared to the intensity of the natural radioactive disintegration series nuclides ^{212}Pb, ^{208}Tl, and ^{212}Bi and the cosmogenic nuclide ^{7}Be. Later, ^{131}I continued to be detected until May 8. The nuclides ^{134}Cs and ^{137}Cs began to be detected on March 30. Figure 2.6b shows the gamma-ray spectrum for April 7, when the radioactivities of these nuclides showed the highest values. Temporal changes in the intensities of these radionuclides transported from the FDNPP are shown in Fig. 2.7. The 30 keV gamma-ray peak of ^{140}Ba, a fission product, is also clearly detected. The presence of the high-boiling point nuclide ^{140}Ba in the atmospheric dust in Higashi-Osaka City also provides evidence that the radioactive dust released from the FDNPP migrated more than 500 km. Higashi-Osaka City is located in the Kinki area, which is the second largest economic zone in Japan. The population of the Kinki area, including Osaka and Kyoto Prefectures, exceeds 20 million. Not only the Tokyo metropolitan area but also the Kinki area was exposed to radioactive contamination due to the FDNPP accident.

Figure 2.7 shows the temporal variations in ^{131}I, ^{134}Cs, and ^{137}Cs radioactivity of atmospheric dust in Higashi-Osaka City. The maximum values of radioactivity occurred on April 6–7 and 17–18, and it is speculated that highly polluted radioactive plumes arrived in the Kinki area on these days. The peak intensity of ^{131}I at this time is almost the same as the atmospheric dust radioactivity in the Tokyo metropolitan area shown in Fig. 2.3. This agreement may imply that the radioactive plume advanced without much dilution in the atmosphere. Although Higashi-Osaka City is far from the FDNPP, the radioactivity peaks of radionuclides on April 6–7 and

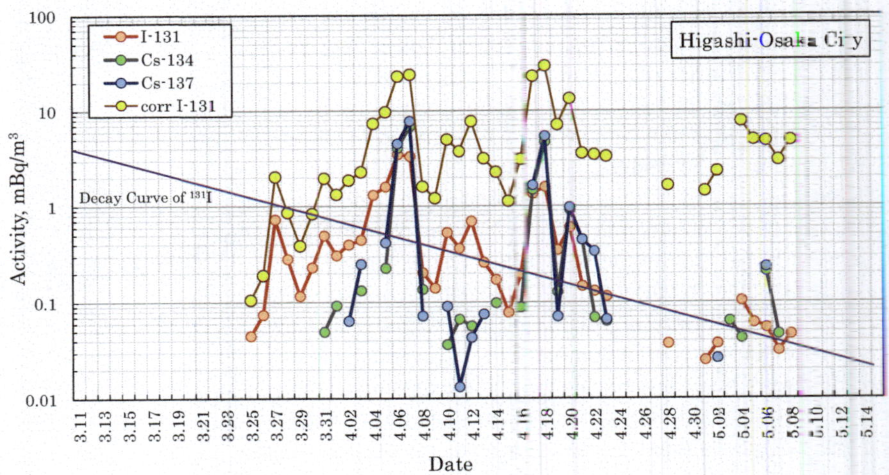

Fig. 2.7 Temporal changes in the activities of radionuclides in atmospheric dust in Higashi-Osaka City, Osaka Prefecture, after the FDNPP accident. Corrected ^{131}I activities are the value decay-corrected to March 16, 2011

17–18 are sharp. This phenomenon also suggests that the radioactive plume moved in the atmosphere without diffusing. The ^{131}I activity in atmospheric dust decreased from April 6, according to its half-life (8.04 days). Most likely, the radionuclides were released continuously from the FDNPP reactors at this time. In the FDNPP accident, the Japanese atmosphere received radioactive contamination by the mixing of a background component with little temporal variation over a wide area and a local high-intensity component transported by the plume. This process may have resulted in the formation of localized highly contaminated radioactive zones called hot spots observed on the surface in various areas of Japan. The activity changes in radionuclides due to the radioactive plume observed in Higashi-Osaka City were also observed in western Japan [20–22], more than 500–1000 km from the FDNPP.

However, as shown in Fig. 2.8, this plume reached Fukuoka City [21], which is far away, 1 day earlier. Higashi-Osaka City and Fukuoka City are in the same southwest direction with respect to the FDNPP, and Higashi-Osaka City and Fukuoka City are 500 km apart. Generally, the radioactive plume released from the FDNPP should reach Fukuoka City via Higashi-Osaka City. On the other hand, the peak of April 17–18 was detected in Higashi-Osaka City but not in Fukuoka City. Applying a simulation based on an advection-diffusion model is suitable for analyzing the dynamics of radioactive plumes. Since the author could not perform such simulations, this analysis cites the results of various atmospheric advection-diffusion simulations published after the FDNPP accident. The simulation performed by NILU (Norsk Institutt for Luftforskning) [23] was in good agreement with the results observed in Higashi-Osaka City and Fukuoka City. The results are shown in Figs. 2.9 and 2.10.

The results of simulations conducted at NILU from April 5 at 15:00 to April 8 at noon in Japan are shown in Fig. 2.9. In the NILU simulation, it is assumed that ^{131}I was released from the FDNPP reactors at a rate of 1×10^{18} Bq/day. The advection and diffusion of the radioactive plume are visualized. The geographic relationship between Higashi-Osaka City and Fukuoka City is shown in the upper left of Figs. 2.9 and 2.10. The atmospheric ^{131}I plume was advected from north to southwest at the initial phase in the NILU simulation. The ^{131}I released from the FDNPP traveled away from the Tokyo metropolitan area and moved south over the Pacific Ocean. On April 6 at 09:00, the radioactive plume turned southwest along the Japanese coast. On the afternoon of April 6, the radioactive plume arrived in Fukuoka City before arriving in Higashi-Osaka City. This plume arrived in Higashi-Osaka City half a day later than in Fukuoka City. The plume then left Fukuoka City on the afternoon of April 7 and Higashi-Osaka City at midnight on the same day.

The result of the NILU simulation from 6:00 on April 17 is shown in Fig. 2.10. In the afternoon of April 17, the radioactive plume reached Higashi-Osaka City. Higashi-Osaka City was covered with this radioactive plume for approximately a day until the afternoon of April 18. However, the radioactive plume did not reach Fukuoka City during this period. The NILU simulations of ^{131}I spread by the advection-diffusion model are in good agreement with the temporal changes in ^{131}I activity in atmospheric dust measured in Higashi-Osaka and Fukuoka cities. Later, both radioactive plumes migrated north over Japan, but there was fortunately

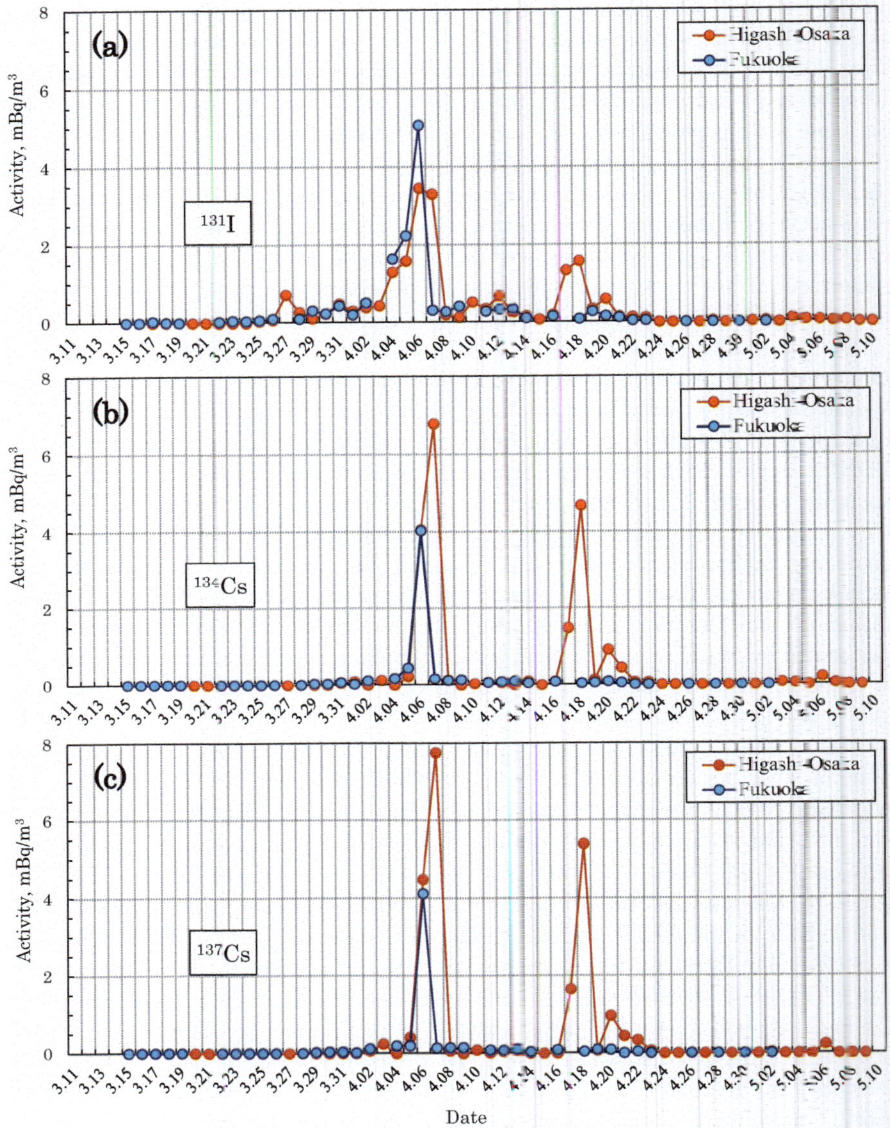

Fig. 2.8 (**a–c**) Temporal changes in the concentration of radionuclides in atmospheric dust transported into Higashi-Osaka City and Fukuoka City. Fukuoka City is located 1000 km southwest from the FDNPP, and Higashi-Osaka City is located 580 km southwest. Radionuclides in atmospheric dust in Fukuoka City were measured by N. Momoshima et al. [21]

no rainfall, and the ground was barely contaminated by radioactivity. However, residents outside breathed the air containing this radioactive dust.

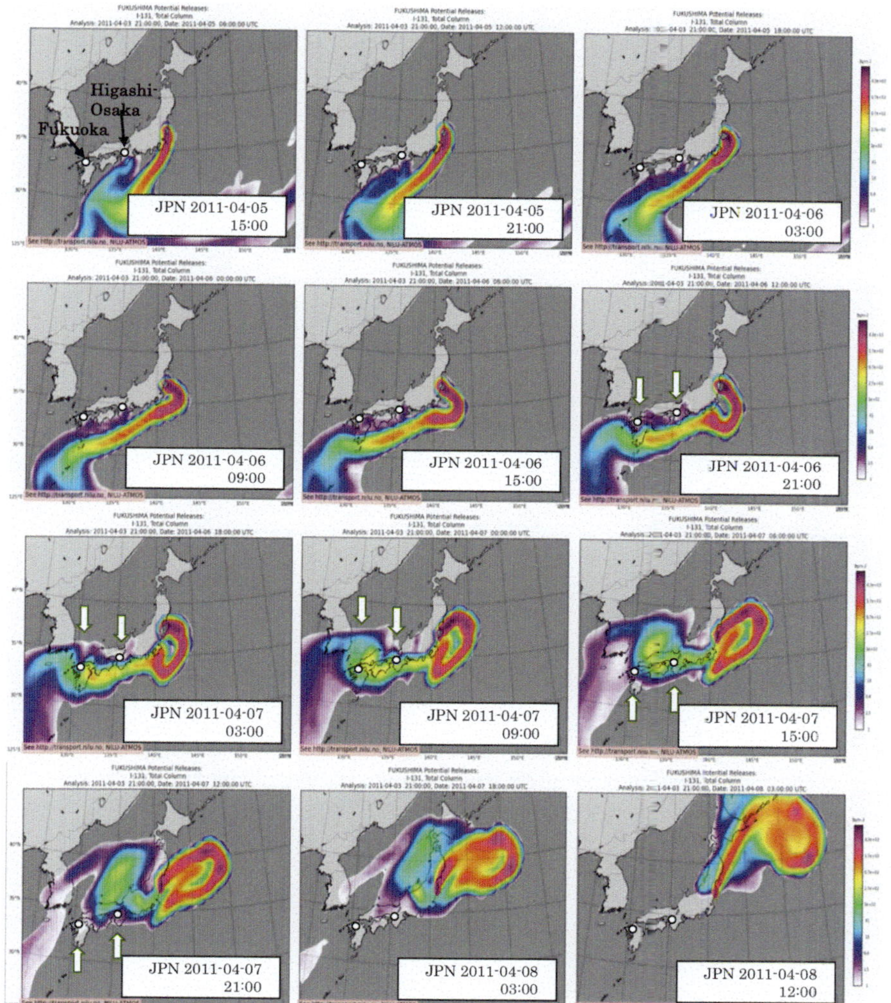

Fig. 2.9 Atmospheric advection-diffusion simulation by the NILU (Norsk Institutt for Luftforskning) for ^{131}I released in the FDNPP accident [23]. As shown in Fig. 2.8, the first peak of radionuclides was found in Fukuoka City on April 6 and Higashi-Osaka City on April 7

To clarify the origin of the radioactive plume that was transported to Higashi-Osaka City, the backward trajectory analysis provided by NOAA's HYSPLIT 4 (National Oceanic and Atmospheric Administration: Hybrid Single-Particle Lagrangian Integrated Trajectory 4) was applied [24]. As shown in Fig. 2.11, the ^{131}I that arrived on April 7 was estimated to have released from the FDNPP in the afternoon of April 4. Similarly, the ^{131}I that arrived on April 18 was released from

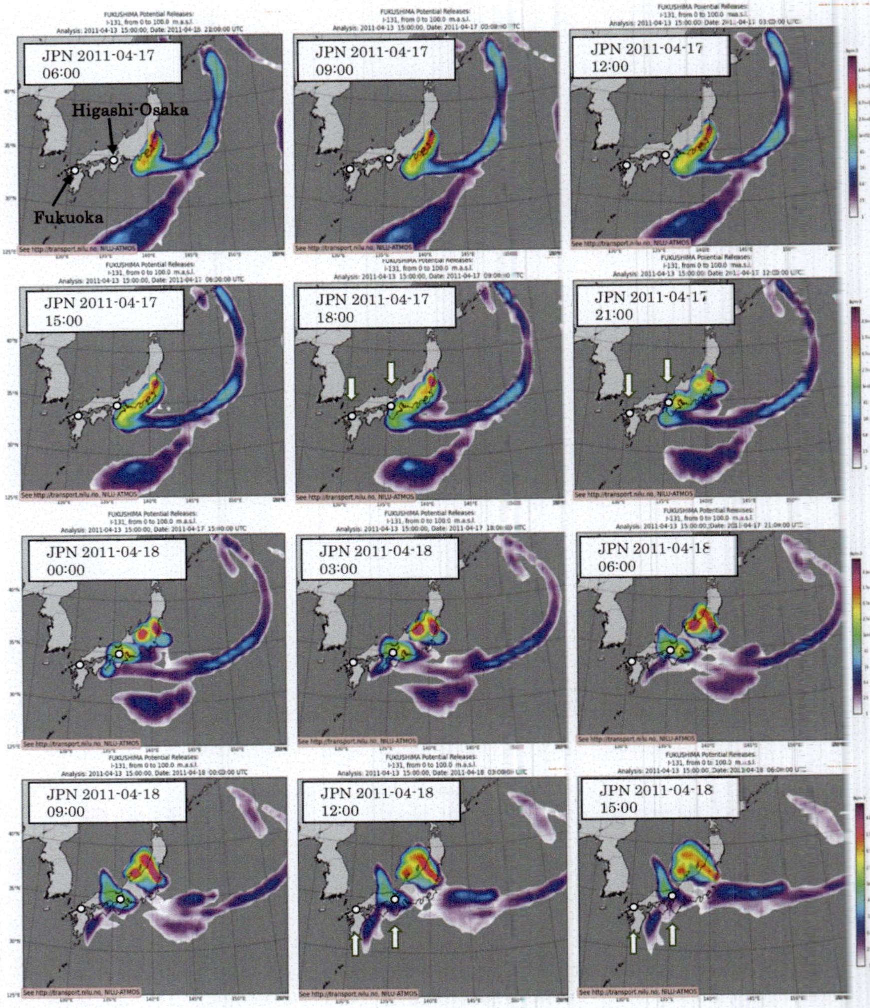

Fig. 2.10 Atmospheric advection-diffusion simulation by the NILU for ^{131}I released in the FDNPP accident [23]. The second peak was found in Higashi-Osaka City on April 17 and 18. This peak was not detected in Fukuoka City (see Fig. 2.8)

the FDNPP in the afternoon of the 16th. A radioactive plume observed on April 5 in Fukuoka City was released from the FDNPP reactor on the morning of April 4. In addition, these analyses obtained almost the same results as the forward trajectory analysis.

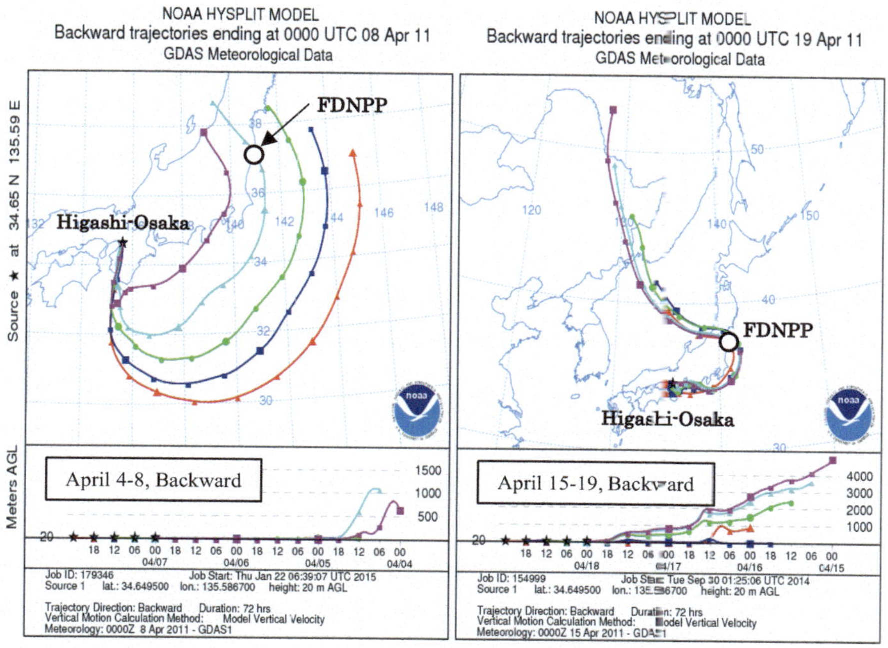

Fig. 2.11 Estimation of the emission source of the radioactively contaminated plume by the backward trajectory analysis provided by the NOAA HYSPLT model. Analysis start time: Left figure is on April 8. Right figure is on April 19. Analysis interval: 6 h. Aqua line in the left figure is shown that the contaminated plume released from the FDNPP on April 4 at 14:00 and reached Higashi-Osaka City on April 7 at 06:00. Magenta line in the right figure is shown that the plume released from the FDNPP on April 16 at 16:00 and reached Higashi-Osaka City on April 18 at 00:00

2.4.3 Quantitative Evaluation of ^{131}I Reaching Western Japan

The dynamics of ^{131}I that reached Higashi-Osaka City and Fukuoka City were quantitatively evaluated. The following parameters were used in the calculations: (1) The assumed ^{131}I release amount from the FDNPP was the primary parameter in the NILU simulation, 1×10^{18} Bq/day. (2) The spread of ^{13} I activity to Higashi-Osaka City and Fukuoka City was read from the NILU simulation map. (3) The ^{131}I emissions from the FDNPP for each event until the end of March 2011 were estimated by the TEPCO values. (4) The ^{131}I activities in atmospheric dust were measured in Higashi-Osaka City and Fukuoka City.

The results estimated from these parameters are shown in Table 2.1. Since the TEPCO's ^{131}I release was estimated separately under steady-state conditions and when some event occurred, the cumulative values for 1 week from March 25–31 were averaged for use in the calculation [25]. The peak intensity of measured ^{131}I was corrected for decay given the elapsed time from the end of March. Specifically,

Table 2.1 Quantitative evaluation of ^{131}I arriving in Higashi-Osaka City and Fukuoka City

NILU simulation parameter: amounts of ^{131}I released from the FDNPP are 1.0×10^{18} Bq/day					
Estimated ^{131}I activity released from the FDNPP reactors by TEPCO [23] (Cumulative value from March 25–31)	Steady state	0.84 PBq	Total released: 4.14×10^{16} Bq		
		0.12 PBq/day	Average released: 5.92×10^{15} Bq/day (X)		
	Events	40.6 PBq			
		5.80 PBq/day			
	Collected date	Higashi-Osaka City		Fukuoka City	
		Detected, Bq/m^3	Estimated release from the FDNPP, Bq	Detected [19], Bq/m^3	Estimated release from the FDNPP, Bq
Measured value (first peak)	Read on NILU's map	200 Bq/m^3 [a]		300 Bq/m^3 [a]	
	April 4	1.2×10^{-3}	0.60×10^{13}	1.6×10^{-3}	0.55×10^{13}
	April 5	1.5×10^{-3}	0.75×10^{13}	2.2×10^{-3}	0.75×10^{13}
	April 6	3.5×10^{-3}	1.8×10^{13}	5.1×10^{-3}	1.8×10^{13}
	April 7	3.2×10^{-3}	1.6×10^{13}	0.32×10^{-3}	0.07×10^{13}
	Total		4.8×10^{13}		3.2×10^{13}
	Daily average		1.2×10^{13}		1.0×10^{13} [b]
	Corrected daily average[c] (Y1)		2.4×10^{13}		2.0×10^{13}
	% Ratio (Y1/X)		**0.41**		**0.34**
Measured value (second peak)	Read on NILU's Map	200 Bq/m^3 [a]		nd[a]	
	April 17	1.2×10^{-3}	6.0×10^{12}	nd	–
	April 18	1.5×10^{-3}	7.5×10^{12}	0.08×10^{-3}	0.0[illegible] $\times 10^{13}$
	Total		1.4×10^{13}		0.0[illegible] $\times 10^{13}$
	Daily average		7.3×10^{12}		0.0[illegible] $\times 10^{13}$
	Corrected daily average[c] (Y2)		2.9×10^{13}		0.08×10^{13}
	% Ratio (Y2/X)		**0.49**		**0.01**

[a] ^{131}I activity estimated by the NILU simulation maps in Figs. 2.9 and 2.10 [21]
[b] Excluding the April 7 value
[c] Decay-corrected value to March 16, 2011

by the decay time of the first peak, one half-life of ^{131}I had elapsed, and by the time of the second peak, two half-lives. As a result, the release of ^{131}I in the plume from the FDNPP reactors was estimated at an average rate of 5.92×10^{15} Bq/day in the last week of March.

On the other hand, the arrival amounts of the first peak of ^{131}I detected in Higashi-Osaka City and Fukuoka City were 2.4×10^{13} Bq/day and 2.0×10^{13} Bq/day, respectively. Therefore, the amounts of ^{131}I that reached Higashi-Osaka City and Fukuoka City were estimated to be 0.41% and 0.34%, respectively, of the amount

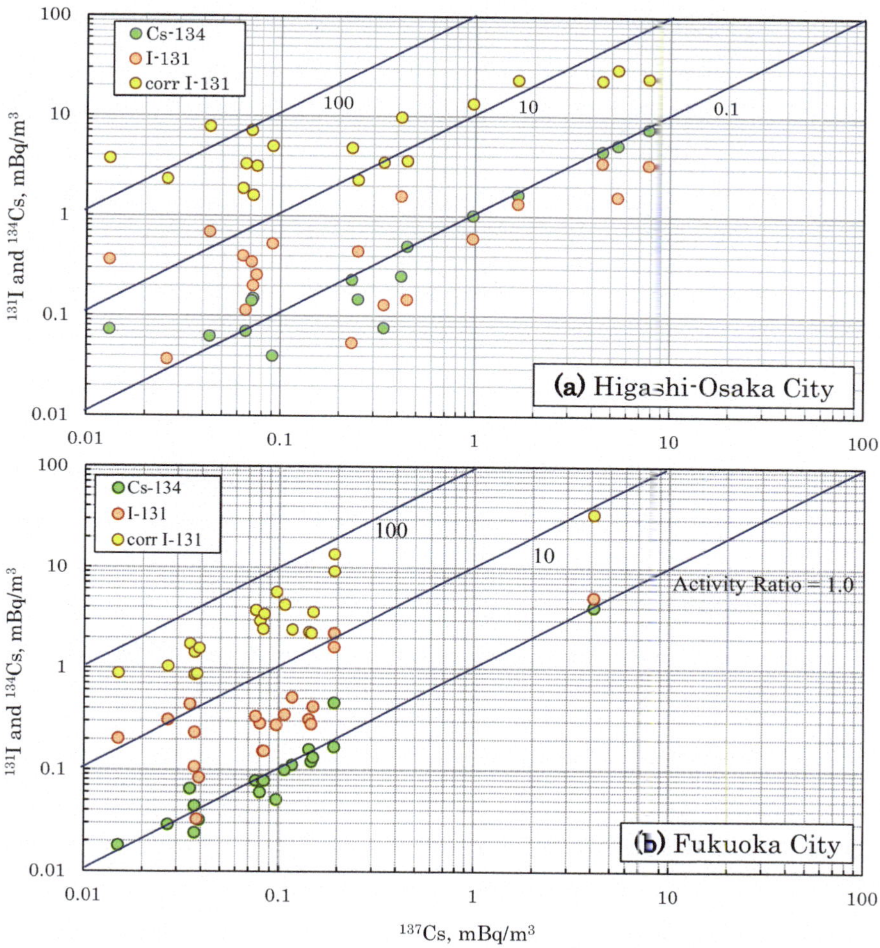

Fig. 2.12 Radioactivity ratio of ^{131}I and ^{134}Cs to ^{137}Cs in atmospheric dust. Yellow circles (corr I^{131}) are the decay-corrected value to March 16, 2011. (**a**): Collected in Higashi-Osaka City. (**b**) Data of Fukuoka City were analyzed by N. Momoshima et al. [21]

released from the FDNPP. Similarly, the value of the second peak in Higashi-Osaka City was 0.49%. Since the estimation value used for this calculation included various error factors, it was difficult to make a precise prediction. That is, approximately 0.3–0.5% of the released amount reached a distance of 500–1000 km from the release source. This information is necessary to consider when developing an exposure management and evacuation plan for residents during a nuclear disaster.

Figure 2.12 plots the activities of ^{134}Cs and ^{131}I against the ^{137}Cs activity detected in atmospheric dust in Higashi-Osaka City and Fukuoka City. The ^{134}Cs/^{137}Cs ratio was approximately 1.0 within the measurement error, showing the same distribution as in Fig. 4.1. The isotopic fractionation of ^{134}Cs and ^{137}Cs was not observed during the process of transporting the contaminated radioactive plume over a long distance. On the other hand, the decay-corrected $^{131}I/^{137}Cs$ ratio showed the same tendency as in Fig. 4.1. This figure plots the soil values in the Tokyo metropolitan area, which are close to those at the FDNPP compared with those in Higashi-Osaka City and Fukushima City. In other words, the similarity in the distributions of the $^{131}I/^{137}Cs$ ratios in Figs. 2.12 and 4.1 means that the behaviors of radioactive iodine and radioactive cesium in the contaminated plume are similar. The significant differences in the half-lives of these nuclides require careful attention, but the similarity in the behavior of these nuclides is also essential information for analyzing the dynamics of radionuclides in the environment in an emergency.

2.5 Evidence for the Meltdown of Nuclear Fuel in the FDNPP Reactors Found in Atmospheric Dust Collected in Higashi-Osaka City, Western Japan

Radioactive plumes from the FDNPP accident were detected on April 6–7 and 18–21, 2011, in Higashi-Osaka City, 580 km southwest of the FDNPP. In these contaminated radioactive plumes, ^{95}Zr (half-life: 64 days) and ^{95}Nb (half-life: 35 days) were found in the atmospheric dust collected on April 18. As described in Sect. 2.4 the radioactive plume containing these nuclides was released from one of the FDNPP reactors on April 16.

Few reports published after the FDNPP accident document that radioactive equilibrium between the nuclides ^{95}Zr and ^{95}Nb was detected from environmental samples. The discovery of such high boiling-point fission nuclides far from the FDNPP is very symbolic. The gamma-ray spectrum is shown in Fig. 2.13, where 724 and 757 keV gamma rays from ^{95}Zr and 766 keV gamma rays from ^{95}Nb are detected. The nuclide ^{95}Zr is a typical fission product and is the parent of ^{95}Nb, and ^{95}Nb is a daughter nuclide in transient equilibrium. ^{95}Zr is produced by (n, p) and (d, p) nuclear reactions of ^{94}Zr and (p, pn) reaction of ^{96}Zr, which are stable nuclides of zirconium in the zircaloy in nuclear fuel cladding. ^{95}Nb is also produced by the (d, n) reaction of ^{94}Zr and (p, 2n) reaction of ^{96}Zr, but its nuclear reaction cross-section is very small, so most of the cases involve ^{95}Zr as a parent nuclide. Zirconium has a melting point of 1588 °C, a boiling point of 4409 °C, and a vapor pressure of 4405 °C at 0.1 MPa.

Therefore, the detection of ^{95}Zr-^{95}Nb in transient equilibrium from environmental samples means that the FDNPP reactor fuel boiled on April 16. Alternatively, the reactor core may not have been sealed entirely at this time, and the dust in the reactor core may have released to the atmosphere [26, 27]. However, April 16, when the plume containing these nuclides was released, was more than a month after the accident, and the reactor at this time could have cooled sufficiently on the outside. However, the inside of the reactor may not have been sufficiently cooled, but the details are unknown. According to TEPCO's report [26, 27], the temperature of the

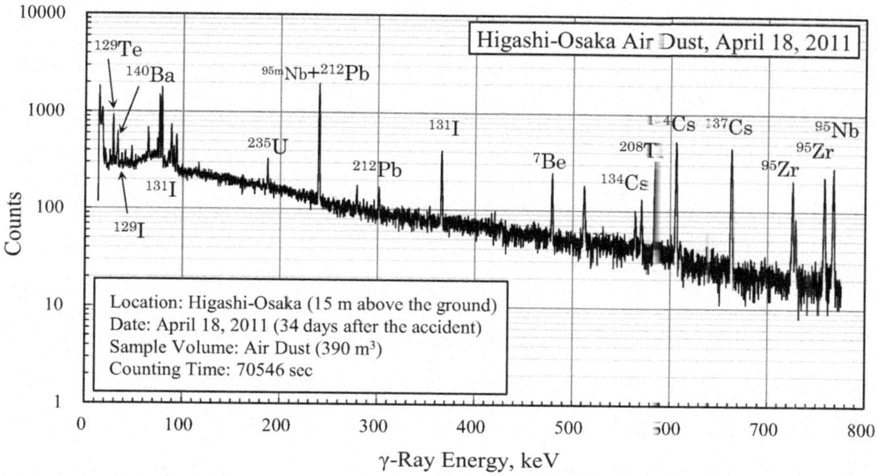

Fig. 2.13 Gamma-ray spectrum of ^{95}Zr-^{95}Nb detected from atmospheric dust in Higashi-Osaka City, Osaka Prefecture

CRD (Control Rod Drive) installed under the pressure vessel of the Unit 2 reactor increased sharply on April 14 and 16. The location of the CRD is where nuclear fuel meltdown debris may be deposited. The temperature of the pressure vessel support skirt at the top of the CRD also increased from April 1–10, before the CRD became hot. However, it is unlikely that the reactor was hot enough to vaporize zirconium. It is reasonable to infer that ^{95}Zr, which was deposited somewhere in the collapsed reactor vessel, nuclear fuel debris, reactor containment building, *etc.*, was scattered during the accident handling process and released to the environment. The spread of such high boiling-point fission nuclides into the environment indicates that the reactor was exposed to considerable temperatures during the accident. This discovery implies that nuclear fuel was reliably meltdown or melt-through in the FDNPP reactors.

The validity of the presence of these nuclides was verified, assuming that ^{95}Zr-^{95}Nb was released on April 16, 2011, and detected in the atmospheric dust in Higashi-Osaka City on April 18. The ^{95}Zr activity detected on the filter paper collected on April 18 was 4.34 mBq/m^3, and that of ^{95}Nb was 2.90 mBq/m^3, so the ^{95}Zr/^{95}Nb activity ratio was 1.50. The three verified assumptions are shown in Table 2.2. Assumptions (1) and (2) cannot explain the ^{95}Zr/^{95}Nb ratio measured on April 18. In other words, the most appropriate method was to estimate the ^{95}Zr/^{95}Nb ratio at the time of the accident from the actual value on April 18. This result is equivalent to assumption (3). As shown in Fig. 2.14, transient equilibrium was not reached on April 18, when these nuclides were detected, and the ^{95}Zr/^{95}Nb ratio at the time of the accident was 8.96. This value can explain the activity even if ^{95}Nb was generated not only by the decay of ^{95}Zr but also by the nuclear reaction of ^{94}Zr or ^{96}Zr. The results of this simulation show that ^{95}Zr-^{95}Nb, which was found in the air dust of Higashi-Osaka, was present in the reactor at the time of the FDNPP accident. It splattered into the atmosphere on April 16, indicating that radioactive

Table 2.2 Radiochemical verification of ^{95}Zr and ^{95}Nb released from the FDNPP reactors

	^{95}**Zr**		^{95}**Nb**		^{95}**Zr/**^{95}**Nb ratio**	
Half-life, day	64.0		35.0		0.44[a]	
Observed, mBq/m^3 (April 18, 2011)	4.34		2.90		1.50	
	Estimated values					
Assumption	^{95}**Zr**		^{95}**Nb**		^{95}**Zr/**^{95}**Nb ratio**	
	March 15	April 18	March 15	April 18	March 15	April 18
(1)	6.72	4.34	0.00	2.52	–	1.72
(2)	6.72	4.34	14.30	9.81	0.47	0.44
(3)	***6.72***	**4.34**	***0.75***	**2.90**	***8.96***	**1.50**

Assumption: (1) Only ^{95}Zr was released from the reactors. (2) ^{95}Zr-^{95}Nb in the transient equilibrium was released from the reactors. (3) ^{95}Nb activity at the accident was estimated from the detected values on April 18, 2011

[a] $^{95}Zr/^{95}Nb$ activity ratio at the transient equilibrium

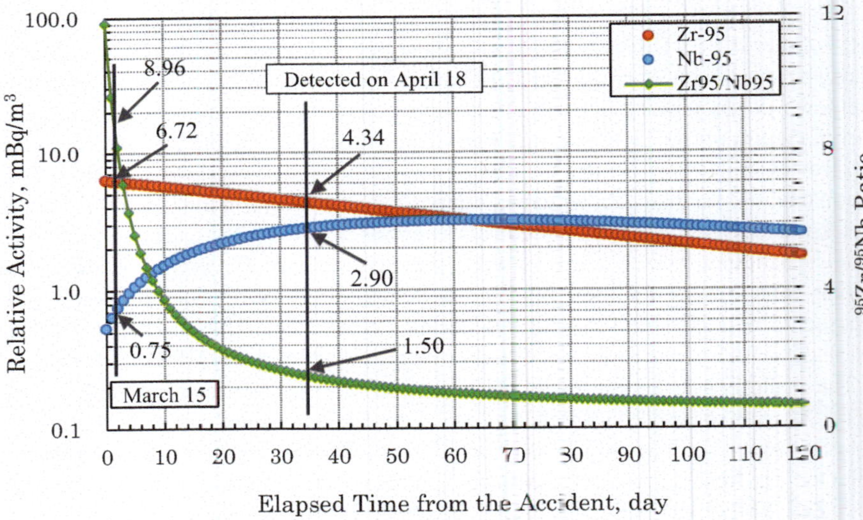

Fig. 2.14 Estimation of the origin of ^{95}Zr-^{95}Nb nuclides detected in atmospheric dust in Higashi-Osaka City, Osaka Prefecture

plumes transported it to Osaka. Nevertheless, from the results of radiation monitoring on the FDNPP premises published by TEPCO [28], no significant changes in air doses were observed between April 12 and 18. These measurements were carried out at 10-min intervals at each monitoring post, as shown in Fig. 1.2. In other words, when the temperature of the CRD in the Unit 2 reactor rose sharply on April 14 and 16, ^{95}Zr-^{95}Nb would have been scattered from the surrounding nuclear fuel debris. Since the air dose on the FDNPP premises did not increase, this phenomenon seems to have occurred locally for a short time. However, if radioactive material was scattered from a reactor in a process that could not be detected by the current monitoring system, it means that the FDNPP accident could not be safely controlled.

As shown in Fig. 2.7, the fission product nuclides ^{131}I, ^{134}Cs, and ^{137}Cs were detected in significant amounts in all atmospheric dust samples collected between April 16 and 23. However, ^{95}Zr-^{95}Nb was detected only in the April 18 sample. This discrepancy suggests that nuclides that vaporize at relatively low temperatures, such as radioactive iodine and radioactive cesium, and the high boiling-point nuclide ^{95}Nb may differ significantly in their processes of release from the collapsed reactor to the atmosphere.

References

1. Energy Environmental Safety Engineering Group (2017) Analysis and evaluation of atmospheric release and transport of radioactive materials in the Fukushima Daiichi Nuclear Power Plant accident. Energy Environmental Safety Engineering Group, Nagoya University [in Japanese]. http://ees.nagoya-u.ac.jp/~env_eng/subjects/subjects/subject_1F.html
2. Steinhauser G, Niisoe T, Harada HK, Shozugawa K, Schneider S, Synal HA et al (2015) Post-accident sporadic releases of airborne radionuclides from the Fukushima Daiichi Nuclear Power Plant site. Environ Sci Technol 49:14028–14035. https://doi.org/10.1021/acs.est.5b03155
3. Morino Y, Ohara T, Watanabe M, Hayashi H, Nishizawa M (2013) Episode analysis of deposition of radiocesium from the Fukushima Daiichi Nuclear Power Plant accident. Environ Sci Technol 47:2314–2322. https://doi.org/10.1021/es304620x
4. Turuta H, Nakajima T (2012) Radioactive materials in the atmosphere released by the accident of the Fukushima Daiichi Nuclear Power Plant. Chikyukagaku (Geochemistry) 46:99–111. [in Japanese]
5. Maruo Y, Ohara T, Nishizawa M (2011) Atmospheric behavior, deposition, and budget of radioactive materials from the Fukushima Daiichi nuclear power plant in March 2011. Geophys Res Lett 38:L00G11. https://doi.org/10.1029/2011GL048689
6. Ohara T, Morino Y, Tanaka A (2011) Atmospheric behavior of radioactive materials from Fukushima Daiichi Nuclear Power Plant. Natl Inst Public Health 60:292–299. [in Japanese]
7. Lujanienė G, Byčenkienė S, Povinec PP, Gera M (2012) Radionuclides from the Fukushima accident in the air over Lithuania: measurement and modelling approaches. Environ Radioact 114:71–80. https://doi.org/10.1016/j.jenvrad.2011.12.004
8. Hayakawa Y (2011) Radiation contour map of the Fukushima Daiichi accident. http://blog-imgs-51-origin.fc2.com/k/i/p/kipuka/0810A.jpg (not available now)
9. Katata G, Chino M (2017) Source term, atmospheric dispersion, and deposition of radionuclides during the Fukushima Daiichi Nuclear Power Station accident. Earozoru Kenkyu 32:237–243. https://doi.org/10.11203/jar.32.237
10. Paatero J, Hämeri K, Jaakkola T, Jantunen M, Koivukoski J et al (2010) Airborne and deposited radioactivity from the Chernobyl accident - a review of investigations in Finland. Boreal Environ Res 15:19–33
11. JAEA (Japan Atomic Energy Agency) (2011) Trial calculation of the atmospheric Cs137 deposition associated with the Fukushima Daiichi Power Plant accident - simulation using World Version of SPEEDI (WSPEEDI) [in Japanese]. https://nsec.jaea.go.jp/fukushima/data/20110906.pdf
12. DOE/NNSA (U.S. Department of Energy/National Nuclear Security Administration) (2011) US DOE/NNSA Response to 2011 Fukushima Incident - data and documentation. https://www.

energy.gov/downloads/us-doennsa-response-2011-fukushima-incident-data-and-documentation (The original version released in 2011 has been reorganized and reuploaded.)
13. NSR (Nuclear Regulation Authority, Japan) (2019) Measurement results of atmospheric dust by NRA and Fukushima Prefecture [in Japanese]. https://radioactivity.nsr.go.jp/ja/list/222/list-1.html
14. Haba H, Kanaya J, Mukai H, Kambara T, Kase M (2012) One-year monitoring of airborne radionuclides in Wako, Japan, after the Fukushima Daiichi nuclear power plant accident in 2011. Geochem J 46:271–278
15. Nakagawa Y, Suzuki T, Kinjo Y, Miyazaki N, Sekiguchi M et al (2011) Measurement of the ambient radioactivity and estimation of human radiation exposure dose in Tokyo with regard to the radioactive substance leakage due to the Fukushima Nuclear Power Plant Accident. Radioisotopes 60:467–472. [in Japanese with English Abstract]
16. Furuta S, Sumiya S, Watanabe H, Nakano M, Imaizumi K et al (2011) Results of the environmental radiation monitoring following the accident at the Fukushima Daiichi Nuclear Power Plant - Interim report (Ambient radiation dose rate, radioactivity concentration in the air and radioactivity concentration in the fallout). JAEA Rev 35:1–89. [in Japanese with English Abstract]. https://jopss.jaea.go.jp/pdfdata/JAEA-Review-2011-035.pdf
17. CPDNP (Center for the Promotion of Disarmament and Non-Proliferation) (2011) Radionuclide detection status at CTBT radionuclide detection observatory installed in Takasaki. https://www.cpdnp.jp/pdf/110729Takasaki_report_Jul26.pdf
18. MEXT (2011) Measurement results of monitoring the concentration of radioactive materials released into the atmosphere by MOD aircraft. https://radioactivity.nsr.go.jp/ja/list/357/list-1.html
19. NIRS (National Institute of Radiological Science) (1963–2010). Radioactivity Survey Data in Japan, No. 1–No. 145. https://kankyo-hoshano.go.jp/07/07.html
20. Sakaguchi Y, Sakama M, Fushimi K, Nakayama S (2011) Measurement of airborne radioactivity from the Fukushima reactor accident in Tokushima, Japan. Nat Sci Res, Tokushima University 25:39–45. [in Japanese with English Abstract]
21. Momoshima N, Sugihara S, Ichikawa R, Yokoyama H (2012) Atmospheric radionuclides transported to Fukuoka. Japan remote from the Fukushima Dai-ichi nuclear power complex following the nuclear accident. Environ Radioact 111:28–32. https://doi.org/10.1016/j.jenvrad.2011.09.001
22. Zhang Z, Ninomiya K, Takahashi N, Shinohara A (2015) Daily variation of I-131, Cs-134 and Cs-137 activity concentrations in the atmosphere in Osaka during the early phase after the FDNPP accident. Radioanal Nucl Chem 303:1527–1531. https://doi.org/10.1007/s10967-014-3752-3
23. NILU (Norwegian Institute for Air Research) (2011) Results of simulation for atmospheric transportation of ^{131}I in the early after of the FDNPP accident. http://transport.nilu.no/products/fukushima (not available now)
24. Draxler RR, Hess GD (2004) NOAA (National Oceanic and Atmospheric Administration). Description of HYSPLIT (Hybrid Single-Particle Lagrangian Integrated Trajectory) 4 Modeling System. NOAA Tech Memo. ERL ARL-224
25. TEPCO (2011) Core condition of the Fukushima Daiichi Nuclear Power Station Units 1 to 3 reactors [in Japanese]. http://www.tepco.co.jp/nu/fukushima_np/images/handouts_111130_09-j.pdf
26. TEPCO (2013) Progress and lessons learned from the Fukushima Daiichi Nuclear Power Plant accident [in Japanese]. http://www.tepco.co.jp/nu/fukushima-np/outline/
27. TEPCO (2012) Estimation of the amount of radioactive material released into the atmosphere at the Fukushima Daiichi Nuclear Power Plant accident [in Japanese]. http://www.tepco.co.jp/cc/press/betu12_j/images/120524j0105.pdf
28. TEPCO (2011) Fukushima Daiichi Nuclear Power Station: fixed-point measurement status by temporary monitoring post [in Japanese]. http://www.tepco.co.jp/decommission/data/monitoring/monitoring_post/index-j.html

Chapter 3
Spatiotemporal Distribution of Radionuclides in Soil in the Tokyo Metropolitan Area

Abstract In the first stage after the FDNPP accident, radioactive contamination on the ground surface in the Tokyo metropolitan area was controlled by weather conditions such as wind direction and rainfall. Many radionuclides in the radioactive plume released from the FDNPP reactors were advected in the atmosphere without falling to the surface. However, some of them precipitated to the ground surface, causing environmental radiation contamination in Tokyo. These radionuclides diffused, which is believed to have caused temporary but significant external and internal exposure to the Japanese people. However, given the destruction of infrastructure due to the massive earthquake, there are almost no measurement data regarding environmental radioactive contamination immediately after the FDNPP nuclear accident. The collapse of infrastructure occurred not only in Fukushima Prefecture where the FDNPP operated but also in the capital Tokyo more than 200 km from the FDNPP. However, the status of environmental radioactive contamination immediately after the accident can be estimated by measuring the contamination by radioactive material recorded in the geosphere or hydrosphere. In this chapter, the geographic distribution of radioactive contamination of soil in the Tokyo metropolitan area during the initial phase of the FDNPP accident is discussed. The results of field sample measurements show that severe environmental radioactive contamination continued for approximately 5 years after the accident. However, after more than 9 years had passed, the radioactive contamination seemed to have changed as radionuclides migrated and were redistributed from the initial contamination situation.

Keywords Tokyo metropolitan area · Radioactive contamination · Soil · Inventory · Spatiotemporal distribution

3.1 Introduction

Immediately after the FDNPP accident, radioactive contamination in the metropolitan area was presumed to have occurred according to the following process. In the first stage, as described in Chap. 2, radioactive contamination on the ground surface

The original version of this chapter was revised (plus a similar explanatory text of the problem as in correction followed by). The correction to this chapter can be found at https://doi.org/10.1007/978-981-15-7368-2_10

H. Yamazaki, *Radioactive Contamination of the Tokyo Metropolitan Area*,
https://doi.org/10.1007/978-981-15-7368-2_3

was controlled by weather conditions such as wind direction and rainfall. Many radionuclides in the radioactive plume released from the FDNPP reactors were advected in the atmosphere without falling to the surface. However, some of them precipitated to the ground surface, causing environmental radiation contamination in Japan. These radionuclides diffused and caused global radioactive contamination, which is believed to have caused temporary but significant external and internal exposure to the Japanese people. However, given the destruction of infrastructure due to the massive earthquake, there are almost no measurement data regarding environmental radioactive contamination immediately after the FDNPP nuclear accident. The collapse of infrastructure occurred not only in Fukushima Prefecture where the FDNPP operated but also in the capital Tokyo more than 200 km from the FDNPP. However, the status of environmental radioactive contamination immediately after the accident can be estimated by measuring the contamination by radioactive material recorded in the geosphere or hydrosphere.

The first phase of environmental radioactive contamination from the Fukushima nuclear accident is considered as follows (References 25–27 in Chap. 2). Initially, hydrogen gas produced by the reaction between the nuclear fuel rod zirconium and the cooling water leaked from a crack in the reactor pressure vessel and accumulated in the reactor building; later, the hydrogen gas exploded, and the reactor building was destroyed. The fission products leaked from the Unit 1 and 3 reactors were released to the environment by the hydrogen explosion in each building. The Unit 4 reactor was not in operation, but the building was destroyed by the hydrogen gas explosion from the Unit 3 reactor. The spent nuclear fuel storage pool was also destroyed, but the nuclear fuel rods were mostly unbroken.

On the other hand, in the Unit 2 reactor, dry venting to prevent the destruction of the reactor pressure vessel was successful, but at this time, massive amounts of fission nuclides that accumulated in this meltdown reactor were released into the atmosphere. Most of these radionuclides are believed to have been released from the exhaust stack of the Unit 2 reactor. Next, these nuclides released from the FDNPP formed radioactive plumes that migrated and diffused into the atmosphere over Japan. At this time, estimates indicate that the people living in the places through which the plume passed received high internal and external exposures to the significant quantity of volatile radionuclides such as radioactive iodine and radioactive cesium, but the actual situation has scarcely been elucidated yet. Third, rainfall caused radionuclides in the plume to precipitate onto the ground, and most radionuclides were adsorbed and fixed to clay minerals in the soil. Alkali earth elements such as radioactive cesium are naturally dissolved in water as soluble ionic species, and radioactive cesium precipitated on concrete or asphalt in urban areas is considered to have been transferred to the ecosystem through rivers as ionic chemical species or to sludge in sewage treatment systems (Chap. 5). In the case of the forests, most radioactive nuclides were retained in litter and plants (Chap. 7). Radionuclides deposited in the hydrosphere were introduced into ecosystems via aquatic organisms and fish (Chap. 8). Radioactive materials released in the FDNPP accident that polluted Japan's environment moved into the hydrosphere such as lakes and oceans, due to rainfall and groundwater. In the Tokyo metropolitan area, Tokyo Bay played an essential role as the final sink for radioactive materials (Chap. 6).

In this chapter, the geographic distribution of environmental radioactive contamination in the Tokyo metropolitan area during the initial phase of the FDNPP accident is discussed. The results of field sample measurements show that severe environmental radioactive contamination continued for approximately 5 years after the accident. However, after more than 9 years had passed, the radioactive contamination seemed to have changed as radionuclides migrated and were redistributed from the initial contamination situation.

3.2 Sample and Radioactivity Measurements

3.2.1 Radioactivity Measurements

The radioactivity of gamma-emitting nuclides in soil, water, aquatic sediments, atmospheric dust, animals, plants, fish, and shellfish described in this book was quantified using the Ge semiconductor gamma-ray measurement system described below.

The activity intensities of radionuclides in the samples were quantified by connecting a 4096 multichannel pulse height analyzer (Labo Equipment, MCA600) to a low-energy HPGe detector (ORTEC, LO-AX/30P) sheathed in 10 cm-thick lead. The specimen was sealed inside a plastic container with a diameter of 5.5 cm and a depth of 2.0 cm in preparation for measurement by gamma-ray spectrometry. The relative geometric efficiency of the Ge detector with respect to the sample volume was calculated using the American NIST Environmental Radioactivity Standards, SRM 4350B (river sediment) and SRM 4354 (freshwater lake sediment). Moreover, the counting efficiency was corrected within the range of sample weight from 2 to 30 g [1]. For samples with a measured weight of less than 2 g, such as air dust samples on the filter, the detection efficiency curve obtained with the NIST SRM standards was extrapolated to correct the counting efficiency. The measurement time was set such that the counting error would be 5% or less, depending on the radioactive intensity of the sample.

In this study, the following nuclides were quantified: ^{95}Zr (γ-ray energy: 757 keV, half-life: 64.0 days), ^{95}Nb (766 keV, 35.0 days), ^{110m}Ag (885 keV, 250 days), ^{131}I (364 keV, 8.04 days), ^{134}Cs (605 keV, 2.06 years), ^{137}Cs (662 keV, 30.1 years), ^{210}Pb (46.5 keV, 22.3 years), and ^{235}U (186 keV, 7×10^8 years). A standard ^{134}Cs solution with known concentration was used to correct the sum peak effect for the ^{134}Cs determination. The detection limits for radionuclides under appropriate conditions were 2.0 Bq/kg for ^{131}I and 0.6 Bq/kg for ^{134}Cs and ^{137}Cs in dry soil samples. The detection limits of these nuclides were almost the same in wet samples, such as fish muscle. When the samples could not be analyzed immediately after sampling, they showed lower measurement precision for ^{131}I than for ^{134}Cs or ^{137}Cs because the activity of ^{131}I decreased by radioactive decay.

The radioactivity was calculated based on a value for the day the sample was obtained. As described in Chap. 1, the Unit 2 reactor melted down late at night on

March 15, 2011, causing the discharge of the most significant quantity of radionuclides during the FDNPP accident [2–6]. Therefore, the radioactivity measured in this study was evaluated based on a value obtained by the radioactive decay corrected to March 16, 2011. Because its half-life is short (8.04 days), ^{131}I was not detected from samples collected after late June 2011.

To evaluate the reliability of the author's radioactivity measurements, they were compared with the quantitative values of the same samples measured at other institutions. Here, in addition to the soil samples, the results of cross-checking are shown for the fish sample discussed in Chap. 8. The soil samples were thoroughly air-dried in an oven at 60 °C after sample collection. To avoid secondary contamination, sieving was not performed, and plant pieces and sand grains with diameters greater than 1 mm were removed mainly with tweezers. To homogenize the sample used for the cross-check, the quartering method was repeated five times and divided into two parts; one part was measured by another institution and the other part by the author's measuring instrument. For cross-checking, the fish samples were frozen after collection. The muscle samples were half-thawed in the laboratory, and muscles were collected on both sides of the spine. One part was measured at another institution, and the other part was measured in the author's laboratory. The fish skin was removed from the muscle to avoid external contamination. Table 3.1 shows the results of cross-checking for these samples. In the sample with low radioactive contamination, the quantification

Table 3.1 Reliability of the radioactivity measurements by gamma-ray spectrometry

Sample	Facility	Decay corrected activity[a], Bq/kg			
		^{134}Cs	^{137}Cs	$^{134+137}$Cs	^{134}Cs/^{137}Cs
Soil 1[b]	Author (A)	13,900 ± 82.2	14,000 ± 84.6	13,950 ± 118	0.993 ± 0.008
	Other (O)	13,500 ± 106	13,400 ± 110	13,450 ± 153	1.007 ± 0.011
	A/O ratio	1.030 ± 0.010	1.045 ± 0.010	1.038 ± 0.014	1.000 ± 0.014
Soil 2[b]	Author (A)	21,500 ± 161	21,800 ± 174	21,650 ± 237	0.986 ± 0.011
	Other (O)	22,400 ± 209	22,200 ± 216	22,300 ± 301	1.009 ± 0.013
	A/O ratio	0.960 ± 0.012	0.982 ± 0.013	0.971 ± 0.018	0.998 ± 0.017
Soil 3[c]	Author (A)	900 ± 17.6	896 ± 18.8	898 ± 25.8	1.004 ± 0.029
	Other (O)	915 ± 20.3	940 ± 22.7	928 ± 30.5	0.973 ± 0.033
	A/O ratio	0.984 ± 0.030	0.953 ± 0.032	0.969 ± 0.044	0.989 ± 0.044
Fish 1[d]	Author (A)	1130 ± 23.1	1110 ± 22.5	1120 ± 32.2	1.018 ± 0.029
	Other (O)	1110 ± 41.2	1070 ± 40.7	1090 ± 57.9	1.037 ± 0.053
	A/O ratio	1.018 ± 0.042	1.037 ± 0.043	1.028 ± 0.060	1.028 ± 0.060
Fish 2[d]	Author (A)	465 ± 13.3	455 ± 11.4	460 ± 17.5	1.022 ± 0.038
	Other (O)	473 ± 18.2	456 ± 17.2	465 ± 25.0	1.037 ± 0.054
	A/O ratio	0.983 ± 0.048	0.998 ± 0.045	0.991 ± 0.066	1.030 ± 0.066

Radioactivities were measured by the method shown in Sect. 3.2. Soil samples bisected after being sufficiently homogenized by quartering method. Fish samples were cut muscles on both sides of the spine, and the frozen muscles sent one side to the cross-checking institutions

[a]Decay-corrected value to March 16, 2011

[b]Collected in Fukushima Prefecture in April 2011

[c]Collected in Tochigi Prefecture in June 2011

[d]Caught in offshore Fukushima Prefecture in December 2011

error is slightly larger, less than approximately ±5%. However, the author's values and other institutions' values are in good agreement within the range of counting error and sample heterogeneity, and the author concludes that the radioactivity values of the environmental samples discussed in this book are sufficiently reliable.

3.2.2 Soil Sampling

Most soil samples were collected from private sites. The owners approved these sample collections. In the case of collection at public places, sampling was carried out with the permission of the administrator when necessary.

For this study, soil samples were collected in the Tokyo metropolitan area and the Kanto district during the 5 years following the FDNPP accident. Figure 3.1 shows the sampling sites. The sites were selected in the Tokyo metropolitan area and the Kanto district approximately 200 km from the FDNPP. Soil sampling was performed by the established IAEA method [7]. In the case of buildings and roads interspersed throughout urban regions, this method could not be applied to radioactive contamination urban areas. In general, the soil samples were obtained from the ground surface along roads where they were not subject to physical perturbation. The sampling sites were located on the flat ground surface by the roadsides. In many cases, they were hardly affected by the flow of rainwater and not covered with plants.

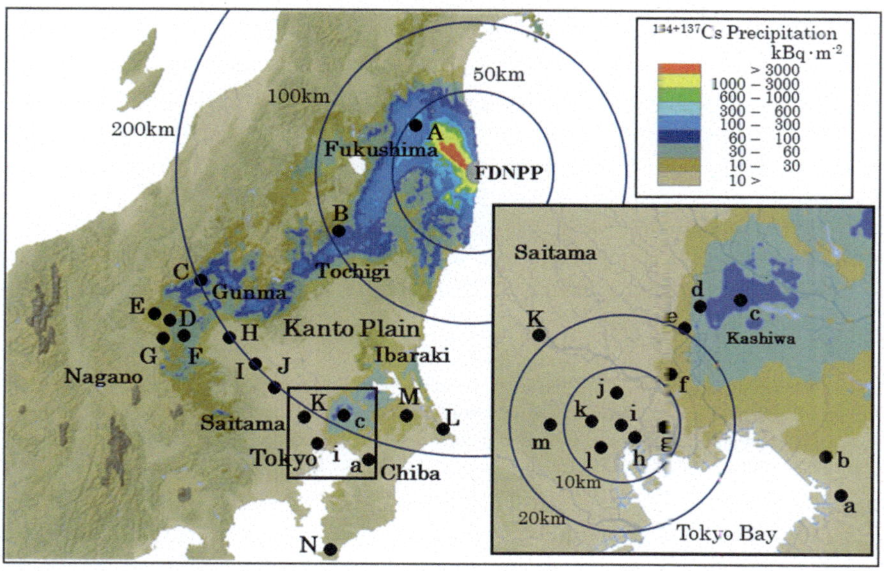

Fig. 3.1 Sampling sites of the soil sample in the Tokyo metropolitan area and Fukushima Prefecture. Geographic distribution of the $^{134+137}$Cs precipitation was cited from the MEXT airborne monitoring map published on November 11, 2011 ([14] a)

Photo 3.1 The sampling Site "i" in the Imperial Palace Outer Garden. A person is sunbathing on the lawn. Probably, he doesn't know that this place is radioactively contaminated (right side in the photograph above)

For example, the soil at Site "i" in Photo 3.1 was sampled in an area of 0.5 m × 1 m and obtained to a depth of 20 cm with the core sample, varying the sampling position slightly at each sampling period in this area. After sampling, the holes were refilled. There was grass, vegetation, etc., at this sampling location, but the soil was exposed at the ground surface. The ground was slightly sloped, so rainwater was assumed to flow gently over the ground surface. Pedestrians could walk on the sampling point. A surface soil up to 1 cm deep below the surface was measured using a ruler and taken from a 10 cm × 10 cm area. The core sediment was collected to a depth of 20 cm using a core sampler with an inner diameter of 5 cm and sliced to an appropriate thickness to prepare an analytical sample. The soil samples were air-dried at 60 °C, and impurities such as vegetation detritus and rocks with diameters of 1 mm or more were removed before their radioactivity was measured.

The inventory of radionuclides in the ground surface was calculated, assuming a soil density of 1.3 g/cm^3 in compliance with the method of the MEXT of Japan [8] because the accurate evaluation of soil density is difficult. The inventories of the radioactively contaminated soil had been estimated in many previous reports using this value. The author measured the density of several soils, and the values were approximately 1.0–1.2 g/cm^3. This difference occurred because soil often contains organic substances such as humus and plant material. If the value of 1.3 g/cm^3 given by MEXT is used for inventory calculation, the soil density is slightly overestimated, but the use of this value is considered safe when radiation exposure from the soil to residents occurs

3.3 Geographic Distribution of Radioactive Cesium and Radiation Dose in Central Japan by MEXT Airborne Monitoring and Typical Gamma-Ray Spectra of Contaminated Soil

To discuss the radioactive contamination of the Tokyo metropolitan area caused by the FDNPP accident, it is necessary to understand the radioactive contamination of the soil by the contaminated radioactive plume released from the FDNPP reactors.

The spread of radioactive plumes in Japan is described in Chap. 2. During the advection-diffusion process of contaminated radioactive plumes, radioactive nuclides precipitated from the plume to the ground surface, which caused radioactive contamination of the soil. Regarding soil contamination in Japan due to the FDNPP accident, monitoring surveys using aircraft were carried out by the MEXT (Ministry of Education, Culture, Sports, Science, and Technology) immediately after the accident. In these monitoring surveys, the air doses 1 m from the surface and the precipitation amounts of ^{134}Cs and ^{137}Cs were estimated. The earliest publicly available air dose measurements were obtained by airborne survey monitoring on March 25, 2011, at a distance of more than 30 km from the FDNPP [9]. Atmospheric ^{131}I and ^{137}Cs were measured by MOD (Ministry of Defense, Japan) aircraft on March 24, 2011, and the results were announced on March 25 [10].

In addition, the monitoring survey was conducted as a joint project between Japan and the United States [11, 12]. The US government's response to the FDNPP accident began with air dose measurements at the Yokota Base in Tokyo and the Embassy in Japan, as shown in Fig. 2.2. However, in Fukushima Prefecture around the FDNPP, before the start of joint airborne monitoring between Japan and the United States, the measurements of air filters (March 18 58 km northwest of FDNPP, 4:15. March 19, 19 km north, 14:59) and soils (March 20, 66 km northwest of FDNPP. March 26, 19 km south) had begun [13]. The DOE (U.S. Department of Energy) participated in the airborne monitoring survey and data analysis from April 6 to July 2, 2011. Immediately after the accident, the monitored area was within 60 km of the FDNPP by DOE and 60–80 km by MEXT.

Over time, the areas of airborne monitoring surveys successively expanded. The radioactive contamination in Tokyo was monitored by an airborne monitoring survey from September 14–18, 2011. A radiation pollution map of the entire eastern part of Japan, including the Tokyo metropolitan area, obtained by this airborne monitoring was published by MEXT on November 11, 2011. Figure 3.2 shows a map of radioactive contamination in eastern Japan ([14] a). The air dose map at 1 m from the ground surface estimated simultaneously with the measurement of radioactive cesium deposition by the MEXT airborne monitoring is shown on the left in Fig. 3.3. Since the survey period varies from region to region, the estimated values of radioactive cesium deposition and air dose shown here have been corrected to the values for October 13, 2011 ([14] a). The right side of Fig 3.3 is the result of a survey in 2018, which is the corrected value for November 15, 2018 ([14] b).

Fig. 3.2 Geographic distribution of the $^{134+137}$Cs precipitation in eastern Japan. This monitoring survey was carried out by the MEXT from April to October 2011, and the results were published in November 2011 ([14] a)

Since the FDNPP accident, an airborne monitoring survey of radioactive contamination on the surface of eastern Japan has been carried out by the NRC (Japanese Nuclear Regulatory Commission), which carried over from the work by

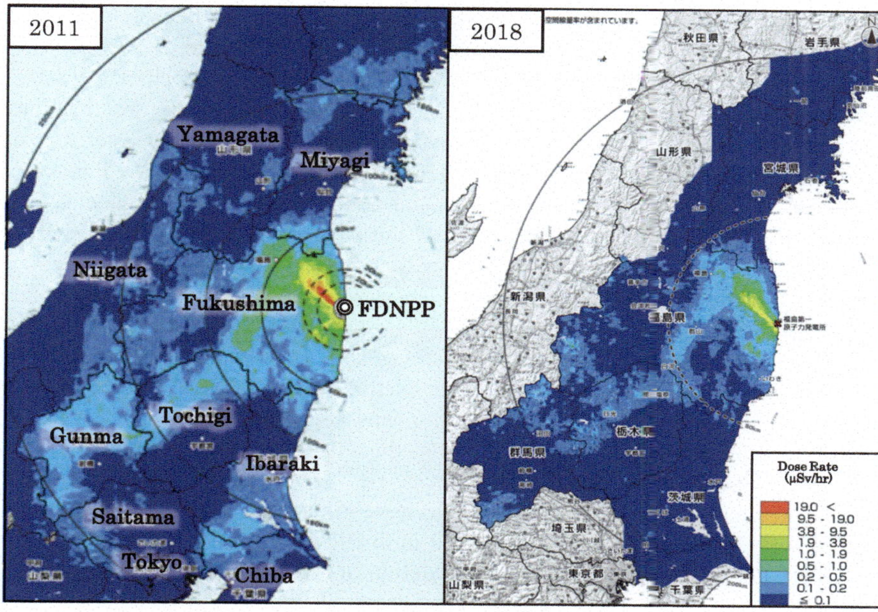

Fig. 3.3 Geographic distributions of the estimated air doses 1 m above the ground surface in eastern Japan. These figures show the results measured by the MEXT airborne monitoring in 2011 ([14] a) and 2018 ([14] b)

the MEXT. However, airborne monitoring surveys have not been conducted since 2012 in Tokyo, Saitama, Chiba, and Kanagawa prefectures, although the population is most concentrated and radioactive contamination is still high in these areas. The response of the Japanese government, which has not continued to monitor the actual state of radioactive contamination in the central area of Japan, is not appropriate from the viewpoint of assessing radiation exposure for residents.

As evident from the geographic distribution of radioactive cesium precipitation and air dose, the contaminated radioactive plumes released from the FDNPP reactors exposed almost all areas within 60 km of the reactor to severe radioactive contamination. In an area approximately 50 km long and 20 km wide from the periphery of the reactor, the precipitation amount of $^{134+137}Cs$ was greater than 1000 kBq/m^2, and the air dose exceeded 9.5 μSv/h. The annual dose in the area shown in red on the left in Fig. 3.3 exceeded 166 mSv. The radioactive plume that caused radioactive contamination in the Tokyo metropolitan area was discharged from the reactor and then moved northwest, causing radioactive contamination of the ground surface. This plume changed direction near the border between Fukushima and Miyagi prefectures and moved southwest along the mountainous area that stretches to the central part of eastern Japan. As a result, large parts of Tochigi and Gunma prefectures were also radioactively contaminated. The radioactive contamination of Tokyo was caused by plumes migrating south from the FDNPP reactors. This radioactive plume also caused radioactive contamination in the densely populated areas of southern Ibaraki Prefecture and northern Chiba Prefecture.

The right side of Fig. 3.3 is the newest airborne monitoring chart available. Although seven and a half years had passed since the accident, the figure illustrates that radioactive contamination continued even then in a wide range of eastern Japan when this map was made. Typical radionuclides released from the FDNPP reactors are ^{131}I, ^{134}Cs, and ^{137}Cs, but ^{131}I disappeared entirely due to radiative decay. The amount of ^{134}Cs decreased to 10% of the initial level, and ^{137}Cs decreased to approximately 85%. Although nuclides with short half-lives have disappeared, long-lived nuclides such as ^{137}Cs remain, so radioactive contamination in eastern Japan is thought to have continued for a while, as shown by this map.

Figure 3.4 presents the gamma-ray spectra of soils in Fukushima Prefecture and Tokyo contaminated by radioactive plumes released from the FDNPP reactors. The earliest soil sample obtained by the author was collected from a schoolyard in Fukushima City approximately 60 km from the FDNPP 8 days after the FDNPP accident and measured on March 22. The peaks of ^{132}Te with a half-life of 3.25 days and its daughter nuclide ^{132}I (half-life: 2.3 h) were still clearly detected. ^{136}Cs, with a half-life of 13.2 days, which is generated in small amounts by fission, was also detected; ^{136}Cs is produced by the (n, p) reaction of ^{138}Ba and the (n, α) reaction of ^{139}La. Although ^{140}Ba with a gamma-ray energy of 30.0 keV was detected, this nuclide is a fission product with a half-life of 12.8 days. At the same site in the schoolyard shown in Fig. 3.4a where decontamination work had not yet been performed, ^{132}Te and ^{132}I were not detected in the soil collected on April 27. Nevertheless, a small peak of ^{136}Cs is still visible in Fig. 3.4c.

Figure 3.4b is a gamma-ray spectrum of the washing sludge from a rescue helicopter collected 11 days after the accident. The flight history of the helicopter is unknown, but the radioactivity intensity is much higher than that of the soil in Fig. 3.4a. The helicopter may have flown over the FDNPP reactors or into a radioactive plume. In other words, this helicopter was highly radioactively contaminated, suggesting that radionuclides were still floating at extremely high concentrations around the DFNPP at this time. Furthermore, these measurements also indicate that radioactive plumes were continuously released from the collapsed FDNPP reactors.

The first soil sample in Tokyo was collected on April 10, 2011. The gamma-ray spectrum of the surface soil from the Imperial Palace Outer Garden in Tokyo is shown in Fig. 3.4d. Short-lived nuclides ^{132}Te and ^{136}Cs were detected even though approximately 1 month had passed since the accident. However, the significant radionuclides contaminating Tokyo soil were ^{131}I, ^{134}Cs, and ^{137}Cs. As shown in Fig. 2.3c, these nuclides carried from the FDNPP by the radioactive plumes that were transported on March 15–16 and March 21–22 were precipitated to the surface by the rain that fell at the time. Thus, the soil in Tokyo is estimated to have been radioactively contaminated because of the washing processes induced by rain. The temporal changes in radioactive cesium intensity in the soil at this site are shown in Fig. 3.6.

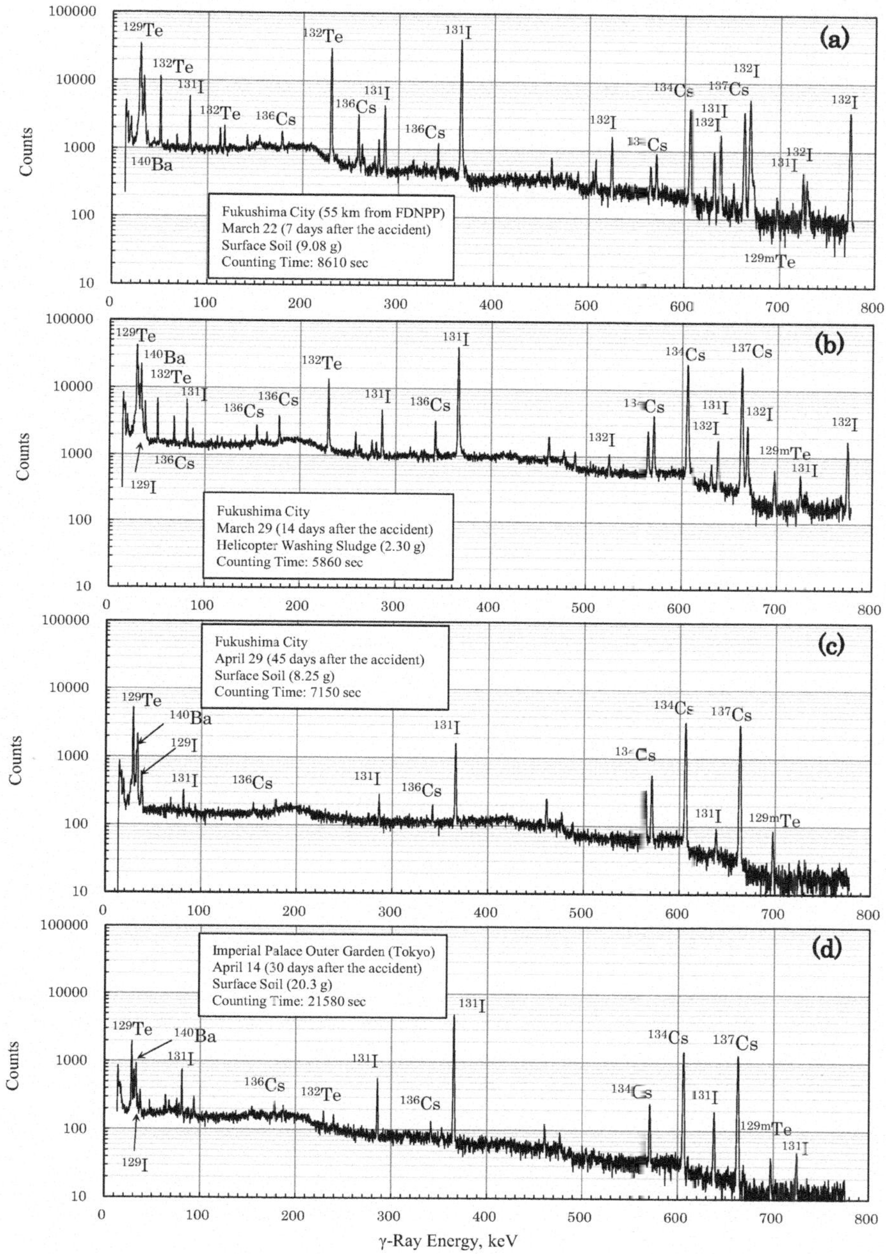

Fig. 3.4 Gamma-ray spectra of the surface soil in Fukushima City and central Tokyo immediately after the FDNPP accident. (**a**) Surface soil in the school garden collected 4 days after the accident. One week has passed before measurement, but the short half-lives fission product ^{132}Te ($T_{1/2}$ = 78 h) and its daughter ^{132}I ($T_{1/2}$ = 2.3 h) found. Neutron capture nuclide ^{136}Cs ($T_{1/2}$ = 13.2 days) with was also detected. (**b**) Collected from the rescue helicopter washing sludge. Probably the helicopter was flying over the FDNPP. The detection of short half-lives nuclides

3.4 Uniformity of Local Distribution of Radioactive Cesium Precipitated on the Ground Surface

In airborne monitoring, the aircraft flew with a trajectory width of approximately 3 km and an altitude of 150–300 m [14]. Therefore, the measured values in the radioactive contamination map of eastern Japan estimated by the MEXT and the NRA (Nuclear Regulation Authority, Japan) must be considered to indicate the average value in a region with a diameter of 300–600 m. In other words, this airborne monitoring has a large blind spot for the monitoring area. Such airborne monitoring is useful for analyzing the distribution and behavior of radioactive contamination in a broad area such as forests and cultivated land. However, cannot be considered suitable for analyzing complex terrains such as urban areas and residential areas. Immediately after the FDNPP accident, simple radiation surveys were performed by citizens in various parts of Japan. As a result, local highly contaminated radioactive areas called hot spots are found in various places throughout Japan. Such hot spots are challenging to detect by examining the local radioactive contamination distribution obtained by airborne monitoring. Therefore, the heterogeneity of radioactive cesium precipitated on the ground surface of Tochigi Prefecture through which the radioactive contamination plume passed is evaluated.

The monitoring site is an uncultivated paddy field at an elevation of 437 m in Nasu-Shiobara City, 110 km west-southwest of the FDNPP. On the north and east sides not far from the monitoring field are mountains. To the west and south are broad plains. Around the soil sampling area are paddy fields and pastures with houses scattered between them. Therefore, the wind is considered unlikely to have been perturbed by the buildings and trees as radioactive cesium precipitated from the plume to the ground surface. If radioactive cesium in the plume is deposited on the surface as soluble cations or their compounds, it is immediately absorbed by clay minerals in the soil. Therefore, if the soil contaminated with radioactive cesium is not physically mixed or moved, the distribution of radioactive cesium in the soil reflects the cesium distribution in the plume.

Figure 3.5 shows the sampling points at the monitoring site. Core samples with a depth of 5 cm were collected from five locations in the middle of an uncultivated paddy field of approximately 1000 m^2. According to the MEXT airborne monitoring, the inventory of $^{134+137}$Cs at this site is estimated to be 100–300 kBq/m^2. Table 3.2 shows the analytical results for ^{134}Cs and ^{137}Cs in the soil collected from each point. The average radioactivity intensity at five points is 1480 ± 391 Bq/kg for ^{134}Cs and 1600 ± 457 Bq/kg for ^{137}Cs. The inventory of $^{134+137}$Cs corrected for decay on October 13, 2011, calculated from this value, is was 189 ± 52 kBq/m^2, which agrees well with the value estimated by the airborne

Fig. 3.4 (continued) means that radionuclides continued to release from the reactors. (**c**) Collected at the same as site (**a**) approximately 6 weeks after the accident. (**d**) Collected at central Tokyo. High activities of ^{131}I were still detected. The date is the measurement date

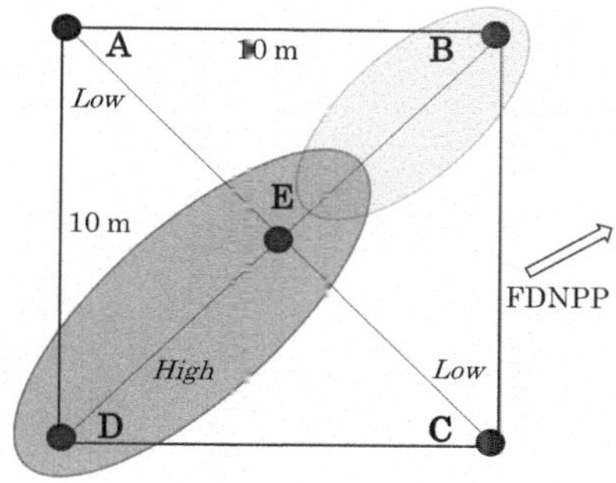

Fig. 3.5 Uniformity of the distribution of radioactive cesium precipitated from the plume. The observed site is the uncultivated paddy field in Nasushiobara City, Tochigi Prefecture. There is 110 km southwest of the FDNPP, and an elevation is 497 m

monitoring. However, the relative standard deviation of the $^{134-137}$Cs activity at these five sites is ±27.6%, which is much larger than the Ge detector counting error of 0.7–1.2% for this soil sample. The distribution of radioactive cesium at Points B, E, and D along the advection course of the radioactive plume is more significant than the average, and it is smaller at Points A and C. In particular, the radioactive cesium intensity at Point A is 40% smaller than the average value. Most likely, the radioactive plume released from FDNPP migrated in the direction from Point B to Point D. There is no clear correlation between the distribution of radioactive cesium and the soil particle size. Such a sizeable non-uniform distribution of radioactive cesium in a narrow area of not more than 10 m^2 suggests non-uniformity in the radioactive plume from which this radioactive cesium was transported. Nasushiobara City is more than 100 km from the FDNPP. However, it is highly likely that radioactive cesium in the plume advancing to this monitoring site retained the non-uniformity distribution released from the reactors.

The formation of hot spots in the radioactively contaminated zone must also take into account the process of radionuclide migration and enrichment after precipitation. Nevertheless, it is unlikely that the radionuclides adsorbed on the soil immediately after contamination can quickly move and be redistributed elsewhere. If the heterogeneity in the advection process of the radioactive plume is preserved in the distribution of radionuclides precipitated on the ground surface, it may be necessary to reconsider the survey method for assessing surface radioactive contamination. The secondary redistribution of radioactive contamination can be predicted by the physical and chemical properties of radionuclides and environmental conditions. Nevertheless, if a radionuclide precipitated from the plume and its distribution was non-uniform, then a sizeable contaminated area needs to be monitored with excellent point spacing. In the FDNPP accident, such a monitoring survey by the Japanese government could not be carried out.

Table 3.2 Uniformity of the distribution of radioactive cesium precipitated on the surface soil in Nasushiobara City, Tochigi Prefecture

Sampling date	Core depth, cm	Sampling point[a]	Average particle size, μm	^{134}Cs Activity, Bq/kg	^{134}Cs Deviation from average, %	^{134}Cs RSD, %	^{137}Cs Activity, Bq/kg	^{137}Cs Deviation from average, %	^{137}Cs RSD, %	$^{134+137}Cs$ Activity, Bq/kg	$^{134+137}Cs$ RSD, %
2011/6/4	0–5	B	9.6	1610	9.2	26.4	1710	6.9	28.6	3320	27.5
		E (Center)	12.2	1740	18.0		1950	21.9		3690	
		D	14.7	1830	24.1		2020	26.3		3850	
		C	8.0	1330	−9.8		1450	−9.4		2780	
		A	9.3	863	−41.5		891	−44.3		1750	
		Average	10.8 ± 2.7	1480 ± 391			1600 ± 457			3080 ± 847	
		Average inventory[b], kBq/m^2		96.2 ± 25.4			104 ± 29.7			200 ± 55.1	

[a]Sampling sites are shown in Fig. 3.5

[b]According to the MEXT airborne monitoring shown in Fig. 3.2 ([14] a), the $^{134+137}Cs$ precipitation at this area 100–300 kBq/m^2

3.5 Vertical Distribution of Radionuclides Precipitated in the Soil

Many papers have been published on the adsorption of radionuclides on soil particles and clay minerals (References 19–30 in Chap. 1) The radionuclides are fixed in the soil and then move in the depositional layer because the clay particles adsorb the radionuclides released to the environment. However, despite many studies on the vertical mixing of aquatic sediments [15–30], research on the mixing of terrestrial soil has not progressed [31–35]. In the FDNPP accident, radioactive nuclides were precipitated onto the ground surface from the advecting radioactive plume in the atmosphere, causing radioactive contamination. To cope with such contamination, it is necessary to understand the adsorption behavior of radionuclides in soil and their migration and mixing processes. Information is also necessary regarding the decontamination of soil containing radionuclides. Furthermore, radioactive cesium deposited on the ground surface in a spike form may be an excellent tracer for elucidating the process of vertical mixing of terrestrial soil. Of course, it is also useful for evaluating the mixing parameters of aquatic sediments as shown in Figs. 2.4 and 6.9.

Chapter 6 discusses the transport and accumulation processes of radionuclides in Tokyo Bay and its inflowing rivers. In this case, it is also essential to understand the vertical mixing of aquatic sediments and the transport of contaminated soil particles by rivers. Table 3.3 shows the vertical distribution of radionuclides in the soil after the FDNPP accident at several points. Site D is a pre-cultivated field of Andosols rich in organic matter on a plateau at an elevation of approximately 1000 m in Gunma Prefecture. The amount of radioactive cesium precipitated is not very large. Since it is a cultivated field, the porosity of the soil is high, but most ^{131}I, ^{134}Cs, and ^{137}Cs was present in the 0–1 cm layer at the surface. For more than 40 days after the radionuclides were deposited here, their vertical distribution was not affected by rainfall infiltration or physical soil mixing. In the forest, approximately 50 m from the cultivated field, radionuclides were adsorbed by fallen leaves and litter that covered the soil surface, and no radioactive contamination of the underlying soil was observed. In this forest, radioactive iodine and radioactive cesium adsorbed on fallen leaves and litter did not easily migrate in the vertical direction. The cultivated land and forest are only 50 m apart, but the concentrations and isotope ratios of radionuclides differed significantly between these areas. This difference may suggest the heterogeneous deposition of these nuclides from the radioactive plume as described in Sect. 3.4. Most of the trees in this forest are deciduous trees, and no new leaves had yet sprouted in mid-March when radionuclides arrived. As discussed in Chap. 7, radionuclides from the radioactive plume that passed through the forest may have been effectively trapped in the bark. Since radioactive cesium adsorbed strongly to the bark, it was expected not to readily move down to the ground surface due to rainfall. However, rain was falling when this area was exposed to the radioactive plume. Some of the radionuclides precipitated on the bark may have been washed away by rainwater and accumulated in fallen leaves.

Table 3.3 Vertical distribution of radionuclides in the soil of the Tokyo metropolitan area

Site	Description	Sampling date	Depth, cm	Activity, Bq/kg						Inventory[a], kBq/m^2		
				^{131}I	^{134}Cs	^{137}Cs	^{131}I	^{134}Cs	^{137}Cs	^{131}I	^{134}Cs	^{137}Cs
				Detected			Decay corrected[b]			Decay corrected[b]		
D	Farmland	2011/4/29	0–1	36.7	176	184	1630	183	184	21.2	2.38	2.39
			1–3	nd	nd	4.5	–	–	4.5	–	–	0.12
			3–5	nd	nd	nd	–	–	–	–	–	–
			5–10	nd	nd	nd	–	–	–	–	–	–
	Forest	2011/4/29	0–5[c]	267	4380	4730	12,900	4570	4740	516[d]	183[d]	190[d]
			6–7[e]	nd	nd	nd	–	–	–	–	–	–
B	Garden	2011/6/13	0–1	nd	6210	6810	–	6920	6860	–	90.0	89.2
			1–5	nd	2800	3120	–	3120	3140	–	162	163
			5–10	nd	80	84.5	–	89.1	85.1	–	5.79	5.53
			10–15	nd	nd	nd	–	–	–	–	–	–
			15–20	nd	nd	nd	–	–	–	–	–	–
i	Park	2011/4/10	0–1	2150	944	961	18,500	966	962	241	12.6	12.5
			1–3	nd	nd	nd	–	–	–	–	–	–
		2011/4/28	0–1	1320	1130	1170	33,700	1180	1180	690	15.3	15.3
			1–3	33.3	33.2	41.9	1360	34.6	42.0	35.4	0.90	1.09
			3–5	nd	nd	14.0	–	–	1.4	–	–	0.04
			5–10	nd	nd	nd	–	–	–	–	–	–
		2011/8/13	0–1	nd	106	121	–	122	122	–	1.59	1.59
			1–3	nd	8.0	14.2	–	9.2	14.3	–	0.24	0.37
			3–5	nd	4.7	5.2	–	5.4	5.2	–	0.14	0.14
			5–10	nd	nd	nd	–	–	–	–	–	–
		2013/5/1	0–2	nd	165	343	–	342	361	–	8.89	9.39
			2–5	nd	33.2	64.1	–	68.7	67.4	–	1.79	1.75
			5–10	nd	nd	5.2	–	–	5.5	–	–	0.14
			10–15	nd	nd	nd	–	–	–	–		

		2013/12/31	0–2	nd	23.8	57.8	–	61.8	61.7	–	1.61	1.60
			2–5	nd	4.0	10.8	–	10.4	11.5	–	0.41	0.45
			5–10	nd	nd	nd	–	–	–	–	–	–
		2014/5/3	0–2	nd	45.7	123	–	133	133	–	3.46	3.46
			2–5	nd	4.0	12.3	–	11.7	13.2	–	0.46	0.52
			5–10	nd	2.1	6.3	–	6.1	6.8	–	0.40	0.44
			10–15	nd	nd	nd	–	–	–	–	–	–
		2014/8/20	0–2	nd	47.4	150	–	152	162	–	3.95	4.21
			2–5	nd	13.2	34.2	–	42.3	37.0	–	1.65	1.44
			5–10	nd	6.8	13.2	–	21.8	14.3	–	1.42	0.93
			10–15	nd	nd	nd	–	nd	nd	–	–	–
		2015/1/3	0–1	nd	215	704	–	787	769	–	10.2	10.0
			1–3	nd	95.2	325	–	349	355	–	9.07	9.23
			3–5	nd	10.7	34.0	–	39.2	37.1	–	1.02	0.97
			5–10	nd	5.3	16.8	–	19.4	18.4	–	1.26	1.20
			10–15	nd	nd	7.2	–	–	7.9	–	–	0.51
			15–20	nd	nd	nd	–	–	–	–	–	–
		2015/5/2	0–1	nd	97.3	354	–	399	390	–	5.19	5.07
			1–3	nd	19.0	68.8	–	77.8	75.7	–	2.02	1.97
			3–5	nd	9.9	38.8	–	40.6	42.7	–	1.06	1.11
			5–10	nd	11.9	46.2		48.8	50.9		3.17	3.31
			10–15	nd	3.5	15.4	–	14.3	17.0	–	0.93	1.11
			15–20	nd	nd	nd	–	–	–	–	–	–
		2016/1/3	0–1	nd	126	608	–	651	680	–	8.46	8.86
			1–3	nd	32.6	149	–	168	166	–	4.37	4.32
			3–5	nd	3.7	19.5	–	19.1	21.8	–	0.50	0.57
			5–10	nd	nd	8.6	–	–	9.6	–	–	0.62
			10–15	nd	nd	7.9	–	–	8.8	–	–	0.57
			15–20	nd	nd	nd	–	–	–	–	–	–

(continued)

Table 3.3 (continued)

Site	Description	Sampling date	Depth, cm	Activity, Bq/kg						Inventory[a], kBq/m^2		
				^{131}I	^{134}Cs	^{137}Cs	^{131}I	^{134}Cs	^{137}Cs	^{131}I	^{134}Cs	^{137}Cs
				Detected			Decay corrected[b]			Decay corrected[b]		
		2016/8/13	0–2	nd	46.3	252	–	295	286	–	7.67	7.44
			2–4	nd	4.2	21.7	–	26.8	24.6	–	0.70	0.64
			4–6	nd	nd	8.3	–	–	9.4	–	–	0.24
			6–10	nd	1.5	9.5	–	9.6	10.8	–	0.50	0.56
			10–15	nd	nd	5.1	–	–	5.8	–	–	0.38
			15–20	nd	nd	nd	–	–	–	–	–	–

[a]Calculated assuming a density of soil is 1300 kg/m^3
[b]Activities were decay corrected to March 16, 2011
[c]Litter and fallen leaves
[d]Calculated assuming a density of litter is 800 kg/m^3
[e]Soil

Site B is a hilly area in Tochigi Prefecture at an elevation of approximately 500 m. Surrounded by paddy fields and pastures are scattered private houses. The soil sampling site is in the yard of a private house where six people live, including older adults and children. The yard has a small flower bed and a place to hang laundry, as well as playground equipment for the children. Since sampling was conducted 3 months after the FDNPP accident, ^{131}I had already disappeared. However, because a small gamma-ray peak of 39.6 keV emitted from the fission product ^{129}I (half-life: 1.57×10^7 years) was observed in the spectrum, the yard was undoubtedly radioactively contaminated with radioactive iodine as well as radioactive cesium. High activities of radioactive cesium were observed in the surface soil at Site B. Unlike Site D, radioactive cesium was also detected in the 1–5 and 5–10 cm layers. When collecting the core samples of the soil, the core sampler was inserted into the ground from the soil surface where radioactive contamination was extreme, so that the upper radionuclides may have adhered to the inner wall of the sampler and been carried to the lower layer. Care must be taken because such a displacement phenomenon gives false results as if there were contamination in the deep layer. However, the inventory of radioactive cesium in the 5–10 cm layer at Site B is approximately 2.2% of the total inventory, much greater than that expected to be carried to the deeper layer by such displacement. Therefore, at Site B, radioactive cesium had penetrated the soil at least deeper than 5 cm. It was not possible to determine whether the process by which radioactive cesium was transported to the depths originated from the soil properties at this site or whether the soil was physically mixed by human activity.

Site “i” is the site discussed in Sect. 3.7. The core sediments were collected at the Imperial Palace Outer Garden in central Tokyo. This site is approximately 1.2 km east-northeast from the Japanese Parliament. Many people visit this park, making it difficult to collect samples. Visitors may walk on the sampling point. A feature of this location is that the amount of radioactive iodine precipitated was much more significant than that of radioactive cesium (see Fig. 4.2). The site was estimated to have been radioactively contaminated on March 21, 2011, but the first core sampling was on April 10. At this time, both radioactive iodine and radioactive cesium were present in the surface 0–1 cm layer on the ground. By the 28th, ^{131}I and ^{134}Cs had penetrated into the 1–3 cm layer and accounted for 5.1% and 5.9% of the total inventory, respectively. However, some of the ^{137}Cs penetrated below the 1–3 cm layer. Perhaps the ^{137}Cs in the deeper layer originated from global fallout. As shown in Fig. 3.6, at Site “i”, the radioactive cesium concentration differed each time sampling was performed. Human activity is thought to have greatly influenced the distribution of radioactive contamination on the ground surface of this site. The core collected in 2015 was only 50 cm from that collected in 2011, but the vertical distributions of the two are very dissimilar. It can be seen that by 2015, the radioactive cesium had diffused to the 10–15 cm layer, but in the case of the 2016 core, most of it remained in the layer shallower than 5 cm (see Table 3.3).

Radionuclides that polluted Japan’s land in the FDNPP accident were precipitated from the atmosphere to the ground surface from radioactive plumes. Then, the radionuclides were adsorbed and retained on the ground surface, but their

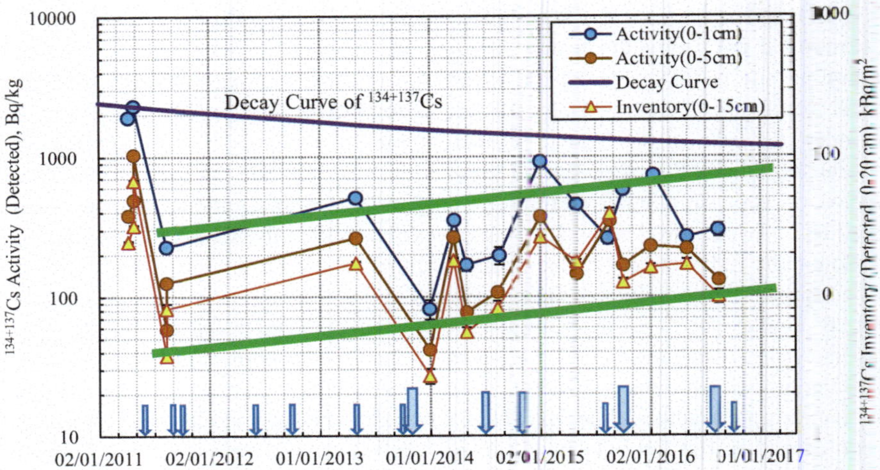

Fig. 3.6 Temporal changes in the concentration and inventory of $^{134+137}$Cs in the surface soil in the Imperial Palace Outer Garden. The radioactive decay curve of $^{134+137}$Cs is shown assuming that the activity of ^{134}Cs and ^{137}Cs are the same immediately after the accident. Vertical arrows showed the monthly rain precipitations. Small: over 200 mm. Medium: over 300 mm. Large: over 400 mm [43]

mechanism and behavior are very complicated. The vertical distributions of radionuclides in the soil and their changes significantly affect the behavior of radionuclides in the environment. Evaluating the vertical distribution of radionuclides in the soil is extremely important for management of radiation exposure and decontamination work for residents. If the inhomogeneity of the radioactive plume reflected the physical and chemical states of the radionuclides, its situation should have been recorded in the soil onto which they were precipitated. The soil itself has a heterogeneous composition and environmental form. Unfortunately, in Fukushima decontamination work has been carried out without such evaluation of the soil characteristics; thus, it is challenging to manage radiation after decontamination work. As of 2020, the contaminated soil recovered from the decontamination work has been stored outdoors without proper protection. This neglect leads to a very high risk of providing a new source of environmental radioactive recontamination.

3.6 Contamination by Radioactive Nuclides in the Surface Soil in the Tokyo Metropolitan Area

Table 3.4 shows the concentrations and inventories of ^{131}I and radioactive cesium in soil samples collected at the sites in the Tokyo metropolitan area shown in Fig. 3.1 within 3 months after the FDNPP accident. Since the sampling dates differ for each site, the detected activities were corrected by radioactive decay to the values for

Table 3.4 Typical analytical results of the concentration and inventory of radionuclides in the soil collected in the Tokyo metropolitan area and Fukushima Prefecture

Prefecture	Site	Description	Sampling date	Taken, cm	Decay corrected activity[a], Bq/kg				Ratio for ^{137}Cs		Inventory[c], kBq/m^2		
											This work[a]		MEXT[b]
					^{131}I	^{134}Cs	^{137}Cs	$^{134+137}$Cs	^{131}I	^{134}Cs	^{131}I	$^{134+137}$Cs	$^{134+137}$Cs
Fukushima	A	Schoolyard	2011/3/19	0–1	122,000	13,900	14,100	28,000	8.65	0.986	1590	363	300–600
			2011/4/27	0–1	142,000	21,400	21,900	43,300	6.48	0.977	1850	563	
Tochigi	B	Pasture[e]	2011/6/4	0–1	43,600	11,100	11,300	22,400	3.86	0.982	566	290	60–100
		Garden	2011/6/13	0–1	24,500	3020	3150	6170	7.78	0.959	318	80.2	
		Farmland	2011/6/13	0–1	91,800	6730	6810	13,500	13.5	0.988	1190	176	
Gunma	C	Paddy field	2011/4/30	0–1	5640	1830	1850	3680	3.05	0.989	73.3	47.9	30–60
	D	Forest	2011/3/26	0–1	582	67.4	70.7	138	8.23	0.953	7.6	1.8	10–30
		Farmland	2011/4/29	0–1	1630	183	184	367	8.85	0.996	21.2	4.8	
				1–3	nd	nd	4.5	4.5	–	–	–	0.1	
	E	Farmland[g]	2011/4/29	0–1	4360	1760	1720	3480	2.53	1.02	56.7	45.1	
	F	Plateau forest[h]	2011/5/1	0–1	2050	824	810	1630	2.53	1.02	26.6	21.2	
	H	Road side	2011/4/28	0–1	3860	740	745	1490	5.18	0.993	50.1	19.3	<10
Nagano	G	Plateau forest[f]	2011/4/28	0–1	513	550	557	1110	0.920	0.986	6.7	14.4	<10
		Hotel garden	2011/4/28	0–1	3350	726	755	1480	4.44	0.962	43.6	19.3	
Ibaraki	L	Garden	2011/4/20	0–1	4850	225	237	462	20.4	0.949	63.1	6.0	<10
	M	Garden	2011/4/20	0–1	8420	105	108	213	78.1	0.977	109	2.8	10–30
Chiba	N	Garden	2011/4/10	0–1	582	59.7	69.2	129	8.41	0.863	7.6	1.7	<10
	a	Road side	2011/4/11	0–1	7020	672	672	1340	10.4	1.00	91.3	17.5	10–30
	b	Road	2011/4/11	0–1	12,100	2460	2510	4970	4.82	0.980	157	64.6	

(continued)

Table 3.4 (continued)

Prefecture	Site	Description	Sampling date	Taken, cm	Decay corrected activity[a], Bq/kg				Ratio for ^{137}Cs		Inventory[c], kBq/m^2		
											This work[a]		MEXT[b]
					^{131}I	^{134}Cs	^{137}Cs	$^{134+137}Cs$	^{131}I	^{134}Cs	^{131}I	$^{134+137}Cs$	$^{134+137}Cs$
	c	Garden	2011/5/19	0–2	40,700	5300	5330	10,600	7.64	0.994	1060	276	60–100
		Road sludge	2011/6/11	0–1	114,000	33,900	33,900	67,800	3.36	1.00	1480	881	
		Road side	2011/6/11	0–1	28,500	17,700	17,700	35,400	1.61	1.00	371	460	
	d	Road side	2011/5/19	0–1	26,100	3210	3190	6400	8.18	1.01	340	83.2	
	e	Garden	2011/5/19	0–2	25,200	1060	1080	2140	23.3	0.981	655	55.6	10–30
Tokyo	f	Road side	2011/5/19	0–1	56,900	3470	3410	6880	16.7	1.02	740	89.4	10–30
	g	Garden (1)	2011/4/16	0–1	25,700	1630	1610	3240	16.0	1.01	334	42.1	
			2011/4/27	0–1	77,500	3560	3600	7160	21.5	0.989	1010	93.1	
		Garden (2)	2011/5/25	0–1	27,700	1710	1740	3450	15.9	0.983	360	44.8	
	h	Road side	2011/4/10	0–1	10,100	588	580	1170	17.4	1.01	131	15.2	<10
	i	Garden[d]	2011/4/10	0–1	18,500	966	962	1930	19.2	1.00	240	25.1	
			2011/4/28	0–1	53,700	1180	1180	2360	45.5	1.00	698	30.6	
				1–3	1360	43.6	62.5	106	21.8	0.698	35.0	3.0	
	j	Garden	2011/4/10	0–1	20,000	644	683	1330	29.3	0.942	259	17.2	
	k	Road side (1)	2011/6/2	0–2	18,200	1780	1760	3540	10.3	1.01	474	92.2	
		Road side (2)	2011/6/2	0–2	7780	475	473	948	16.4	1.00	202	24.6	
		Road side (3)	2011/6/2	0–2	11,600	514	536	1050	21.7	0.959	302	27.2	
	l	Road side	2011/6/2	0–2	11,100	1230	1220	2450	9.10	1.01	290	63.8	
	m	Garden	2011/5/19	0–2	6740	432	442	874	13.2	0.976	175	22.0	

Saitama	I	Rode side	2011/4/28	0–1	nd	151	154	305	–	0.979	–	4.0	<10
	J	Road side	2011/4/28	0–1	3130	204	207	412	15.1	0.986	40.7	5.4	
	K	River bank	2011/4/10	0–1	6130	248	252	500	24.3	0.981	79.6	6.5	
							Average		14.3 ± 13.9	0.978 ± 0.053			

Detection limit: ^{131}I <2 Bq/kg, ^{134}Cs and ^{137}Cs <0.6 Bq/kg

nd: not detected

[a]Decay-corrected value to March 16, 2011

[b]Airborne monitoring results by the MEXT ([14] a)

[c]Calculated assuming 1300 kg/m^3 for the soil density

[d]Measurements continued from April 10, 2011, to August 16, 2016. The results are shown in Fig. 3.6

Elevation [e]500 m, [f]1430 m, [g]1100 m, [h]1390 m

March 16, 2011. The background of Fig. 3.1 is the geographic distribution of $^{134+137}$Cs precipitation obtained by airborne monitoring. Figure 3.2 has the same background as Fig. 3.1. This map was released on November 11, 2011, by the MEXT. The distribution map of the pollution by radioactive cesium shown here is called a revised version. The MEXT released a pollution map of the same area on October 21, 2011, but a municipality where the nuclear plant was installed protested that the pollution level was too high. Therefore, the Japanese government changed the processing method for the background radioactivity in creating an airborne monitoring map and reprocessed the data to reduce the apparent contamination concentration. The original contamination map released on October 21 is shown in Fig. 9.5. The change in the background processing method decreased the radioactive cesium inventory by dozens of kBq/m^2. This treatment also significantly reduced the reliability of the airborne monitoring data.

The author's measurement sites do not cover the entire Tokyo metropolitan area. However, the measured data showed that the geographic distribution of radioactive cesium contamination was in good agreement with the MEXT pollution map published at a later date. Radioactive cesium was detected in surface soil at all stations sampled between March 19 and June 13, 2011. Despite a very short half-life of 8.04 days, ^{131}I was also discovered in all surface soils except at Site "I" in Saitama Prefecture (see Fig. 3.1). This result suggests that the radioactive contamination of the ground due to the FDNPP accident spread widely throughout the Tokyo metropolitan area at the beginning of the accident.

The inventories of samples collected to a surface depth of only 1 cm may have been estimated slightly lower than the actual values, considering the vertical distribution of radionuclides in soil shown in Table 3.4. However, it is believed that the difference is at most within 5%. In many areas shown in Table 3.4, the $^{134+137}$Cs inventory estimated by MEXT from airborne monitoring is lower than the value measured in this study. In particular, the tendency is remarkable in the urban areas of Chiba Prefecture and Tokyo. This difference in the inventories suggests that the estimation of radioactive contamination levels by airborne monitoring is limited in densely populated areas. The likely underestimation of the inventories in an urban area by airborne monitoring affects local and long-term estimates of external exposure for the inhabitants.

3.6.1 *Inventory of ^{131}I in the Tokyo Metropolitan Area*

The ^{131}I inventory in the Tokyo metropolitan area converted to the value on March 16, 2011, was nd (not detected: <0.03 kBq/m^2)—1480 kBq/m^2. Since the inventory of Fukushima City (Site A) was 1590–1850 kBq/m^2, ^{131}I deposition was smaller in all areas of the Tokyo metropolitan area than in Fukushima City. However, the ^{131}I inventories at Site B in Tochigi Prefecture, Site "c" in Chiba Prefecture, and Site "g" in Tokyo exceeded 1000 kBq/m^2. Site B is close to Fukushima, where significant plumes released from the FDNPP have passed, and it is easy to deduce that large amounts of radioactive material were deposited. Site “c”, on the other hand, features an area where hot spots were formed immediately after the accident. Although this site is approximately 200 km from the FDNPP, its inventory is higher than that of the surrounding area. Later, MEXT airborne monitoring (Fig. 3.2) confirmed that the area was a highly radioactively contaminated zone. In Tokyo, Site “g” in the east and Sites “I” and “j” in the city center showed high inventories. The inventories in central Tokyo were higher than those at other sampling sites in the surrounding area. This geographic distribution of ^{131}I in the Tokyo metropolitan area provides evidence that the FDNPP accident caused high concentrations of ^{131}I to develop central Tokyo.

The radioactive iodine detected in tap water supplied from the Kanamachi water purification plant in Tokyo on March 23, 2011, was the first event of radioactive iodine contamination in Tokyo from the FDNPP accident. In Japan at this time, the standard for radioactive iodine in drinking water was less than 300 Bq/kg based on the food and drink intake limit set by the Japan Nuclear Safety Commission (currently the NRA: Nuclear Regulatory Authority, Japan). In addition, the Food Sanitation Law of the Ministry of Health, Labor, and Welfare specified a provisional standard value of 100 Bq/kg in beverages for infants. The radioactive iodine concentrations of tap water at this water purification plant greatly exceeded this value. The radioactive iodine concentration of tap water at the Kanamachi water purification plant, announced by the Tokyo metropolitan government, were 210 Bq/kg on March 22 ([36] a) and 190 Bq/kg on March 23. No radioactive cesium values were listed in these announcements ([36] b). The Edogawa River water is used by the Kanamachi water purification plant to produce tap water. The Edogawa River is discussed in Chap. 6 of this book as a source of radioactive contamination for Tokyo Bay. Rain fell on the 21st in the Tokyo area. As predicted by Dr. Ishida in Fig. 2.1, radioactive plumes with extremely high ^{131}I concentrations had invaded eastern Tokyo and northern Chiba Prefecture on the 21st. Most likely, the Tokyo Metropolitan Government Bureau of Waterworks used the polluted raw water from the Edogawa River at this time. The radioactive materials released in the FDNPP accident were being transported to Tokyo, and the people of Tokyo learned this fact for the first time through the press. Site “f” was located approximately 1 km from the Kanamachi water purification plant. Its ^{131}I inventory was 740 kBq/m^2. This value is approximately half that in Fukushima City, but it is estimated that ^{131}I at the same level as Fukushima was deposited in the catchment area of the Kanamachi water purification plant. This information raised worries about the effects of radioactive contamination of tap water due to the FDNPP accident. Since March 28, ^{131}I has not been detected in the tap water at this water treatment plant (nd: <20 Bq/kg) ([36] c), but the concern about radioactive contamination of tap water due to the FDNPP accident has not disappeared.

3.6.2 Inventories of ^{134}Cs and ^{137}Cs in the Tokyo Metropolitan Area

The geographic distribution of radioactive cesium inventories also depended heavily on the soil sampling site. As with ^{131}I, its broad distribution was similar to that determined by the MEXT airborne monitoring. However, the ratio of ^{131}I to radioactive cesium inventories varied significantly from sampling point to sampling point. This difference is discussed in detail in Chap. 4.

The $^{134+137}$Cs inventory at Site A in Fukushima City showed good agreement between the MEXT estimates and the measured values. However, farther south in Tochigi Prefecture, the measured value was significantly higher than the estimated MEXT value. In Gunma, Nagano, Ibaraki, and Saitama prefectures, the estimated MEXT values, and the measured values were almost the same. Many of these sampling areas were in the countryside. On the other hand, in Chiba Prefecture, the measured values at all sites except Site N were much higher than the MEXT estimates. The measured values at all sites in Tokyo were also greater than the estimated MEXT values.

The $^{134+137}$Cs inventory at site “c” in Kashiwa City, Chiba Prefecture was 276–881 kBq/m^2, which was almost the same level as 363–563 kBq/m^2 in Fukushima City. Site “c” was located on the side of the road next to a park so that it may have been affected by the trees planted in the park. It is highly probable that rainwater containing massive amounts of radionuclides that had been transported down tree trunks was flowing into the road sludge with the highest inventory of $^{134+137}$Cs at 881 kBq/m^2. This site also had an extremely high ^{131}I inventory. However, in the roadside soils approximately 3 m away, the $^{134+137}$Cs inventory was only halved, despite ^{131}I decreasing to one-quarter of that in the road sludge. In the garden soil in the park, ^{131}I decreased to approximately 70% of the value in the road sludge, while $^{134+137}$Cs decreased significantly to approximately 30% of the value in the road sludge.

Thus, the concentrations and inventories of radioactive iodine and radioactive cesium fluctuated in a complicated manner, even at very closely spaced sampling points. Since various factors are presumed to have been involved in the behavior of these radionuclides precipitated from plumes to the ground surface, it would be challenging to generalize their behavior for each radionuclide. However, estimating the behavior of radionuclides from the environmental conditions at each sampling point is imperative, and determining the environmental conditions at the sampling points and collecting samples that meet the objectives of the analysis are also essential. Unnecessarily collecting and analyzing samples is not scientific.

The $^{134+137}$Cs inventory in central Tokyo was lower than that in northern Chiba. However, it a value of approximately 100 kBq/m^2 was observed in eastern Tokyo adjacent to this highly polluted area. Sites “g”–“l” are in not only the center of Tokyo but also the center of legislation, administration, and economics in Japan. It is incredibly alarming that such important areas of Japan are exposed to radioactive contamination. The high contamination by ^{131}I attenuates rapidly over time.

However, radioactive cesium, especially ^{137}Cs, has a long half-life of 30.1 years and will remain in this area for a long time. Radioactive cesium deposited on office buildings and infrastructure is likely to be washed away by rain and re-adsorbed and fixed in the surrounding soil. Attention must be paid to the behavior of radioactive cesium derived from urban structures.

In the immediate aftermath of the FDNPP accident, ^{131}I, ^{134}Cs, and ^{137}Cs were detected in the soil of the Tokyo metropolitan area. High activities and inventories of the radionuclides were found in eastern Tokyo and northern Chiba. The decay-corrected inventories of ^{131}I and $^{134+137}$Cs on March 16, 2011, in these areas, were 35–1480 kBq/m^2 and 15.2–881 kBq/m^2, respectively, which are almost the same levels as those in Fukushima City. As described in Chap. 5, the radioactive cesium concentration in the sludge incineration ash was higher in the urbanized eastern part of Tokyo and was a correlated with changes in rainfall in this area. It is possible that the radioactive cesium precipitated on concrete and asphalt is more likely to be discharged than that on the soil. These results suggest that the Tokyo metropolitan area continues to be affected by radioactive contamination caused by the FDNPP accident. The densely populated central regions of Japan have been exposed to massive radiation contamination. It is unprecedented worldwide that the FDNPP accident has caused large-scale radiation exposure to the general public. One hopes that long-term health impact assessments and monitoring will be conducted not only for Fukushima residents but also for residents of the Tokyo metropolitan area.

3.7 Temporal Changes in Radioactive Cesium in Park Soil from Central Tokyo

The soil at Site “i”, located in the Imperial Palace Outer Garden in the center of Tokyo, was continuously surveyed for radioactivity concentrations and inventories. As shown in Photo 3.1, the soil collection site is located in a part of the park where people can freely move around. The first sampling took place on April 10, 2011, when ^{131}I (2150 Bq/kg) and $^{134+137}$Cs (1910 Bq/kg) were detected with high radioactivity, but ^{131}I disappeared quickly due to radioactive decay. However, as shown in Table 3.3, most of the radioactive cesium accumulated on the soil surface throughout the monitoring period and was not detected in layers deeper than 15 cm. Figure 3.6 shows the temporal changes in the average $^{134+137}$Cs concentrations in the layers 1 and 5 cm below the surface. The inventory of $^{134+137}$Cs shows the cumulative values for layers up to 15 cm deep where $^{134+137}$Cs was detected. The decay curve of radioactive cesium, assuming that the amounts of ^{134}Cs and ^{137}Cs deposited immediately after the accident were the same, is also shown in Fig. 3.6. On April 28 and August 13, 2011, 2 cores were collected from the same site, approximately 10 cm apart, to verify the accuracy of the inventory. One of these cores was used to measure the vertical distribution, and the analytical results were expressed as vertically cumulative values to calculate the inventory. The other core was sliced to a thickness of 5 cm and analyzed for the inventory. The $^{134+137}Cs$ inventories of these 2 cores, standardized on the date of collection, were 32.6 and 67.5 kBq/m^2 on April 28 and 4.1 and 8.2 kBq/m^2 on August 13. These differences in inventories for adjacent core samples are attributed to the fact that the horizontal distribution of radioactive cesium infiltrated into the soil results in highly localized contamination, rather than to differences in sampling methods. Most likely, this speculation seems reasonable. Furthermore, in the cores collected on March 23, 2014, on August 14, 2015, and on May 1, 2016, the concentrations of $^{134+137}Cs$ in the 0–1 cm layer and the 0–5 cm layer were close or reversed. In these cores, radioactive cesium penetrated deep into the soil without being retained on the soil surface. It was not possible to determine whether such distributions of radioactive cesium were due to local physical vertical mixing of the soil at the sampling point, or whether ultrafine contaminated particles in the sediment had advected deeper. The spatial-temporal distribution of radioactive cesium in the soil of urban areas is discussed below.

From April 10–28, 2011, after the FDNPP accident, the activities and inventories of ^{131}I and $^{134+137}Cs$ increased sharply. If this increase was not the result of heterogeneous distribution of the radionuclides in the soil, the diffusion of radionuclides from the FDNPP to the Tokyo area was probably ongoing [37]. However, on August 13, the $^{134+137}Cs$ radioactivity decreased to approximately 10% of its initial value, and ^{131}I was not found. That is, it can be assumed that the new emission of radionuclides from the FDNPP reactors to the Tokyo area had subsided by that time. Chapters 5 and 7 emphasize the phenomenon that the concentrations of radioactive cesium that had polluted the environment suddenly decreased immediately after the FDNPP accident. The changes in radioactive cesium concentrations in the sludge incineration ash at the Tokyo Water Reclamation Center, which are described in Chap. 5, also decreased sharply to approximately one-tenth of the initial value in the 9 months after the accident [38]. However, the concentration of $^{134+137}Cs$ in the park soil decreased sharply and seemed to have increased gradually since August 2011. As shown in Table 3.3, no significant penetration of radioactive cesium into the subsurface was observed over time. Presumably, radioactive cesium that was initially lost from the park soil in Tokyo was not an ionic species that is absorbed by clay minerals and fixed to the soil. Perhaps it was present as an ultrafine particle species that could move through soil voids with rainwater, which can be inferred from various situational evidence obtained by analyzing many samples [39–47].

The adsorption behavior of radionuclides on clay minerals in the soil has been studied for many years, and it is well known that radioactive cesium penetrates clay mineral layers where it undergoes strong ionic adsorption and binds to soil particles (References 19–30 in Chap. 1). Heavy rain associated with a category five super typhoon drenched the Tokyo metropolitan area from May 28–30, 2011, totaling 108 mm of rain in central Tokyo [48]. The radioactive cesium on the soil of the Imperial Palace Outer Garden might have been washed out in this heavy rain. Perhaps, the ultrafine radioactive cesium particles called hot particles and the radioactive cesium adsorbed on fine clay particles or plant litter and humus were washed away from the soil surface. As a result, there is no denying that the concentration and inventory of radioactive cesium may have decreased. However, as shown by the vertical arrows in Fig. 3.6, torrential rains have occurred in the Tokyo metropolitan area many times since then. In other words, soil runoff due to rainfall alone cannot explain the rapid decrease in radioactivity immediately after the accident. This fact also implies that radioactive cesium deposited immediately after the FDNPP accident may have contained a large amount of mobile cesium in a highly migratory form. In other words, radioactive cesium that disappeared from the soil early in the accident is thought to have existed as a chemical species that was not easily retained in the soil. The concentration and inventory of radioactive cesium in the soil of the park at Site "i" seem to have fluctuated significantly since the second year of the accident, but compared to the expected decrease in radioactivity intensity due to the radioactive decay of $^{134+137}$Cs, its value appears to be steadily increasing over time. These spatiotemporal variations in radioactive cesium in soil, as shown in Fig. 3.6, indicate that the recycling system for the transport and supply of radioactive cesium at this location may not have been balanced. It is possible that the supply was much lower than the amount lost due to movement on the ground surface. The Tokyo metropolitan area is surrounded by many highly contaminated areas, not just that around the FDNPP reactors. Thus, the possibility that radioactive cesium continued to enter Tokyo with atmospheric dust cannot be denied.

Most likely, the temporal changes in radioactivity intensity in the Imperial Palace Outer Garden soil may reflect the effect of uneven distribution depending on the sampling area. To overcome this difficulty, Chap. 5 investigates the radioactive contamination of sludge incineration ash at a sewage treatment plant, which captured temporal fluctuations in radioactive contamination in Tokyo.

Many of the author's data discussed in this chapter have been published in PLOS ONE [49]. The data were revised and recalculated as needed before inclusion in this book.

References

1. Yasuhara M, Yamazaki H (2002) The impact of 150 years of anthropogenic pollution on the shallow marine ostracode fauna, Osaka Bay, Japan. Mar Micropaleontol 55:63–74. https://doi.org/10.1016/j.marmicro.2005.02.005
2. NA Independent (The Independent Investigation on the Fukushima Nuclear Accident) (2014) The Fukushima Daiichi Nuclear Power Station Disaster: investigating the myth and reality. Routledge, London
3. TEPCO (Tokyo Electric Power Company) (2012) The investigation reports of the Fukushima Nuclear accident [in Japanese]. http://www.tepco.co.jp/cc/press/betu12_j/images/120620j0303.pdf
4. NISA (Nuclear and Industrial Safety Agency) (2011) Evaluation on the state of core in units 1, 2 and 3 related to the accident of TEPCO's Fukushima Daiichi Nuclear Power Station [in Japanese]. http://warp.da.ndl.go.jp/info:ndljp/pid/3491887/www.meti.go.jp/earthquake/nuclear/pdf/20110606-1nisa.pdf
5. TEPCO (2012) Estimation of the amount of radioactive material released to the atmosphere from the Fukushima Daiichi Nuclear Power Plant accident [in Japanese]. http://www.tepco.co.jp/cc/press/betu12_j/images/120524j0101.pdf

6. Chino M, Nakayama H, Nagai H, Terada H, Katata G et al (2011) Preliminary estimation of release amounts of ^{131}I and ^{137}Cs accidentally discharged from the Fukushima Daiichi Nuclear Power Plant into the atmosphere. J Nucl Sci Technol 48:1129–1134. [in Japanese]. https://doi.org/10.1080/18811248.2011.9711799
7. IAEA (International Atomic Energy Agency) (2004) Soil sampling for environmental contaminants. IAEA-TECDOC-1415
8. MEXT (Ministry of Education, Culture, Sports, Science and Technology of Japan) (2011) Calculation results and reasoning for internal exposure considered in compilation of "provisionally approach" [in Japanese]. http://radioactivity.nsr.go.jp/ja/contents/4000/3912/24/1305995_0512_1.pdf
9. MEXT (2011) Airborne monitoring results around the Fukushima Daiichi Nuclear Power Plant [in Japanese]. https://radioactivity.nsr.go.jp/ja/contents/4000/3783/24/1304280_2610.pdf
10. MEXT (2011) Measurement results of monitoring the concentration of radioactive materials released into the atmosphere by MOD aircraft [in Japanese]. https://radioactivity.nsr.go.jp/ja/list/357/list-1.html
11. MEXT (2011) Measurement results of airborne monitoring by MEXT and US Department of Energy aircraft [in Japanese]. https://radioactivity.nsr.go.jp/ja/contents/4000/3710/24/1305820_20110506.pdf
12. MEXT (2011) Measurement results of secondary airborne monitoring by MEXT and US Department of Energy aircraft [in Japanese]. https://radioactivity.nsr.go.jp/ja/contents/5000/8969/24/110616_2nd.pdf
13. DOE/NNSA (U.S. Department of Energy/National Nuclear Security Administration) (2011) US DOE/NNSA Response to 2011 Fukushima Incident - data and documentation. https://www.energy.gov/downloads/us-doennsa-response-2011-fukushima-incident-data-and-documentation (The original version released in 2011 has been reorganized and reuploaded.)
14. MEXT (2011) MEXT's airborne monitoring results in Iwate, Shizuoka, Nagano, Yamanashi, Gifu, and Toyama prefectures and revisions to airborne monitoring that take into account the effects of natural nuclides. (a) https://radioactivity.nsr.go.jp/ja/contents/5000/4899/24/1910_111112nd.pdf, (b) https://radioactivity.nsr.go.jp/ja/contents/15000/14214/24/190308_13th_air.pdf
15. Goldberg ED, Koide M (1962) Geochronological studies of deep sea sediments by the ionium/thorium method. Geochim Cosmochim Acta 26:417–450
16. Berger WH, Heath GR (1968) Vertical mixing in pelagic sediments. J Mar Res 26:134–143
17. Bonatti E, Fisher DE, Joensuu O, Rydell HS (1971) Postdepositional mobility of some transition elements, phosphorus, uranium and thorium in deep sea sediments. Geochim Cosmochim Acta 35:189–201
18. Robbins JA, Edgington DN (1975) Determination of recent sedimentation rates in Lake Michigan using Pb-210 and Cs-137. Geochim Cosmochim Acta 39:285–304
19. Guinasso NL Jr, Schink DR (1975) Quantitative estimates of biological mixing rates in abyssal sediment. J Geophys Res 80:3032–3043
20. Cochran JK, Allen RC (1979) Particle reworking in sediments from the New York Bight apex: evidence from Th-234/U-238 disequilibrium. Estuar Coast Mar Sci 9:739–747
21. Olsen CR, Simpson HJ, Peng TH, Bopp RF, Trier RM (1981) Sediment mixing and accumulation rate effects on radionuclide depth profile in Hudson estuary sediments. J Geophys Res 86:11020–11028
22. McKee BA, DeMaster DJ, Nittrouer CM (1982) Rate pf particle mixing and sediment accumulation on the continental shelf near the mouth of the Yangtse River. Trans Am Geophys Union 63:65–75
23. Beasley TM, Carpenter R, Jennings CD (1982) Plutonium, Am-241 and Cs-137 ratios, inventories and vertical profiles in Washington and Oregon continental shelf sediments. Geochim Cosmochim Acta 46:1931–1946
24. DeMaster DJ, Cochan JK (1982) Particle mixing rates in deep-sea sediments determined from excess Pb-210 and Si-32: profiles. Earth Planet Sci Lett 61:257–271

25. Evans DE, Alberts JJ, Clark RA III (1983) Reversible ion-exchange fixation of cesium-137 leading to mobilization from reservoir sediments. Geochim Cosmochim Acta 47:1041–1049. https://doi.org/10.1016/0016-7037(83)90234-X
26. Nittrouer CA, DeMaster DJ, McKee BA, Cutshall NH, Larsen IL (1983/1984) The effect of sediment mixing on Pb-210 accumulation rates for the Washington Continental Shelf. Mar Geol 54:201–221
27. Heit M, Miller KM (1987) Cesium-137 sediment depth profiles and inventories in Adrondack Lake sediments. Biogeochemistry 3:243–265
28. MacKenzie AB, Scott RD (1993) Sellafield waste radionuclides in Irish Sea intertidal and salt marsh sediments. Environ Geochem Health 15:173–184
29. Golosov VN, Belyaev VR, Markelov MV (2013) Application of Chernobyl-derived ^{137}Cs fallout for sediment redistribution studies: lessons from European Russia. Hydrol Process 27:781–794. https://doi.org/10.1002/hyp.9470
30. Yamada S, Kitamura A, Kurikami H, Yamaguchi M, Malins A et al (2015) Sediment and ^{137}Cs transport and accumulation in the Ogaki Dam of eastern Fukushima. Environ Res Lett 10:014013. https://doi.org/10.1088/1748-9326/10/1/014013
31. Almgren S, Isaksson M (2006) Vertical migration studies of ^{137}Cs from nuclear weapons fallout and the Chernobyl accident. J Environ Radioact 91:90–102. https://doi.org/10.1016/j.jenvrad.2006.08.008
32. Rosén K, Öborn I, Lönsjö H (1999) Migration of radiocaesium in Swedish soil profiles after the Chernobyl accident, 1987-1995. J Environ Radioact 46:45–66. S0265-931X(99)00040-5 [pii]
33. Kaste JM, Heimsath AM, Hohmann (2006) Quantifying sediment transport across an undisturbed prairie landscape using cesium-137 and high resolution topography. Geomorphology 76:430–440. https://doi.org/10.1016/j.geomorph.2005.12.007
34. Szabó KZ, Udvardi B, Horváth A, Bakacsi Z, Pásztor L et al (2012) Cesium-137 concentration of soils in Pest County, Hungary. J Environ Radioact 110:38–45. https://doi.org/10.1016/j.jenvrad.2012.01.023
35. Koarashi J, Atarashi AJ, Matsunaga T, Sato T, Nagao S et al (2012) Factors affecting vertical distribution of Fukushima accident-derived radiocesium in soil under different land-use conditions. Sci Total Environ 431:392–401. https://doi.org/10.1016/j.scitotenv.2012.05.041
36. Tokyo Metropolitan Government Bureau of Waterworks (2011) About result of a measurement of radioactivity of tap water [in Japanese]. (a) https://www.waterworks.metro.tokyo.jp/press/h22/press110323-01.html, (b) https://www.waterworks.metro.tokyo.jp/press/h22/press110323-02.html, (c) https://www.waterworks.metro.tokyo.jp/press/h22/press110328-01.html
37. NILU (Norsk Institute for Luftforskning) (2011) Results of simulation for atmospheric transportation of ^{131}I in the early after the FDNPP accident. http://transport.nilu.no/products/fukushima (not available now)
38. Bureau of Sewerage, Tokyo Metropolitan Government (2017) Measurement result of radiation level around water reclamation centers and radioactivity through wastewater treatment. http://www.gesui.metro.tokyo.jp/english/oshi/
39. Matsunaga T, Ueno T, Amano H, Tkatchenko Y, Kovalyov A et al (1998) Characteristics of Chernobyl-derived radionuclides in particulate from in surface waters in the exclusion zone around the Chernobyl Nuclear Power Plant. J Contam Hydrol 35:101–113. S0169-7722(98)00119-3 [pii]
40. Adachi K, Kajino M, Zaizen Y, Igarashi Y (2013) Emission of spherical cesium-bearing particles from an early stage of the Fukushima nuclear accident. Sci Rep 3:2554. https://doi.org/10.1038/srep02554
41. Itoh S, Eguchi T, Kato N, Takahashi S (2014) Radioactive particles in soil, plant, and dust samples after the Fukushima nuclear accident. Soil Sci Plant Nutr 60:540–550. https://doi.org/10.1080/00380768.2014.907735
42. Abe Y, Iizawa Y, Terada Y, Adachi K, Igarashi Y et al (2014) Detection of Uranium and chemical state analysis of individual radioactive microparticles emitted from the Fukushima

Nuclear accident using multiple synchrotron radiation X-ray analyses. Anal Chem 86:8521–8525. https://doi.org/10.2021/ac501998d

43. Martin PG, Griffiths I, Jones CP, Stitt CA, Davies-Milner M et al (2016) In-situ removal and characterization of uranium-containing particles from sediments surrounding the Fukushima Daiichi Nuclear Power Plant. Spectrochim Acta B 117:1–7. https://doi.org/10.1016/j.sab.2015.12.010
44. Yamaguchi N, Mitome M, Kotone AH, Asano M, Adachi K et al (2016) Internal structure of cesium-bearing radioactive microparticles released from Fukushima nuclear power plant. Sci Rep 6:20548. https://doi.org/10.1038/srep20548
45. Furuki G, Imoto J, Ochiai A, Yamasaki S, Nanba K et al (2017) Caesium-rich micro-particles: a window into the meltdown events at the Fukushima Daiichi Nuclear Power Plant. Sci Rep 7:42731. https://doi.org/10.1038/srep42731
46. Miura H, Kurihara Y, Sakaguchi A, Tanaka K, Yamaguchi N et al (2018) Discovery of radiocesium-bearing microparticles in river water and their influence on the solid-water distribution coefficient (Kd) of radiocesium in Kuchibuto River in Fukushima. Geochem J 52:1–1■. https://doi.org/10.2343/geochemj.2.0517
47. Martin PG, Louvel M, Cipiccia S, Jones CP, Batey DJ et al (2019) Provenance of uranium particulate contained within Fukushima Daiichi Nuclear Power Plant Unit 1 ejecta material. Nat Commun 10:2801. https://doi.org/10.1038/s41467-019-10937-z
48. Japan Meteorological Agency (2020) Search past weather data. https://www.data.jma.go.jp/obd/stats/etrn/select/prefecture00
49. Ishida M, Yamazaki H (2017) Radioactive contamination in the Tokyo metropolitan area in the early stage of the Fukushima Daiichi Nuclear Power Plant (FDNPP) accident and its fluctuation over five years. PLoS One 12:e0187687. https://doi.org/10.1371/journal.pone.0187687

Chapter 4
Characteristics of the Distributions of ^{131}I and Radioactive Cesium in the Soil of the Tokyo Metropolitan Area After the FDNPP Accident

Abstract The national and local governments measured the temporal changes in ^{131}I in the atmosphere after the FDNPP accident. These results indicated that the radioactive plume released from the FDNPP spread from Fukushima to the Tokyo metropolitan area and Tokyo itself. The Tokyo metropolitan area was exposed in highly contaminated plumes on March 15–16 and 20–22 immediately after the accident. In central Tokyo, there was a single day with 0.0 mm of light rain on March 16 and a reasonably heavy rainfall of 33.5 mm on March 21–22. Coincidentally, on these rainy days, highly contaminated radioactive plumes had reached central Tokyo, causing radionuclides in the plumes to be washed out and precipitated and leading to radioactive contamination of the ground surface. The radioactive plume spread over in the Tokyo metropolitan area due to the FDNPP accident, significant heterogeneity in radioactive iodine and radioactive cesium was observed on the ground surface in a narrow area. Such non uniformity of the radioactive nuclides precipitated at the surface indicates that the plume, including radionuclides, was advancing with this uneven distribution. In this chapter, the characteristics of radioactive plumes that were transported into the Tokyo metropolitan area are evaluated from the geographic distribution of ^{131}I and radioactive cesium.

Keywords Central Tokyo · Radioactive iodine · High contamination · Soil · $^{131}I/^{137}Cs$ ratio

4.1 Introduction

The national and local governments measured the temporal changes in ^{131}I in the atmosphere after the FDNPP accident, as shown in Fig. 2.3. This figure indicates that the radioactive plume released from the FDNPP spread from Fukushima to the Tokyo metropolitan area and Tokyo itself (References 1, 3–6, 9 in Chap. 2). These results suggest to imagine that the Tokyo metropolitan area was exposed in highly contaminated plumes on March 15–16 and 20–22 immediately after the accident. Furthermore, the results are also in good agreement with the estimation of the radioactive plume route map shown in Fig. 2.1. In central Tokyo, there was a single

H. Yamazaki, *Radioactive Contamination of the Tokyo Metropolitan Area*,
https://doi.org/10.1007/978-981-15-7368-2_4

day with 0.0 mm of light rain on March 16 and a reasonably heavy rainfall of 33.5 mm on March 21–22 [1]. Coincidentally, on these rainy days, highly contaminated radioactive plumes had reached central Tokyo causing radionuclides in the plumes to be washed out and precipitated and leading to radioactive contamination of the ground surface. The radioactive plume spread over in the Tokyo metropolitan area due to the FDNPP accident, as shown in Fig. 3.5 and Table 3.2, significant heterogeneity in radioactive iodine and radioactive cesium was observed on the ground surface in a narrow area. Such non uniformity of the radioactive nuclides precipitated at the surface indicates that the plume, including radionuclides, was advancing with this uneven distribution. In this chapter, the characteristics of radioactive plumes that were transported into the Tokyo metropolitan area are evaluated from the geographic distribution of ^{131}I and radioactive cesium. Excellent papers on the isotope ratios of radionuclides released by the FDNPP accident have been published [2–9], but the details not discussed here.

4.2 Concentrations and Inventories of ^{131}I, ^{134}Cs, and ^{137}Cs in the Soil of Tokyo

Figure 4.1 shows the relationship among ^{131}I, ^{134}Cs, and ^{137}Cs concentrations in soils collected from the Tokyo metropolitan area and Fukushima Prefecture, as listed in Table 3.4. To describe the relative relationship, the concentrations of these nuclides were corrected with radioactive decay values to March 16, 2011. The

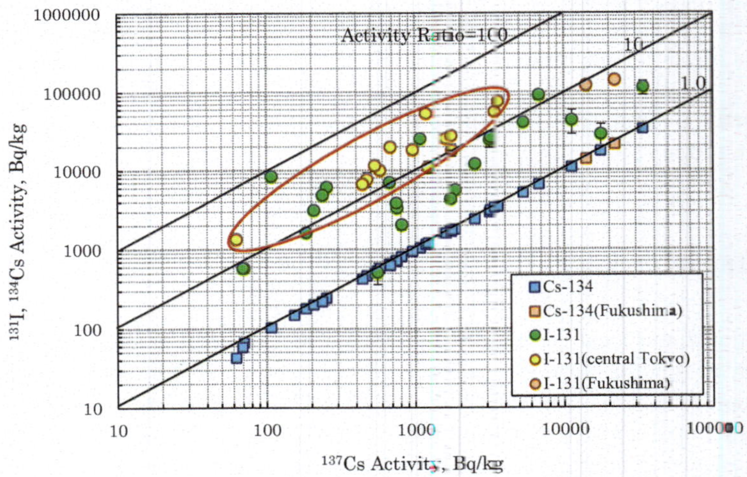

Fig. 4.1 Concentration correlation between ^{131}I and ^{134}Cs for ^{137}Cs in the soil collected from the Tokyo metropolitan area and Fukushima Prefecture. Radioactivities were decay-corrected to March 16, 2011. Activity ratios of ^{131}I to ^{137}Cs in central Tokyo surrounded by the red circle were high. The ^{134}Cs/^{137}Cs ratios were almost 1.0 in every site

Table 4.1 Average values of the ^{131}I/^{137}Cs and ^{134}Cs/^{137}Cs ratios by the region in the Tokyo metropolitan area

Area[a]	Prefecture and region[a]	n	^{131}I/^{137}Cs[b]	^{134}Cs/^{137}Cs[b]
Near the FDNPP (Area F)	Fukushima Tochigi	5	8.05 ± 3.53 (7.64 ± 0.07)	0.978 ± 0.012
Far from the FDNPP (Area G)	Gunma Nagano Southern Chiba	9	4.91 ± 2.95 (3.48 ± 0.09)	0.984 ± 0.038
Around Tokyo (Area H)	Ibaragi Saitama Northern Chiba Tokyo	25	18.9 ± 15.4 (9.67 ± 0.07)	0.971 ± 0.070
	Average (weighted average)	39	14.3 ± 13.9 (7.55 ± 0.04)	0.978 ± 0.053 (0.992 ± 0.002)

The weighted average value in parentheses
[a]See Fig. S2.1 in Chap. 2 for the region of each prefecture
[b]Decay-corrected value to March 16, 2011

^{134}Cs/^{137}Cs ratio was almost 1.0 at every sampling site. This isotope ratio means that radioactive cesium released from the FDNPP reactor was not affected by the isotope fractionation effect in the process of migrating and precipitating into the Tokyo area. On the other hand, the ^{131}I/^{137}Cs ratio fluctuated greatly between 0.92 and 78.1. As shown in Chap. 3, ^{131}I precipitated on the ground surface is strongly adsorbed and fixed on soil particles. Its behavior is similar to that of radioactive cesium. It is conceivable that the significant fluctuations in the ^{131}I/^{137}Cs ratio shown in Fig. 4.1 reflect differences in the concentration distributions of ^{131}I and ^{137}Cs in the plume or differences in the behaviors of ^{131}I and ^{137}Cs when precipitating from the plume to the ground surface.

Table 4.1 shows the ^{131}I/^{137}Cs and ^{134}Cs/^{137}Cs radioactivity ratios of three regions along the path of the radioactive plume from the FDNPP to Tokyo. For ^{131}I/^{137}Cs, the weighted average result is also shown because a significant attenuation correction for accompanying radioactive decay was performed in the measurement of ^{131}I radioactivity. The ^{134}Cs/^{137}Cs ratio released in the FDNPP accident was estimated to be approximately 1.2–1.25 [10, 11]. The weighted average ^{134}Cs/^{137}Cs ratio of the soil collected at 39 sites in the Tokyo metropolitan area and Fukushima City was 0.992 ± 0.002. No regional differences in this value were observed. However, there was a significant regional difference in the ^{131}I/^{137}Cs ratios. In Fukushima and Tochigi prefectures (Area F), which are close to the FDNPP, the value was 7.64 ± 0.07, and in Gunma and Nagano prefectures (Area G), which are farther away, the value was 3.48 ± 0.09. These ratios are smaller than the estimated ^{131}I/^{137}Cs ratio of 11–50 released from the FDNPP [10, 11]. However, the ^{131}I/^{137}Cs ratio in Tokyo and the suburbs of Tokyo (Area H) was 9.67 ± 0.07, which was in good agreement with the value estimated by NISA [10]. A Comparison with Fig. 2.1 illustrates the following points; the plume that polluted Tochigi, Gunma, and Nagano prefectures was transported on March 15, and the plume that contaminated

Tokyo reached there on March 21. Although the plume on the 15th passed through Tokyo, soil contamination in Tokyo was avoided because little rain fell on that day. However, since Areas F and G had significant rainfall on March 15–16, it is estimated that ^{131}I, ^{134}Cs, and ^{137}Cs precipitated from plumes of different origins from central Tokyo. The differences in the geographic distributions of the ^{131}I/^{137}Cs ratios in the Tokyo metropolitan area are thought to result from the differences in the arrival date of the plume that caused soil pollution. Presumably, the ratio of ^{131}I/^{137}Cs in the plume released from the FDNPP reactor varied on different days.

The concentrations and inventories of ^{131}I, ^{134}Cs, and ^{137}Cs were analyzed in the central part and suburb of Tokyo (Area H), where the ^{131}I/^{137}Cs ratios were high. The results with the values corrected for decay on March 16, 2011, are shown in Tables 4.2 and 4.3. If ^{131}I precipitated from a plume that reached Tokyo on March 21, the concentration and inventory of ^{131}I listed in these tables would have declined to approximately 65% of the initial values by the 21st.

The $^{134+137}$Cs inventory in central Tokyo was 17.3–92.1 kBq/m^2, approximately one-tenth of that in Fukushima City at 363–563 kBq/m^2 (Table 3.4). Fukushima City is located approximately 60 km northeast of the FDNPP site. The $^{134+137}$Cs inventory was 13.7–277 kBq/km^2 in the area around Tokyo, but it was particularly high in Chiba and Kashiwa cities in northern Chiba Prefecture. On the other hand, the inventories of ^{131}I were 202–698 kBq/m^2 in central Tokyo and 175–1050 kBq/m^2 in the surrounding area, slightly lower than 1590–1850 kBq/m^2 in Fukushima City. The inventory of ^{131}I was also high in northern Chiba Prefecture and eastern Tokyo. Kashiwa City showed a high amount of $^{134+137}$Cs deposition according to MEXT's airborne monitoring [12], but Chiba City was not included in the highly contaminated area. This distribution of radioactive iodine and radioactive cesium implies that so-called hot spots where radioactive contamination was locally concentrated formed in Chiba City.

4.3 Geographic Distribution of the ^{131}I/^{137}Cs Radioactivity Ratio in Central Tokyo

There are two crucial estimates for the amounts of radioactive iodine and radioactive cesium released during the early days of the FDNPP accident. One is from TEPCO [11], and the released amounts of ^{131}I, ^{134}Cs, and ^{137}Cs are 400 PBq, 10 PBq, and 8 PBq, respectively. On the other hand, the NISA values are 160 PBq, 18 PBq, and 15 PBq, respectively [10]. The government of Japan has adopted the NISA values in its report [13] to the IAEA Ministerial Meeting. Many researchers and institutions have estimated the amounts of radionuclides released from the FDNPP, but the estimates vary widely. In such a nuclear disaster, the process of the accident is simulated using many hypothetical parameters, and naturally, the results differ depending on the parameters used. Performing a simulation is essential, but there is no point in performing a simulation that is inconsistent with field measurements.

Table 4.2 Concentrations and inventories of ^{131}I, ^{134}Cs, and ^{137}Cs in central Tokyo

Site	Sampling date	Depth, cm	Detected, Bq/kg			Corrected[a], Bq/kg			Inventory[a, b], kBq/m²			Ratio to ^{137}Cs	
			^{131}I	^{134}Cs	^{137}Cs	^{131}I	^{134}Cs	^{137}Cs	^{131}I	^{134}Cs	^{137}Cs	^{131}I	^{134}Cs
Imperial Palace East Garden	2011/4/10	0–1	2130	629	682	20,000	644	683	259	8.37	8.88	29.3	0.943
Imperial Palace Outer Garden	2011/4/28	0–1	1210	1130	1180	53,700	1180	1180	698	15.3	15.3	45.5	1.000
In front of the Cabinet Office	2011/6/2	0–2	8.60	448	471	7780	475	473	202	12.4	12.3	16.4	1.004
In front of the Ministry of Finance			12.8	484	534	11,600	514	536	302	13.4	13.9	21.7	0.959
Tameike Intersection			20.1	1680	1750	18,200	1780	1760	473	46.3	45.8	10.3	1.011
Roppongi Intersection			12.2	1160	1210	11,100	1210	1230	289	31.5	32.0	9.10	0.984

The geographic distributions of the $^{131}I/^{137}Cs$ ratios are shown in Fig. 4.2

[a]Decay-corrected value to March 16, 2011

[b]Calculated assuming a soil density of 1300 kg/m^3

Table 4.3 Concentrations and inventories of ^{131}I, ^{134}Cs, and ^{137}Cs in a residential garden around Tokyo

Site	Sampling date	Depth, cm	Detected, Bq/kg			Corrected[a], Bq/kg			Inventory[a, b], kBq/m^2			Ratio to ^{137}Cs	
			^{131}I	^{134}Cs	^{137}Cs	^{131}I	^{134}Cs	^{137}Cs	^{131}I	^{134}Cs	^{137}Cs	^{131}I	^{134}Cs
Asaka City	2011/5/19	0–2	99.3	1000	1076	25,200	1060	1080	655	27.6	28.1	23.3	0.981
Suginami Ward			24.9	407	440	6740	432	442	175	11.2	11.5	15.2	0.977
Misato City			54.7	988	1070	14,800	1050	1080	385	27.3	28.1	13.7	0.972
Katsushika Ward			150	3090	3220	40,500	3280	3240	1050	85.3	84.2	12.5	1.012
Nagareyama City			61.9	2420	2560	16,800	2560	2570	437	66.6	66.8	6.54	0.996
Kashiwa City			97.8	4970	5310	26,500	5270	5330	689	137	139	4.97	0.989
Chiba City	2011/4/28		842	5240	5320	40,700	5300	5330	1060	138	139	7.64	0.994
Nagareyama City[c]	2011/6/11	0–5	nd	95.5	106	–	104	106	–	6.76	6.89	–	0.977
Kashiwa City[c]			nd	558	594	–	605	597	–	39.3	38.8	–	1.013

The geographic distributions of the ^{131}I/^{137}Cs ratios are shown in Fig. 4.2
[a]Decay-corrected value to March 16, 2011
[b]Calculated assuming a soil density of 1300 kg/m^3
[c]Collected at a place different from May 19

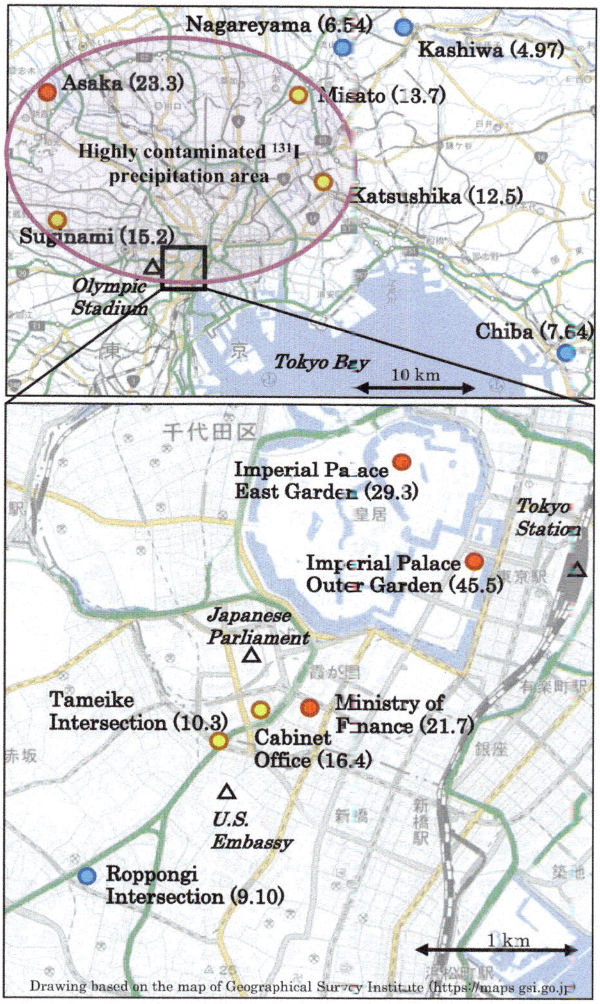

Fig. 4.2 $^{131}I/^{137}Cs$ ratios observed in the central area of Tokyo. The concentrations are shown in Tables 4.2 and 4.3. The activity ratios are decay-corrected to March 16, 2011. Approximately 500 km^2 area surrounded by an ellipse was highly contaminated by ^{131}I

For cesium nuclides, which have the same physical and chemical properties, the estimated $^{134}Cs/^{137}Cs$ activity ratios are 1.25 for TEPCO and 1.20 for NISA. These values are significantly different from the actual measured values of 1.0 at the field sites. This difference means that there are still challenges in estimating emissions.

Figure 4.2 shows the geographic relationship among the soil sampling points in Tables 4.2 and 4.3. Central Tokyo is located at the center of Japan's administration and contains embassies from many countries, including the US Embassy. The Imperial Palace, Tokyo Station, and a busy downtown area are also located in this region. In this figure, the $^{131}I/^{137}Cs$ ratio corrected for radioactive decay to March 16, 2011, is shown for each sampling point.

In central Tokyo, the highest ratios of ^{131}I/^{137}Cs were 29.3–45.5 in parks in the Imperial Palace. The concentration of radioactive cesium at this location was at the same level as general sites in central Tokyo and suburbs, clearly indicating that a relatively large amount of ^{131}I had precipitated. However, at the Tameike intersection and Roppongi intersection several kilometers southwest of the Imperial Palace, the ^{131}I/^{137}Cs ratios decreased to 9.10–10.3. The ^{131}I inventory at this location was 289–473 kBq/m^2, which was not unusually small compared to 259–698 kBq/m^2 at the Imperial Palace park. Nevertheless, $^{134+137}$Cs inventories at the Tameike intersection and Roppongi intersection were 63.5–92.1 kBq/m^2, which were higher than those at the Imperial Palace, as shown in Table 4.2. In other words, the ^{131}I/^{137}Cs ratios were lower due to relatively large amounts of radioactive cesium precipitating at these sites. The depositional behaviors of radioactive iodine and radioactive cesium released during the FDNPP accident are complicated even in such a small area of several kilometers square at most. As noted in Chap. 3, the distribution of radionuclides in the radioactive plume released from the FDNPP reactor was extremely heterogeneous, and these nuclides migrated to Tokyo without mixing even during the process of advection in the atmosphere. Presumably, this caused the variations in the geographic distribution of ^{131}I, ^{134}Cs, and ^{137}Cs in the soil.

In the suburbs of Tokyo, the ^{131}I inventories showed high values of 655–1050 kBq/m^2 in Katsushika Ward, Kashiwa City, and Asaka City. However, in Katsushika and Kashiwa, the $^{134+137}$Cs inventory showed extremely high values of 170–276 kBq/m^2, which resulted in relative decreases in ^{131}I/^{137}Cs. This situation was similar to that observed in central Tokyo. According to the MEXT airborne monitoring (Fig. 3.2), the amount of $^{134+137}$Cs deposition in Chiba City was less than 10 kBq/m^2, but the inventory of $^{134+137}$Cs detected by the author was 277 kBq/m^2, which is much larger. At this point, the ^{131}I inventory also showed a significant value of 1060 kBq/m^2. The data in this area not listed in Table 4.3 were also high. However, the ^{131}I/^{137}Cs ratio was 7.64, and it was believed that the plume that passed through the highly contaminated area in northern Chiba Prefecture also reached Chiba City.

As shown by the red circle in Fig. 4.2, it was presumed that ^{131}I, which was relatively high in concentration compared to radioactive cesium, spread over region of approximately 500 km^2, including the central Tokyo area, and a massive amount of ^{131}I precipitated onto the ground. Naturally, during this process, the people of Tokyo were breathing air containing high concentrations of radioactive iodine.

References

1. Japan Meteorological Agency (2020) Search past weather data [in Japanese]. https://www.data.jma.go.jp/obd/stats/etrn/select/prefecture00
2. Ohara T, Morino Y, Tanaka A (2011) Atmospheric behavior of radioactive materials from Fukushima Daiichi Nuclear Power Plant. J Natl Inst Public Health 60:292–299. [in Japanese with English Abstract]

3. Chino M, Nakayama H, Nagai H, Terada H, Katata G et al (2011) Preliminary estimation of release amounts of ^{131}I and ^{137}Cs accidentally discharged from the Fukushima Daiichi Nuclear Power Plant into the atmosphere. J Nucl Sci Technol 48:1129–1134. https://doi.org/10.1080/18811248.2011.9711799
4. Amano H, Akiyama M, Chunlei B, Kawanura T, Kishimoto T et al (2012) Radiation measurements in the Chiba Metropolitan Area and radiological aspects of fallout from the Fukushima Dai-ichi Nuclear Plants accident. J Environ Radioact 111:42–52. https://doi.org/10.1016/j.jenvrad.2011.10.019
5. Kawada Y, Yamada T (2012) ^{134}Cs/^{137}Cs activity ratio of radioactive cesium released by nuclear accident. Isotope News 697:16–20. [in Japanese]
6. Haba H, Kanaya J, Mukai H, Kanbara T, Kase K (2012) One-year monitoring of airborne radionuclides in Wako, Japan, after the Fukushima Dai-ichi nuclear power plant accident in 2011. Geochem J 46:271–278
7. Komori M, Shozugawa K, Nogawa N, Matsuo M (2013) Evaluation of radioactive contamination caused by each plant of Fukushima Daiichi Nuclear Power Station. BUNSEKI KAGAKU 62:475–483. [in Japanese with English Abstract]
8. Chaisan K, Smith JT, Bossew P, Kirchner G, Laptev GV (2013) Worldwide isotope ratios of the Fukushima release and early-phase external dose reconstruction. Sci Rep 3:2520. https://doi.org/10.1038/srep02520
9. Nishizawa Y, Yoshida M, Sanada Y, Torii T (2015) Distribution of the ^{134}Cs/^{137}Cs ratio around the Fukushima Daiichi nuclear power plant using an unmanned helicopter radiation monitoring system. J Nucl Sci Technol 53:468–474. https://doi.org/10.1080/00223131.2015.1071721
10. NISA (Nuclear and Industrial Safety Agency) (2011) Evaluation on the state of core in units 1, 2 and 3 related to the accident of TEPCO's Fukushima Daiichi Nuclear Power Station [in Japanese] http://warp.da.ndl.go.jp/info:ndljp/pid/3491887/www.meti.go.jp/earthquake/nuclear/pdf/20110606-1nisa.pdf
11. TEPCO (2012) Estimation of the amount of radioactive material released to the atmosphere from the Fukushima Daiichi Nuclear Power Plant accident [in Japanese]. http://www.tepco.co.jp/cc/press/betu12_j/images/120524j0101.pdf
12. MEXT (2011) MEXT's airborne monitoring results in Iwate, Shizuoka, Nagano, Yamanashi, Gifu, and Toyama prefectures and revisions to airborne monitoring that take into account the effects of natural nuclides [in Japanese]. https://radioactivity.nsr.go.jp/ja/contents/5000/4899/24/1910_111112nd.pdf
13. Japanese Government (2011) Report of Japanese Government to the IAEA Ministerial Conference on Nuclear Safety - The Accident at TEPCO's Fukushima Nuclear Power Stations. https://japan.kantei.go.jp/kan/topics/201106/iaea_houkokusho_e.html

Chapter 5
Role of the Wastewater Treatment System in the Behavior of Radioactive Cesium Precipitated in the Urban Area of Tokyo

Abstract The risk of radioactive contamination of densely populated urban areas is not due only to the leakage of radioactive materials from nuclear power plants and nuclear fuel reprocessing plants. Nuclear batteries and depleted uranium loaded on satellites and aircraft can be sources of radioactive contamination in urban areas. The risk of terrorism using nuclear materials cannot be ignored. Furthermore, radioactive materials used in medicine are also released daily into the environment along with patient excrement. Thus, not only the Tokyo metropolitan area but also many large cities around the world are exposed to these risks of radioactive contamination. Tokyo is likely the first metropolis in the world to be highly contaminated by radioactive materials. Radioactive nuclides deposited on the ground surface are known to be adsorbed and retained by clay minerals in the soil, but the behaviors of radioactive nuclides deposited on concrete and asphalt in urban areas are often unknown. The Tokyo special ward, which has a sewage treatment system that mixes domestic wastewater and rainwater for each region, is the best place to analyze the dynamics of radioactive materials in urban areas. In this chapter, the temporal changes in the radioactive cesium concentration in the sludge incineration ash of each sewage treatment plant, which were released on the Tokyo homepage after the Fukushima nuclear disaster, were analyzed. As a result, the spatiotemporal dynamics of radioactive cesium deposited in urbanized areas became apparent. The behavior of radioactive cesium deposited in the Tokyo special ward is strongly influenced by weather conditions such as rainfall.

Keywords Tokyo · Wastewater treatment system · Radioactive cesium contamination · Incineration ash · Periodicity

5.1 Introduction

Globally, nuclear power plants currently in operation will be decommissioned one after another as the use of renewable energy such as wind power and solar power is promoted. However, the risk of radioactive contamination of densely populated urban areas is not due only to the leakage of radioactive materials from nuclear

H. Yamazaki, *Radioactive Contamination of the Tokyo Metropolitan Area*,
https://doi.org/10.1007/978-981-15-7368-2_5

power plants and nuclear fuel reprocessing plants. Nuclear batteries and depleted uranium loaded on satellites and aircraft can be sources of radioactive contamination in urban areas. The risk of terrorism using nuclear materials cannot be ignored. Furthermore, radioactive materials used in medicine are also released daily into the environment along with patient excrement. Thus, not only the Tokyo metropolitan area but also many large cities around the world are exposed to these risks of radioactive contamination.

An important issue regarding radioactive contamination in the Tokyo metropolitan area due to the FDNPP accident was that the Japanese government did not conduct a survey of radioactive contamination or simulate the spatiotemporal variations in contamination. Additionally, if such work was performed, it was not announced at the appropriate time immediately after the accident. As mentioned in Chap. 2, Japan established the SPEEDI system for predicting radiation dose due to environmental radioactive contamination in an emergency, but this system was utterly useless in the event of the FDNPP accident. On September 2, 2014, the Science Council of Japan published a report entitled "Comparison of model calculation results on transport and deposition processes of radioactive materials released into the environment by TEPCO's Fukushima Daiichi Nuclear Power Plant accident" [1]. Highly academic intercomparisons of many atmospheric transport models of radioactive materials were conducted. Nevertheless, this report did not address the dynamics of radioactive contamination in the Tokyo metropolitan area, the most important populated area in Japan, or the transport and depositional processes of radioactive materials in Tokyo. The purpose of this committee was to evaluate simulations to elucidate the process of radioactive transport and deposition. Such verification is essential to prepare for a possible nuclear disaster. However, the Commission's report did not discuss any issues regarding radioactive contamination in the Tokyo metropolitan area, the center of Japan. Did many people desire to see such a report?

As sewage treatment systems are complete in most of these cities, the role of sewage treatment systems in removing radioactive material from radioactively contaminated cities is significant [2–4]. The Tokyo metropolitan area is a large urban area that has just been exposed to radioactive contamination, and it is an excellent experimental field to elucidate the relationship between the behavior of radioactive materials deposited there and the sewage treatment system. The capital Tokyo has an area of 2188 km^2 and a population of 13.9 million; its 23 special wards have an area of 619 km^2, and 9.6 million people live there. Most of the area except for parks and shrines is urbanized, and the Tokyo special ward is a town covered with concrete and asphalt, such as roads and buildings. In most of the Tokyo special ward, urban drainage is treated by a sewage system that combines domestic drainage and rainwater. Urban wastewater is treated at sewage treatment plants installed in various places, where waste is separated into sludge and purified water. The sludge is dried and incinerated, and the sludge incinerated ash is processed into flakes with cement and buried in a landfill called the Central Breakwater in Tokyo Bay. Purified water is released into the river and flows into Tokyo Bay.

Tokyo is likely the first metropolis in the world to be highly contaminated by radioactive materials. Radioactive nuclides deposited on the ground surface are known to be adsorbed and retained by clay minerals in the soil, but the behaviors of radioactive nuclides deposited on concrete and asphalt in urban areas are often unknown. The Tokyo special ward, which has a sewage treatment system that mixes domestic wastewater and rainwater for each region, is the best place to analyze the dynamics of radioactive materials in urban areas. In this chapter, the temporal changes in the radioactive cesium concentration in the sludge incineration ash of each sewage treatment plant, which were released on the Tokyo homepage after the Fukushima nuclear disaster, were analyzed. As a result, the spatiotemporal dynamics of radioactive cesium deposited in urbanized areas became apparent. The behavior of radioactive cesium deposited in the Tokyo special ward is strongly influenced by weather conditions such as rainfall.

5.2 Wastewater Treatment System in Tokyo

The radioactivity of incinerated sludge ash discharged from the Tokyo wastewater treatment center has been monitored by Tokyo since May 2011, and the results have been provided on the Tokyo website ([5] a, b). In most parts of Tokyo, domestic wastewater from homes and offices is sent to sewage treatment centers through sewers. During rainfall, rainwater also flows into these sewer pipes and is mixed with domestic wastewater, sent to a sewage treatment center, and purified. In Tokyo, such a sewage treatment plant is called a water reclamation center. Radioactive materials such as radioactive iodine and radioactive cesium are captured very efficiently in the sewage treatment system at the water reclamation center and become part of the sludge. The water reclamation center burns this sludge to make incinerated ash. At the time of the FDNPP accident, this sludge incineration ash was mixed with cement, processed into flakes, and buried in a landfill called the Central Breakwater in Tokyo Bay.

It is undeniable that volatile radioactive iodine and radioactive cesium could have been scattered into the atmosphere during the incineration of contaminated sludge in the early stages of the FDNPP accident. The Tokyo government did not expect that the Tokyo wastewater treatment system would be exposed to radioactive contamination due to a nuclear disaster. In other words, there is a strong possibility that radiation monitoring of the wastewater treatment system, such as radioactivity in sludge, sludge incineration ash, fly ash, and combustion exhaust gas, was not carried out during March 2011, when the most highly radioactive plume hit Tokyo. At least, such data cannot be found on the website of the Tokyo government.

The Tokyo government is equipped with domestic wastewater-rainwater purification treatment systems in various places in the metropolitan area [6]. Therefore, the time-series fluctuations in the radioactivity of the incineration ash discharged from the water reclamation center are considered to reflect the amount and behavior of radioactive material deposited in the catchment area of each center. Figure 5.1

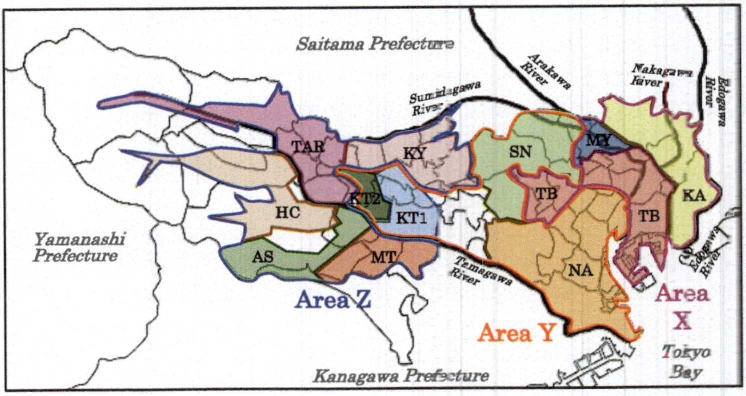

Fig. 5.1 Catchment areas of the Tokyo water reclamation centers [6]. Area classification of the water reclamation centers is shown in Tables 5.1 and 5.2. Area X: Eastern Tokyo. Highly contaminated by radionuclides. This area is mainly a residential downtown. Area Y: Central Tokyo. This region is the center of Japan's politics and economy. Area Z: Western Tokyo. The radioactive contamination level is low. It is mainly a suburban residential area. The original of this figure was published in PLOS ONE (M. Ishida and H. Yamazaki, November 14, 2017) [7]. However, it was misidentified the information on the watershed of the water reclamation centers published by the Tokyo government. This map is a revised version of the catchment areas

shows the catchment area of each water reclamation center. This map indicates the water reclamation center that purifies most of the mixed wastewater discharged from the Tokyo special ward. Many water reclamation centers have sludge incinerators installed in their facilities. However, NA, TB, and KA shown in Fig. 5.1 are sludge treatment plants. These sludge treatment plants receive sludge from several water reclamation centers and incinerate it. The catchment area of each facility has been announced by Tokyo, and the annual sludge discharges ([5] a, b, [6]) are shown in Tables 5.1 and 5.2.

Therefore, the time-series fluctuations in the radioactive cesium concentration of the sludge incineration ash of the water reclamation center published by the Tokyo metropolitan government were analyzed for 5 years after the FDNPP accident. Moreover, the distribution and behavior of radioactive cesium in the urban area of the Tokyo metropolitan area were evaluated.

5.3 Temporal Changes in the Radioactive Cesium Concentration of Sludge Incineration Ash Released from the Water Reclamation Centers of Tokyo

Regarding the processes that transport radioactive material from the catchment to the environment via rainfall and rivers, various measurement data and simulation results based on these data have been reported [9–16]. A detailed comparison of Swedish

Table 5.1 Annual changes in the concentrations and amounts of radioactive cesium contained in the incineration ash discharged from the wastewater treatment plants

Tokyo Water Reclamation Center			$^{134+137}$Cs in Ash, Bq/kg (annual average)					Annual ash discharged, ton/year						Annual discharged amounts of $^{134+137}$Cs by Ash, TBq ($\times 10^{12}$ Bq)					
Area	Code	Name	2011	2012	2013	2014	2015	2011	2012	2013	2014	2015	Total	2011	2012	2013	2014	2015	Total
X	KA	Kasai	26,159	8770	4202	2490	1606	5260	6597	5720	6444	6159	30,180	0.138	0.058	0.024	0.016	0.010	0.245
	MY	Miyagi	9776	3488	2060	1099	842	1320	1543	1095	1466	1319	6743	0.013	0.005	0.002	0.002	0.001	0.023
	TB	Tobu	9649	4146	2082	1271	690	6814	8515	8003	8535	8259	40,126	0.066	0.035	0.017	0.011	0.006	0.134
	Average or Total		15,195	5468	2781	1620	1046	13,394	16,655	14,819	16,444	15,737	77,048	0.216	0.099	0.043	0.028	0.017	0.403
Y	SN	Shingashi	8108	3262	1671	988	559	2850	3539	3666	3668	3603	17,326	0.023	0.012	0.006	0.004	0.002	0.046
	NA	Nanbu	5965	2410	1098	716	461	10,218	13,295	12,685	13,201	12,990	62,389	0.061	0.032	0.014	0.009	0.006	0.122
	KT1	Kitatama 1	6419	2413	1166	737	400	982	1332	1323	1321	1328	6286	0.006	0.003	0.002	0.001	0.001	0.013
	KT2	Kitatama 2	8551	2997	1217	818	528	319	302	375	368	339	1703	0.003	0.001	0.000	0.000	0.000	0.005
	Average or Total		7260	2770	1288	815	487	14,369	18,468	18,049	18,559	18,259	87,704	0.093	0.048	0.022	0.014	0.009	0.186
Z	KY	Kiyose	1829	751	351	242	135	1268	1608	1606	1635	1607	7722	0.002	0.001	0.001	0.000	0.000	0.005
	MT	Minamitama	633	607	296	192	102	519	676	639	669	657	3160	0.000	0.000	0.000	0.000	0.000	0.001
	TAR	Tamagawa	2217	759	358	221	107	892	1183	1288	1220	1236	5818	0.002	0.001	0.000	0.000	0.000	0.004
	HC	Hachioji	2538	956	414	302	164	446	516	287	466	401	2115	0.001	0.000	0.000	0.000	0.000	0.002
	AS	Asagawa	1940	673	245	131	72	213	302	489	358	396	1758	0.000	0.000	0.000	0.000	0.000	0.001
	Average or Total		1832	749	333	218	116	3337	4285	4308	4347	4297	20,574	0.006	0.003	0.001	0.001	0.001	0.012
Whole average or total			6982	2603	1263	767	472	31,099	39,409	37,176	39,350	38,292	185,327	0.316	0.149	0.066	0.044	0.026	0.601

These values estimated in data published by the Tokyo water reclamation centers ([5] a, b). The values of 2011 are cumulative from April to December

Table 5.2 Estimating the radioactive cesium balance in Tokyo by the wastewater treatment system

Tokyo Water Reclamation Center			Catchment area	Radioactive cesium discharged as ash[a]	Estimated inventory[b]				Estimated outflow for 5 years after the accident, %	
					MEXT airborne monitoring		Soil measurement			
Area	Code	Name	km^2 [s]	TBq [t]	kBq/m^2	TBq [u]	kBq/m^2	TBq [v]	[t/u]	[t/v]
X	KA	Kasai	110	0.280	25	2.75	50	5.50	10.2	5.1
	MY	Miyagi	17	0.027	6	0.10	20	0.34	27.0	7.9
	TB	Tobu	136	0.157	7	0.90	40	5.44	16.5	2.9
	Total or Average[c]		(263)	(0.464)	(14.3)	(3.75)	(43.7)	(11.3)	(12.4)	(4.1)
Y	SN	Shingashi	104	0.054	5	0.52	7	0.73	10.4	7.4
	NA	Nanbu	211	0.143	5	1.06	10	2.11	13.5	6.8
	TK1	Kitatama 1	51	0.015	5	0.26	10	0.51	5.8	2.9
	TK2	Kitatama 2	27	0.005	5	0.14	–	–	3.6	–
	Total or average[c]		366 (393)	0.212 (0.217)	5.0 (5.0)	1.84 (1.98)	9.2	3.35	11.5 (11.0)	6.3
Z	KY	Kiyose	80	0.006	1	0.08	5	0.40	7.5	1.6
	MT	Minamitama	59	0.001	1	0.06	–	–	1.7	–
	TAR	Tamagawa	94	0.004	1	0.09	–	–	4.4	–
	HC	Hachioji	85	0.002	1	0.09	–	–	2.2	–
	AS	Asagawa	39	0.001	1	0.04	–	–	2.5	–
	Total or average[c]		80 (356)	0.006 (0.014)	1.0 (1.0)	0.08 (0.36)	5.0	0.40	7.5 (3.9)	1.5
Whole total or average[c]			709 (1012)	0.682 (0.695)	8.0 (6.0) [u/s]	5.67 (6.09)	21.6 [v/s]	15.1	**12.0** (11.4)	**4.5**

The original of this table published in PLOS ONE (M. Ishida and H. Yamazaki, November 14, 2017) [7] was incorrect data entered due to insufficient analysis. The values shown here are revised values

[a]Amounts of radioactive cesium discharged as the ash was obtained decay-corrected value to March 16, 2011. $^{134}Cs/^{137}Cs$ ratio immediately after the accident assumed to be 1.0

[b]The $^{134+137}Cs$ precipitations by the MEXT airborne monitoring [8] in the areas Z and Y are estimated values

[c]The values in parentheses are the total or average for all the centers

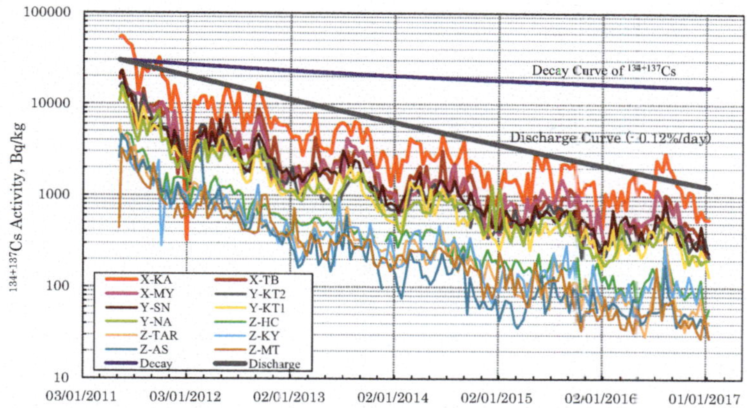

Fig. 5.2 Temporal changes in $^{134+137}$Cs activities in the incineration ash discharged from the Tokyo wastewater reclamation centers. The data are quoted from the Tokyo government website ([5] a, b)

model predictions and actual observations following the Chernobyl accident was also performed to examine the migration process of ^{137}Cs through sewage treatment systems in urban environments [2]. However, the behavior of radionuclides in large urban areas remains almost entirely unknown. Radioactive cesium in sludge incineration ash discharged from Tokyo water reclamation centers can be used as a tracer to evaluate the behavior of radioactive cesium in urban environments.

In this study, the temporal changes in the radioactive cesium concentrations of sludge incineration ash discharged from 12 water reclamation centers in Tokyo shown in Figs. 5.1 and 5.2 from May 2011 to January 2017 were investigated. The radioactive cesium concentrations in the sludge incineration ash from these 12 water reclamation centers were listed on the Tokyo website [6]. Tables 5.1 and 5.2 show the names of each water reclamation center.

5.3.1 Temporal Changes in Radioactive Cesium Concentration of Sludge Incineration Ash

Figure 5.2 shows the changes over time in the radioactive cesium radioactivity of sludge incineration ash discharged from each wastewater treatment center. The first measurements were published on May 10, 2011. The Kasai Center (KA), which has the highest concentration in the eastern part of Tokyo, showed 24,000 kBq/kg for ^{134}Cs, 29,100 Bq/kg for ^{137}Cs, and 53,200 Bq/kg for $^{134+137}$Cs. On the other hand, the Minami Tama Center (MT), a watershed in the suburbs of western Tokyo, showed the lowest concentrations of 195 Bq/kg for ^{134}Cs, 241 Bq/kg for ^{137}Cs, and 436 Bq/kg for $^{134+137}$Cs. Nevertheless, the concentration of $^{134+137}$Cs at this center increased to 3200 Bq/kg on May 18.

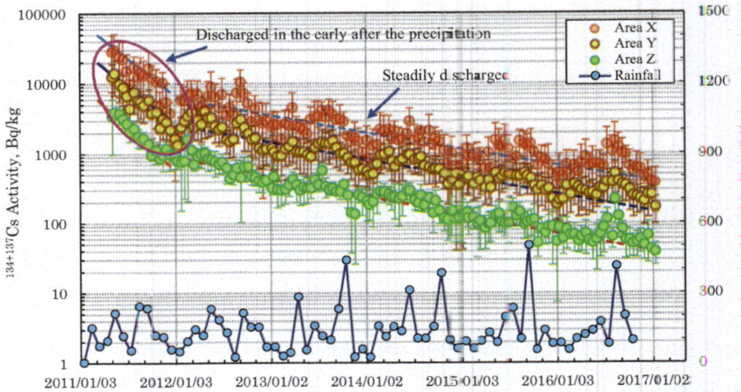

Fig. 5.3 Relationship between temporal changes in $^{134+137}$Cs activities in incineration ash and monthly rainfall in Tokyo [17]

Table 5.3 Concentrations of radioactive cesium discharged by incineration ash after the FDNPP accident

Area	Initial concentration of $^{134+137}$Cs[a], Bq/kg	$^{134+137}$Cs in the sharp decrease period		
		End of the sharp decrease[b], Bq/kg	Easy discharged, Bq/kg	Discharged[c], %
X	50,000	8000	42,000	84
Y	20,000	3000	17,000	85
Z	6000	1100	4900	82

The amounts of released ^{137}Cs were estimated from Fig. 5.3. The ^{137}Cs activities were decay-corrected value to March 16, 2011

[a]Estimated from the concentrations in May 2011

[b]From immediately after the accident to December 2011

[c]The discharged proportion in the initial 10 months to the total discharge. The total discharge was cumulative from October 2011 to December 2016

As shown in Fig. 5.3 and Table 5.3, the radioactive cesium concentrations in the sludge incineration ash decreased almost exponentially with time. The discharge rates were faster than the radioactive decay rate of $^{134+137}$Cs released by the FDNPP accident, and radioactive cesium deposited on the surface was forcefully removed by the sewage treatment system and recovered at water reclamation centers. The half-life of radioactive cesium removed from the ground surface was estimated to be 1.13–1.24 years after 2012. The half-life of this decline is much shorter than the ^{134}Cs half-life of 2.06 years, suggesting that radioactive cesium is relatively easily removed in urban areas. Not all of the radioactive cesium deposited in Tokyo was discharged from the wastewater treatment system, but in central Tokyo, 0.15–0.17% of the radioactive cesium precipitated on the ground surface was removed daily.

5.3.2 *Geographic Distribution and Existing Species of Radioactive Cesium in Tokyo*

Figure 5.2 shows that the amounts of radioactive cesium deposited vary by region. Its three regions are Area X (KA, TB, and MY), Area Y (KT2, SN, KT1, and NA), and Area Z (HC, TAR, KY, AS, and MT), as shown in Fig. 5.1. Area X is mainly a downtown residential zone, Area Y contain mostly company and government offices, and Area Z is a residential suburban region. This variation is in agreement with the results of the MEXT airborne monitoring and the soil measurements by the author, indicating that the radioactive cesium activity in the sludge incineration ash reflects the activity levels in the catchment area of each water reclamation center. Figure 5.3 shows the arithmetic mean values of radioactive cesium concentrations in sludge incineration ash discharged from each water reclamation center for Areas X, Y, and Z. In each area, the concentration of radioactive cesium in the sludge incineration ash decreased rapidly in the first 8 months after the accident. As shown in Table 5.3, 82–85% of the radioactive cesium moved from the surface to the water reclamation center during the first phase of the accident. Such a rapid decrease in the concentration of radioactive cesium deposited in the environment immediately after the FDNPP accident is also noted in Chaps. 3 and 7.

This phenomenon for a rapid decline in radioactive cesium concentrations in the early stages of the accident did not occur only in urban areas covered with concrete or asphalt. It is speculated that highly mobile radioactive cesium exists as nonionic ultrafine particles and is therefore difficult to absorb and retain in the soil. As pointed out in Chap. 3, the presence of ultrafine particles containing high concentrations of radionuclides called hot particles (References 40–47 in Chap. 3) may have contributed to the mobility of radioactive cesium in the early stages of the accident. In the Tokyo sewage treatment system, radioactive cesium that was easily washed away in the first stage was also estimated to have been deposited as non-adsorbable species in the catchment areas of the water reclamation centers. Perhaps this radioactive cesium existed as hot fine particles.

It was assumed that the radioactive cesium deposited in Tokyo would quickly move to the water reclamation center without being fixed in the soils because they were covered with concrete and asphalt. However, this assumption was a mistake. As clearly shown in Fig. 5.3 and Table 5.3, the release rate of radioactive cesium hardly changed, even in areas with different urban environments. Instead, Area Z, which has a low proportion of concrete and asphalt, appears to have a shorter half-life for the decrease in radioactive cesium and greater mobility than Areas X and Y. Most of the radioactive cesium flowing into the Tokyo wastewater treatment system is estimated to have been washed away by rainwater from the catchment area of the water reclamation centers. These results suggest that the processes by which the wastewater treatment system recovered the radioactive cesium that contaminated Tokyo due to the FDNPP accident were very complicated. In particular, the fact that concrete and asphalt may have a higher ability to retain radioactive cesium than soil is an essential finding for assessing the dynamics of radioactive cesium contamination in urban areas.

5.4 Frequency Analysis Between the Radioactive Cesium Concentration in Sludge Incineration Ash and Monthly Rainfall

Rainwater, as well as domestic wastewater, flows into the Tokyo water reclamation centers. Rainwater is thought to have the effect of washing away radioactive cesium deposited on the ground surface and transporting it to the water reclamation centers. Therefore, the relationship between the concentration of radioactive cesium in sludge incineration ash and the amount of rainfall was investigated. The relationship between the radioactive cesium activity in the sludge incineration ash and the monthly rainfall in the Tokyo metropolitan area was analyzed by frequency analysis using the MATLAB software.

As already stated, the radioactive cesium activity in the soil decreased abruptly immediately after the FDNPP accident in the garden soil of Site "i" (Sect. 3.7) and the forest environments (Chap. 7). This phenomenon was also found in the school ground soils of Koto Ward in eastern Tokyo [18]. Even in the case of sludge incineration ash, as shown in Fig. 5.3, the rapid decrease in the radioactive cesium activity in the immediate aftermath of the accident has been confirmed. This figure also shows monthly rainfall fluctuations in Tokyo [17]. However, these phenomena with rapid changes cannot be easily explained based on run-off via rainwater for dynamic precipitation levels in Tokyo [19]. The rapid decrease in the radioactive cesium activity in sludge incineration ash in the immediate aftermath of the accident can be explained by hypothesizing that highly mobile chemical species of radioactive cesium that were deposited immediately after the accident flowed into and were discharged from the wastewater treatment system. The author could not find these radioactive cesium species directly, but cesium-bearing microparticles alluded to by many researchers [References 40–47 in Chap. 2] are considered the native mobile species of radioactive cesium estimated in this book. As shown in Table 5.3, 82–85% of the radioactive cesium from wastewater treatment systems was discharged during the 8 months when the radioactivity decreased rapidly in the immediate aftermath of the accident.

Regarding the steady-state period from October 2011 to January 2017, the periodicity between rainfall and radioactive cesium activity in the sludge incineration ash was examined with the MATLAB software. The results are presented in Fig. 5.4 and show that in the urbanized Areas X and Y, the rainfall and radioactive cesium activity conform in 4-, 8-, and 12-month cycles. This frequency matching suggests that in these districts, radioactive cesium flowed into the wastewater treatment systems carried by rainwater. However, in Area Z, which has a different urban structure, such periodicity is not found. This result suggests that the decrease

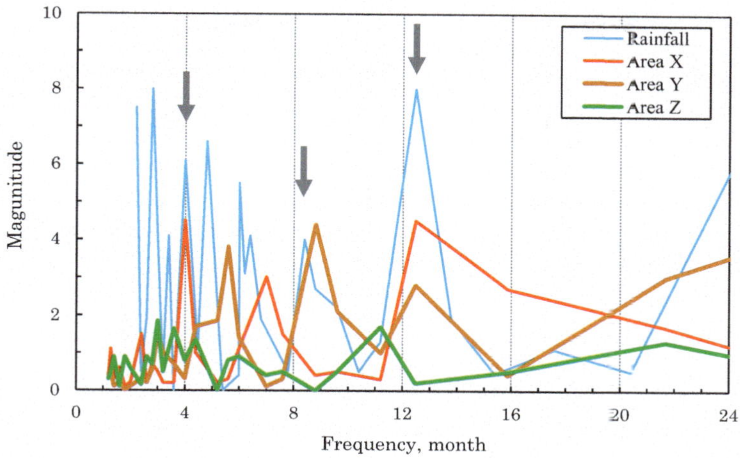

Fig. 5.4 Periodicity between $^{134+137}Cs$ contamination in incineration ash and monthly rainfall. Frequency analysis was performed using MATLAB software. Vertical arrows indicate matching periods

in the radioactive cesium concentration observed in the sludge incineration ash in the suburb of Tokyo was not dependent on rainfall. The movement of radioactive cesium in large urban areas such as Tokyo is strongly influenced not only by the state of radioactive cesium but also by the urban environment.

The adsorption behavior of cesium to suspended particles has already been mentioned [10, 20]. Recently, ultrafine particles with high activity of radioactive cesium discharged by the FDNPP accident have been detected in different regions [References 40–47 in Chap. 3]. Presumably, radioactive cesium on such ultrafine particles does not exist as an ionic species; hence, it likely cannot be maintained adsorbed in clay layers. If ultrafine particles containing radioactive cesium are assumed to have been deposited on the ground surface in the immediate aftermath of the accident, they were probably more easily transported by falling rain or blowing wind than ionic cesium held in the soil. Furthermore, radioactive cesium contained in such ultrafine particles is insoluble in water, so presumably, when radioactive cesium contained in these microparticles is deposited on concrete or asphalt, it is selectively washed away by rainfall.

Quantitative data could not be obtained for the coverage of buildings, concrete, and on to the ground in Areas X, Y, and Z. However, the period of decrease in radioactive cesium in the sludge incineration ash in Areas X and Y is consistent with the fluctuating cycle of rainfall amounts. This result implies that radioactive cesium deposited in the central area of Tokyo is effectively removed by rainfall. On the other hand, the radioactive cesium activity in the garden soil of Site "i" located in Area X dose not fluctuate significantly due to rainfall after the accident. However, in the case of Area Z, despite the shorter half-life of decrease than in Areas X and Y, periodicity is not observed between rainfall and the radioactive cesium outflow. Elucidation of such difficult-to-understand phenomena is an issue for the future.

These results indicate that not only the physical and chemical states of radionuclides but also the urban structure may have significant influences on the behaviors of radionuclides when urban areas, such as the Tokyo metropolitan area, receive radioactive contamination. The tremendous discharge rate of radioactive cesium from the wastewater treatment system in the early stages of the accident is an essential parameter that must be considered when planning for radioactive material decontamination in a contaminated urban area during a nuclear disaster.

5.5 Radioactive Cesium Balance in Tokyo by Sludge Incineration Ash from the Wastewater Treatment System

Table 5.1 shows the annual amount of sludge incineration ash discharged from the Tokyo water reclamation centers and the annual average concentration of radioactive cesium (average arithmetic value). These data were calculated based on figures published on the Tokyo government website ([5] a, b, [6]). This table also shows the annual amounts of radioactive cesium released based on the amounts of incinerated ash and its radioactivity concentrations. The amount of sludge incineration ash emissions is thought to depend on the area and population in the catchment region of the water reclamation center, although the Nanbu Center (NB) with the largest area has had the greatest emissions. The total emissions of sludge incineration ash over 5 years exceeded 60,000 ton. Compared to suburban Area Z, incinerated ash emissions in densely populated Areas X and Y were approximately four times higher. Table 5.2 shows the area of the catchment region of each center. These data have already been published in PLOS ONE [7], but the author's aggregation of the catchment areas was incorrect. Table 5.1 shows the corrected values. Therefore the values in Table 5.2 have also been revised from those of the original paper.

Radioactive contamination levels in the Tokyo metropolitan area were classified into Areas X, Y, and Z based on the radioactive cesium concentration in the sludge incineration ash and the radioactive contamination levels in the soil, as shown in Fig. 5.1. The amounts of radioactive cesium discharged as sludge incineration ash from immediately after the accident until December 2015 were estimated to be 0.403 TBq in Area X, 0.186 TBq in Area Y, and 0.012 TBq in Area Z in terms of radioactivity without correction for radioactive decay (TBq: $\times 10^{12}$ Bq). The total amount of radioactive cesium transferred from these areas to the incineration ash via the water reclamation centers exceeded 6 TBq. The incinerated ash contaminated with radioactivity is solidified by mixing it with cement and then buried in a landfill called the Central Breakwater in Tokyo Bay. Since cesium, which is an alkali element, is highly likely to be ionized during solidification of cement, the radioactive cesium in the solidified debris may leach into the ground by the action of groundwater after burial.

The role of the Tokyo wastewater treatment system in the decontamination of radioactively contaminated urban areas was evaluated. Table 5.2 shows the results estimated using the values corrected for radioactive decay to March 16, 2011. The inventories of $^{134+137}$Cs in Areas X, Y, and Z are estimated from the MEXT airborne monitoring and soil measurements. Although the uncertainty of the measurements suggests that these values are not very reliable, there is no other suitable estimation method. In addition, radioactivity measurements for the soils in the catchments of TK2, MT, TAR, HC, and AS could not be implemented.

Although the amount of $^{134+137}$Cs deposited in the study area was estimated to be 6.09 TBq from the MEXT airborne monitoring performed in October 2011 [8, 21], its value was determined to be 15.1 TBq from the soil measurements. The inventory estimated from the airborne monitoring in the soil measurement areas was 5.67 TBq. Although there are considerable differences between these values, the results indicate that it is difficult to determine the amount of radioactive material deposition in urban areas. However, both methods reveal that 60–75% of the radioactive cesium deposited in the study area accumulated in Area X and that 22–33% accumulated in Area Y. Area Z accounts for one-third of the total area, but the radioactive cesium inventory is less than a few percent.

On the other hand, the amounts of $^{134+137}$Cs discharged as incinerated ash decay-corrected to the values on March 16, 2011, were estimated to be 0.646 TBq in Area X, 0.393 TBq in Area Y, and 0.014 TBq in Area Z. Therefore, the proportion of radioactive cesium that was washed out and discharged by the wastewater treatment system in Tokyo by the end of 2015 was estimated to be 11.4% by airborne monitoring and 4.5% by soil measurements. Additionally, relatively high outflow rates were observed in urbanized Areas X and Y. This distribution means that mobile radioactive cesium species deposited on concrete and asphalt, as discussed in Sect. 5.3, were quickly transferred to wastewater treatment systems soon after the accident. However, this result also suggests that approximately 90% of the deposited cesium may remain in Tokyo. The fact that most of the radioactive cesium remains in the urban environment more than 5 years after the accident is a concern from the viewpoint of radiation protection for residents.

5.6 Temporal Changes in Radioactive Cesium Concentration in Incinerated Ash at a Waste Incineration Plant in the Suburbs of Tokyo

There are two levels of legal standards in Japan for waste contaminated with radioactive material [22]. The designated standard for disposal of waste by separation, incineration, landfill, *etc.*, is 8000 Bq/kg. The clearance level for reusing waste among the general public is 100 Bq/kg. Various debates continue in Japan about the suitability of these standards, and not all Japanese people have accepted these standards. Nevertheless, the IAEA has evaluated the Japanese standard that landfilling of contaminated waste with radioactive cesium concentrations of less than 8000 Bq/kg in managed landfills is consistent with international methodologies [23].

Monitoring the radioactive contamination of household and industrial wastes in areas affected by the FDNPP accident is important not only from the viewpoint of controlling the radiation exposure of residents but also from that of the disposal of contaminated wastes. Matsudo City and Kashiwa City in Chiba Prefecture, which are suburbs of Tokyo, are believed to be highly contaminated areas in the Tokyo metropolitan area. Since immediately after the FDNPP accident, these cities have been monitoring radioactive contamination at waste treatment plants and have published the results [24, 25]. Matsudo City is located approximately 15 km northeast of central Tokyo, with a population of 490,000 and an area of 61 km^2, and many areas are residential regions. Kashiwa City is approximately 30 km northeast from Tokyo and has a population of 430,000 and an area of 115 km^2. In addition to housing, Kashiwa City has many educational and research facilities. These cities are adjacent. In these areas, $^{134+137}$Cs deposition according to the MEXT airborne monitoring was 10–100 kBq/m^2. The concentration of $^{134+137}$Cs on the soil surface immediately after the FDNPP accident were 965–5250 Bq/kg ($n = 6$) in Matsudo city and 6400–67,800 Bq/kg ($n = 4$) in Kashiwa city.

The dynamics of radioactive contamination of domestic and industrial waste discharged from a typical suburban area in Japan that has received radioactive contamination from the FDNPP accident, are of great interest. Section 5.3 evaluates the dynamics of radioactive contamination in Tokyo's wastewater treatment systems but does not discuss solid waste such as garbage. In Matsudo City, the concentrations of incinerated ash and fly ash solidified from waste treatment facilities in the city have been maintained since July 4, 2011, and the data are available on the city's website [24]. Kashiwa City also released data from August 17, 2011, but it seems impossible to see past data at present [25]. The results are shown in Fig. 5.5.

Chiba Prefecture also received substantial damage on March 11, 2011, due to the Tohoku-Pacific Ocean Earthquake. The earthquake entirely or partially destroyed 17 buildings in Kashiwa City, and one person died. In Matsudo City, 140 buildings were entirely or partially destroyed. These areas were affected by contaminated plumes containing radioactive materials derived from the FDNPP disaster. As a result, citizens brought large amounts of radioactively contaminated debris to the waste treatment facilities of these cities, which was incinerated. Waste treatment facilities in Japan differ in the types of waste received (household waste, industrial waste, sewage sludge, *etc.*), and pretreatment (whether separated or mixed for each material) is different in each facility. The type of incinerator, the method of collecting fly ash, the method of treating main ash and fly ash, and the method of solidification also vary depending on the facility. Therefore, it is difficult to evaluate the status of radioactive contamination from the radioactive concentrations of incinerated ash discharged from a waste treatment site. Nevertheless, in Japan, municipalities generally collect their local waste, so the temporal changes in the

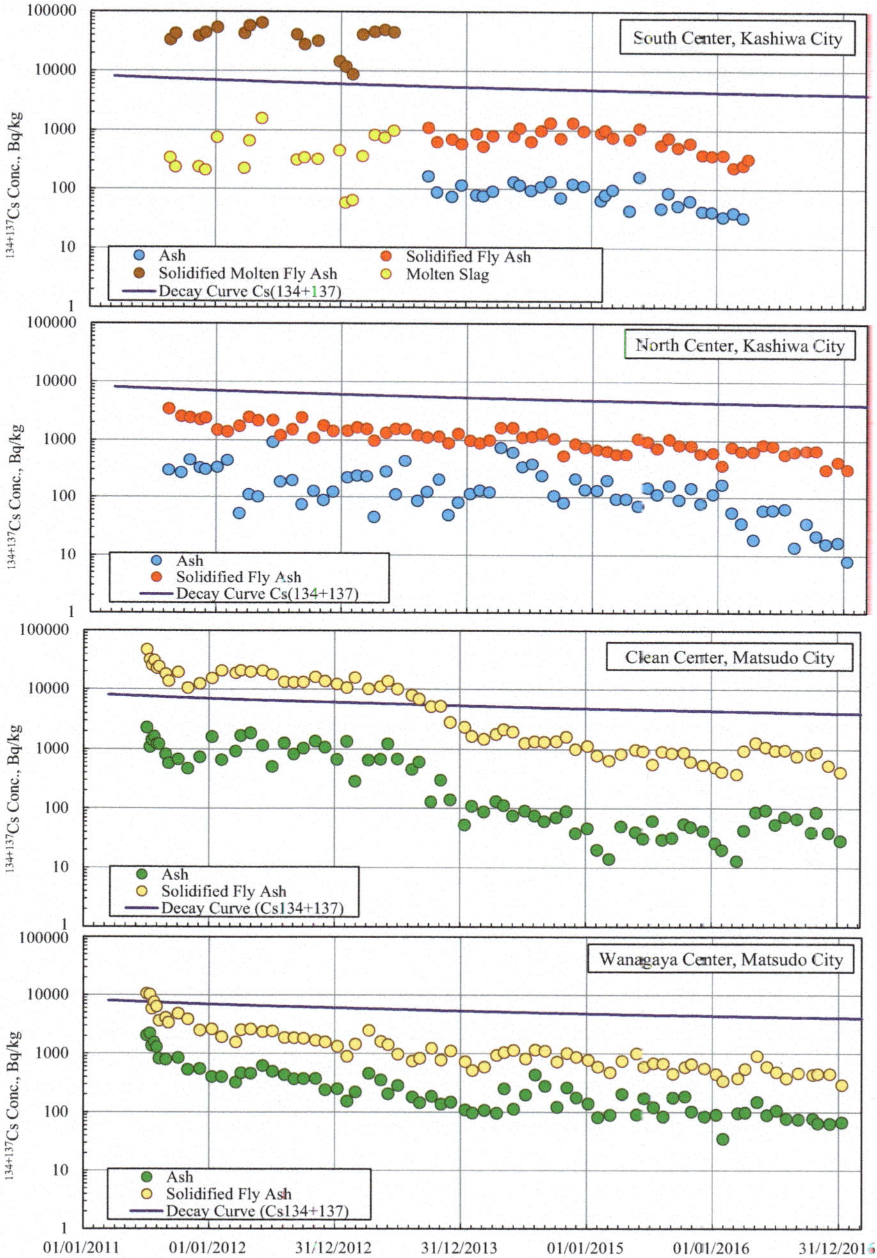

Fig. 5.5 Temporal changes in $^{134+137}$Cs activity in incineration ash at the garbage treatment plants [24, 25]. Precipitations of $^{134+137}$Cs in these cities were 30–300 kBq/m^2, according to the MEXT airborne monitoring [8]

radioactive concentrations of incinerated ash should reflect the fluctuations in radioactive contamination in an area.

Figure 5.5 shows the temporal changes in the $^{134+137}$Cs concentrations of the incinerated ash from waste treatment plants obtained from the websites of Kashiwa City [25] and Matsudo City [24]. Table 5.4 shows the concentrations at the appropriate times. The South Center in Kashiwa City seems to have changed its waste disposal method in the summer of 2013. Until then, the radioactive concentrations of the solidified molten fly ash and molten slag were measured, but after this time the main ash (Ash) and the solidified fly ash were measured. Although no measurement data were available for this center from spring 2016 to the end of 2018, recent data can be seen. The $^{134+137}$Cs concentrations of the solidified fly ash discharged from this center during the 2 years after the accident were high at 38,600 ± 15,300 Bq/kg (n = 18). The radioactive decay of ^{134}Cs should have reduced the level of radioactive contamination by $^{134+137}$Cs to approximately 70% of the initial value, but as seen from Fig. 5.5, this radioactivity concentration hardly decreases. However, data from the North Center, which has continuously measured the $^{134+137}$Cs concentration of solidified fly ash since immediately after the accident, show that the concentration decay is consistent with the radioactive decay rate of these nuclides. The tendency of the $^{134+137}$Cs concentrations in the incinerated ash of the waste collected from Kashiwa City in the South and North centers to change over time may be due to different incineration methods. Since the treatment method of the South Center changed in the fall of 2013, the radioactive cesium concentrations of ash and solidified fly ash have reached almost the same levels as those of the North Center. Usually, a fly ash melting furnace has a high temperature of 1300 °C or more, so there is concern that radioactive cesium may be vaporized into the atmosphere but no information is available on this process. If radioactive cesium vaporizes from the waste combustion furnace, the surrounding environment would be polluted again.

Regarding the incinerated ash at the waste disposal site in Matsudo City the concentration of $^{134+137}$Cs from July 2011 to the present (2020) has been announced [24]. Figure 5.5 shows the data up to January 2017. The average concentrations for October to December 2019 are shown in Table 5.4. Although the radioactive cesium contamination of the soil in Matsudo City was lower than that in Kashiwa City, the concentration of $^{134-137}$Cs in the incinerated ash was much higher in Matsudo City. In the Clean Center and Wanagaya Center, the concentration ratios between the main ash (Ash) and the solidified fly ash were significantly different. As shown in Table 5.4, the rates were 14–31 at the Clean Center and 4.8–6.5 at the Wanagaya Center. This contrast is considered to reflect the differences between the incinerators and the method for solidifying fly ash. The concentration factor is an essential factor for the operation of a waste treatment system because the incineration of waste is intended for volume reduction.

A sharp decrease in $^{134+137}$Cs concentration was observed in Matsudo city's waste incineration ash after the FDNPP accident. This trend was the same as that in sludge incineration ash in Tokyo (Sect. 5.3). As shown in Table 5.4, $^{134+137}$Cs decreased by approximately 60% in 4 months. The radioactive contamination level was not predetermined for the waste coming into the treatment facility. Therefore,

Table 5.4 Changes in $^{134+137}$Cs contamination in waste incineration ash in the Kashiwa and Matsudo cities

Kashiwa City									
South Center					**North Center**				
Date	**n**	**Molten slag [a], Bq/kg**	**Solidified molten fly ash [b], Bq/kg**	**Concentration factor, [b]/[a]**	**Date**	**n**	**Ash [c], Bq/kg**	**Solidified fly ash [d], Bq/kg**	**Concentration factor, [d]/[c]**
Aug–Sep, 2011	3	287 ± 73	37,900 ± 6440	130	Aug–Oct, 2011	3	336 ± 97	2800 ± 543	8.3
Apr–Jun, 2013	3	853 ± 118 (2.79)	46,600 ± 2440 (1.23)	55					
		Ash [c], Bq/kg	Solidified fly ash [d], Bq/kg	Concentration factor, [d]/[c]					
Sep–Nov, 2013	3	109 ± 48	810 ± 254	7.4					
Feb–Apr, 2016	3	39.3 ± 6.0 (0.36)	271 ± 49 (0.34)	6.9	Oct–Dec, 2016	3	18.3 ± 3.2 (0.055)	467 ± 186 (0.17)	26
Jan, Jun, 2019	2	66.9 ± 18.2 (0.61)	391 ± 172 (0.48)	5.9	Oct–Dec, 2019	3	47.3 ± 12.7 (0.14)	376 ± 30 (0.13)	8.0
Matsudo City									
Clean Center					**Wanagaya Center**				
Date	**n**	**Ash [c], Bq/kg**	**Solidified fly ash [d], Bq/kg**	**Concentration factor, [d]/[c]**	**Date**	**n**	**Ash [c], Bq/kg**	**Solidified fly ash [d], Bq/kg**	**Concentration factor, [d]/[c]**
July, 2011	3	1610 ± 510	34,000 ± 9340	21	July, 2011	3	1780 ± 395	8510 ± 2230	4.8
Oct–Nov, 2011	3	630 ± 139 (0.39)	14,200 ± 4770 (0.42)	23	Oct–Nov, 2011	3	639 ± 173 (0.36)	3720 ± 1180 (0.44)	5.8
Oct–Nov, 2016	3	55.2 ± 27.0 (0.034)	761 ± 197 (0.022)	14	Oct–Nov, 2016	3	68.7 ± 7.6 (0.039)	446 ± 8 (0.052)	6.5
Oct–Nov, 2019	3	13.2 ± 3.3 (0.008)	413 ± 125 (0.012)	31	Oct–Nov, 2019	3	66.7 ± 25.5 (0.038)	344 ± 59 (0.040)	5.2

Values in parentheses are the ratio to the initial concentration at each facility this reduction in concentration suggests that the highly mobile species of $^{134+137}$Cs deposited on outdoor rubble were selectively washed away by rainfall. Kashiwa City did not show such a sharp decrease in the concentration. Kashiwa City is the most heavily polluted area on the outskirts of Tokyo, and decontamination work was carried out in significant areas of the city from October 2011 to December 2012 (see Fig. 6.6) [26]. However, no trace of this decontamination work is found in the $^{134+137}$Cs concentrations of the waste incineration ash.

As shown in Table 5.4, the radioactive cesium concentrations in the main ash (Ash) discharged from the waste treatment facilities in Kashiwa and Matsudo cities are lower than the clearance level for radioactive waste in Japan, which is less than 100 Bq/kg. However, the concentrations in the solidified fly ash still approximately 300–500 Bq/kg, which are much higher than the clearance level. Nearly 10 years after the accident, radioactive contamination of waste incineration ash is found in cities more than 200 km from the FDNPP. This situation indicates that the living environment of these cities is also polluted. Environmental radioactive contamination due to the FDNPP accident has continued to seriously affect residents.

Many of the author's data discussed in this chapter have been published in PLOS ONE [7]. The data were revised and recalculated as needed before inclusion in this book.

References

1. SCJ (Science Council of Japan) (2014) Comparison of model calculation results on transport and deposition processes of radioactive materials released into the environment by TEPCO's Fukushima Daiichi Nuclear Power Plant accident [in Japanese]. https://www.scj.go.jp/ja/info/kohyo/pdf/kohyo-22-h140902-j1.pdf
2. Gallego E (2006) MUD: a Model to investigate the migration of ^{137}Cs in the Urban environment and Drainage and sewage treatment systems. J Environ Radioact 85:247–264. https://doi.org/10.1016/j.jenvrad.2004.10.017. PMID: 16095773
3. Pratama MA, Yoneda M, Yamashiki Y, Shimada Y, Matsui Y (2014) Modeling migration of Cs-137 in sewer system of Fukushima City using urban environment and drainage system (MUD). Int J Eng Tech 6:402–408. https://doi.org/10.7763/IJET.2014V6.732
4. Nishio T, Kitano M, Sakai M, Takakura A, Katahira K et al (2015) Radioactive caesium contamination due to Fukushima Daiichi Nuclear Power Plants accident in Osaka City: evaluation of accumulation and decontamination of radioactive materials via reverse logistics function of a city. Radioisotopes 64:229–244. [in Japanese]
5. Bureau of Sewerage, Tokyo Metropolitan Government (2017) Measurement result of radiation level around water reclamation centers and radioactivity through wastewater treatment [in Japanese]. (a) http://www.gesui.metro.tokyo.jp/english/oshi/, (b) https://www.gesui.metro.tokyo.lg.jp/business/technology-statistics/radiological info/2019/index.html
6. Bureau of Sewerage, Tokyo Metropolitan Government (2017) List of Tokyo's water reclamation centers. http://www.gesui.metro.tokyo.jp/english/aboutus/ourprofile/15/
7. Ishida M, Yamazaki H (2017) Radioactive contamination in the Tokyo metropolitan area in the early stage of the Fukushima Daiichi Nuclear Power Plant (FDNPP) accident and its fluctuation over five years. PLoS One 12:e0187687. https://doi.org/10.1371/journal.pone.0187687
8. MEXT (2011) MEXT's airborne monitoring results in Iwate, Shizuoka, Nagano, Yamanashi, Gifu, and Toyama prefectures and revisions to airborne monitoring that take into account the

effects of natural nuclides [in Japanese]. https://radioactivity.nsr.go.jp/ja/contents/5000/4899/24/1910_111112nd.pdf

9. Comans RNJ, Hilton J, Voitsekhovitch O, Laptev G, Popov V et al (1998) A comparative study of radiocesium mobility measurements in soils and sediments from the catchment of a small upland oligotrophic lake (Devoke Water, UK). Water Res 32:2846–2855. https://doi.org/10.1016/S0043-1354(98)00038-4
10. Matsunaga T, Ueno T, Amano H, Tkatchenko Y, Kovalyov A et al (1998) Characteristics of Chernobyl-derived radionuclides in particulate form in surface waters in the exclusion zone around the Chernobyl Nuclear Power Plant. J Contam Hydrol 35:101–113. https://doi.org/10.1016/S0169-7722(98)00119-3
11. Monte L, Brittain JE, Håkanson L, Smith JT, van der Perk M (2004) Review and assessment of models for predicting the migration of radionuclides from catchments. J Environ Radioact 75:83–103. https://doi.org/10.1016/j.jenvrad.2003.11.004
12. Håkanson L (2004) A new generic sub-model for radionuclide fixation in large catchments from continuous and single-pulse fallouts, as used in a river model. J Environ Radioact 77:247–273. https://doi.org/10.1016/j.jenvrad.2004.03.010
13. Ya I (2007) Chernobyl radionuclide distribution and migration. Health Phys 93:410–417. https://doi.org/10.1097/01.HP.0000285092.10598.41
14. Garcia-Sanchez L, Konoplev AV (2009) Watershed wash-off of atmospherically deposited radionuclides: a review of normalized entrainment coefficients. J Environ Radioact 100:774–778. https://doi.org/10.1016/j.jenvrad.2008.08.005
15. Evrard O, Chartin C, Onda Y, Lepage H, Cerdan O et al (2014) Renewed soil erosion and remobilization of radioactive sediment in Fukushima coastal rivers after the 2013 typhoons. Sci Rep 4:4574. https://doi.org/10.1038/srep04574
16. Wakiyama Y, Onda Y, Yoshimura K, Kato H, Konoplev A et al (2014) Radiocesium wash-off associated with soil erosion from various land uses after the Fukushima Dai-ichi NPP accident. Geophys Res Abstr 16:EGU2014–EG10185
17. Japan Meteorological Agency (2020) Search past weather data [in Japanese]. https://www.data.jma.go.jp/obd/stats/etrn/select/prefecture00
18. Koto Ward Government (2020) Analytical results of radionuclides activities in the soil of school yard in Koto Ward, Tokyo [in Japanese]. https://www.city.koto.lg.jp/380303/bosai/daishinsai/hoshano/sokute/61429.html
19. Japan Meteorological Agency (2020) Search past weather data [in Japanese] http://www.data.jma.go.jp/obd/stats/etrn/select/prefecture008
20. Gossuin Y, Colet JM, Roch A, Muller RN, Gillis P (2002) Cesium adsorption in hydrated iron oxide particles suspensions: an NMR study. J Magn Reson 157:132–136. https://doi.org/10.1006/jmre.2002.2581
21. GSI (Geospatial Information Authority of Japan) (2014) Extension site of distribution map of radiation dose, etc. [in Japanese]. http://ramap.jmc.or.jp/map/eng
22. MOE (Ministry of Environment, Japan) (2014) Difference between two standards of 100 Bq/kg and 8,000 Bq/kg [in Japanese]. https://www.env.go.jp/jishin/attach/waste_100-8000.pdf
23. IAEA (2013) Final report. The follow-up IAEA international mission on remediation of large contaminated areas off-site the Fukushima Daiichi Nuclear Power Plant. https://www.iaea.org/sites/default/files/final_report230114.pdf
24. Matsudo City (2020) Measurement result of radioactive material concentration such as incineration ash [in Japanese]. https://www.city.matsudo.chiba.jp/chuumoku/houshasen/taisaku_taiou/gomi_seisou/shoukyakubai.html
25. Kashiwa City (2020) Radioactivity measurement result of incineration ash discharged from Kashiwa City Incineration Plant [in Japanese]. https://www.city.kashiwa.lg.jp/houshasenkanren/3313/3322/index.html
26. Kashiwa City (2012) Decontamination implementation plan, 2nd edn. [in Japanese]. https://www.city.kashiwa.lg.jp/houshasenkanren/3331/3332/p032036_d/fil/jissikeikaku250613.pdf

Chapter 6
The Importance of Tokyo Bay as a Reservoir for Radioactive Materials Precipitated in the Tokyo Metropolitan Area

Abstract The Tokyo metropolitan area is located in the Kanto Plain, which is found in roughly the center of Japan. The area of high-concentration pollution surrounding Tokyo are the catchments of rivers flowing into Tokyo Bay. Radioactive nuclides deposited there are carried by rainfall and river water and finally flow into Tokyo Bay. That is, Tokyo Bay serves as a reservoir for radioactive nuclides deposited in the Tokyo metropolitan area. Tokyo Bay functions in two ways as a reservoir for radionuclides. One function is to deposit radioactive material that has flowed into Tokyo Bay within sediments and to act as a natural storage facility. The other function is to work as a storage facility for decontaminated radioactive material, which is artificially collected from the land in the metropolitan area. The radioactive materials collected at wastewater treatment facilities in Tokyo are buried as sludge incineration ash at the Central Breakwater under construction in the northwestern part of Tokyo Bay. By the end of 2016, approximately 0.68 TBq of radioactive cesium had been buried in this landfill. In any case, clarifying the behavior of radioactive nuclides in Tokyo Bay is crucial. Therefore, to elucidate the dynamics of radioactive nuclides in the Tokyo metropolitan area, it is necessary to analyze the migration process of radionuclides in these watersheds and their accumulation mechanism in Tokyo Bay. Because the Tokyo metropolitan area has mountains, forests, rural areas, and urban areas, it is difficult to evaluate the routes by which radioactive nuclides deposited on the ground surface flow into Tokyo Bay. However, elucidating the influx processes of radioactive nuclides into Tokyo Bay and the advection, diffusion, and deposition processes into the sediment helps in understanding the role of Tokyo Bay as a reservoir for radioactive nuclides. The revelation of these dynamics is essential information to estimate the future fate of radioactive nuclides deposited in the Tokyo metropolitan area.

Keywords Tokyo Bay · Radioactive cesium · Spatiotemporal change · Sediment · Flux · Inventory · Pollution sources

H. Yamazaki, *Radioactive Contamination of the Tokyo Metropolitan Area*
https://doi.org/10.1007/978-981-15-7368-2_6

6.1 Introduction

The Tokyo metropolitan area is located in the Kanto Plain, which is found in roughly the center of Japan. The Tokyo metropolitan area administratively includes Tokyo, Kanagawa, Chiba, Saitama, Ibaraki, Tochigi, Gunma, and Yamanashi prefectures. Ibaraki, Tochigi, and Gunma prefectures border Fukushima Prefecture (See Figs. S2.1 and S2.2 in Chap. 2). To the east is the Pacific Ocean, and to the north and west are large forested areas. In the Fukushima disaster, the entire Tokyo metropolitan area was contaminated with radioactive cesium ($^{134+137}$Cs) at approximately 10–300 kBq/m^2 or more, but in the northern part of Chiba Prefecture, approximately 30 km northeast of the center of Tokyo, was particularly contaminated with high concentrations (Fig. 3.2). The highest value shown in Table 3.4 was 881 kBq/m^2. Mountainous forested areas, approximately 100 km farther north, also suffered large-scale radioactive contamination. According to the MEXT airborne monitoring [1], the precipitation of $^{134+137}$Cs in many forested areas may have exceeded 100 kBq/m^2. These areas of high-concentration pollution surrounding Tokyo are the catchments of rivers flowing into Tokyo Bay. Radioactive nuclides deposited there are carried by rainfall and river water and finally flow into Tokyo Bay. That is, Tokyo Bay serves as a reservoir for radioactive nuclides deposited in the Tokyo metropolitan area.

Tokyo Bay is a closed inner bay, but there are many ports, and the bay is congested by navigating ships. In addition, many high-rise buildings have been constructed coastal landfills, and many residents live there. Many factories and commercial facilities are situated along the coast, and approximately ten million people live around Tokyo Bay. If radioactive material from the catchment regions of the Tokyo metropolitan area flows into Tokyo Bay, the radioactive contamination of Tokyo Bay will directly affect the lives of the area residents. Additionally, many inhabitants pursue fishing in this sea since high-quality fish have long been caught in Tokyo Bay.

Tokyo Bay functions in two ways as a reservoir for radionuclides. One function is to deposit radioactive material that has flowed into Tokyo Bay within sediments and to act as a natural storage facility. The other function is to work as a storage facility for decontaminated radioactive material, which is artificially collected from the land in the metropolitan area. As shown in Chap. 5, the radioactive materials collected at wastewater treatment facilities in Tokyo are buried as sludge incineration ash at the Central Breakwater under construction in the northwestern part of Tokyo Bay. By the end of 2016, approximately 0.68 TBq of radioactive cesium had been buried in this landfill.

In any case, clarifying the behavior of radioactive nuclides in Tokyo Bay is crucial. Therefore, to elucidate the dynamics of radioactive nuclides in the Tokyo metropolitan area, it is necessary to analyze the migration process of radionuclides in these watersheds and their accumulation mechanism in Tokyo Bay. Because the Tokyo metropolitan area has mountains, forests, rural areas, and urban areas, it is difficult to evaluate the routes by which radioactive nuclides deposited on the ground surface flow into Tokyo Bay. However, elucidating the influx processes of radioactive nuclides into Tokyo Bay and the advection, diffusion, and deposition processes into the sediment helps in understanding the role of Tokyo Bay as a reservoir for radioactive nuclides. The revelation of these dynamics is essential information to estimate the future fate of radioactive nuclides deposited in the Tokyo metropolitan area.

Many of the data presented in this chapter have already been published by PLOS ONE in the authors [2].

6.2 Oceanographic Structure of Tokyo Bay

Tokyo Bay is a closed bay 70 km from north to south and 20 km from east to west, with a total area of 1380 km^2 and an average depth of 15 m. It is connected to the Pacific Ocean by the southernmost strait, and its bay mouth is 21 km wide and 700 m deep. The water depth near the Aqua Line shown in Fig. 6.1 is approximately 25–30 m, but the water depth increases rapidly from this point to offshore areas. A submarine valley formed in the center of the strait connects to the open ocean. The closure index determined by the MOE (Ministry of the Environment, Japan) is 1.78 [3]; thus, it is a closed bay where the Japanese Water Pollution Control Law regulates the emissions of nutrients. The closure index is determined as follows: $\{(\sqrt{S}/W) \times (D_1/D_2)\}$, where S is the area of the bay, W is the width of the bay mouth, D_1 is the maximum depth of the bay and D_2 is the depth of the bay mouth. Sea areas with an index of 1.0 or higher are very likely to be eutrophic and are therefore subject to drainage regulations by the MOE. In areas Y and Z shown in Fig. 6.1, DO (dissolved oxygen) in the bottom water is depleted in the summer, and anoxic sea areas appear [4, 5]. The average residence time of seawater varies by season but is reported to be approximately 31 days. Tokyo Bay is closed, but the flow of seawater is complicated. In addition to tidal currents, permanent currents flow throughout the bay, and circular drifts dominate surface water movements: clockwise in winter and counterclockwise in summer [6]. The bottom water moves in a direction opposite to that of the surface water flow. Ocean water from the Pacific flows north down the bottom of the bay until it reaches the innermost part of the bay [4, 5].

Central Tokyo is located on the western side of the bay, surrounded by the metropolitan zones that form the heart of Japan, with a total population of 38 million. The basins of the rivers that flow into Tokyo Bay from the Tokyo metropolitan area occupy a land area of 9100 km^2, and the total amount of flowing river water varies greatly depending on weather conditions, but the average is approximately 1.4×10^7 m^3/day. The principal rivers are the Edogawa, Old Edogawa, Arakawa, Tamagawa, Sumidagawa, and Tsurumigawa rivers [4, 5].

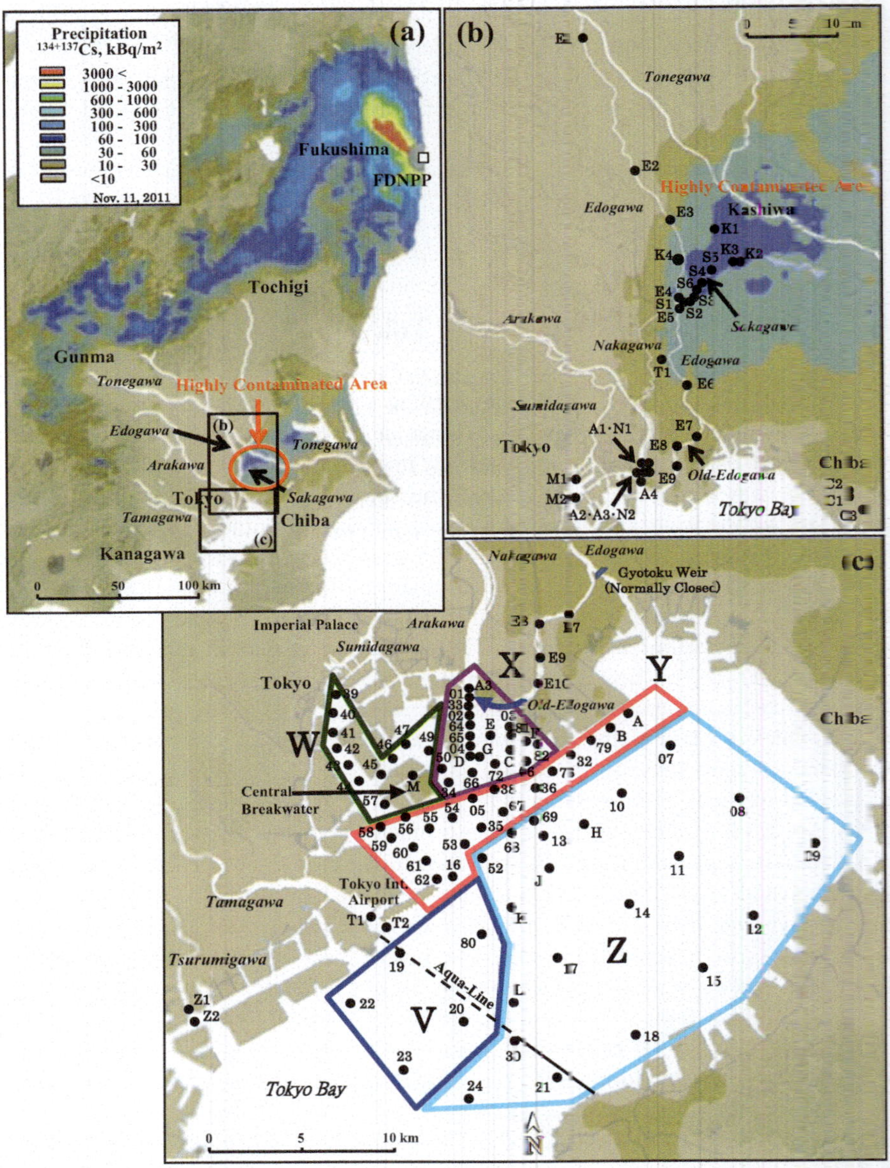

Fig. 6.1 Study areas and sampling sites. Geographic distribution of the radioactive cesium precipitation is quoted from the MEXT airborne monitoring [1]. (**a**) Study area. (**b**) Sampling sites in the Edogawa river system. (**c**) Sampling sites in Tokyo Bay. V: Tamagawa River estuary, W: Sumidagawa River estuary, X: Old Edogawa River estuary, Y: Offshore the Old Edogawa and Sumidagawa estuaries, Z: Central Tokyo Bay. Aqua-Line: Tokyo Bay crossing road. Water in Old Edogawa flows in the direction of the blue arrow in (**c**)

6.3 Status of Radioactive Contamination Around Tokyo Bay and the FDNPP Disaster

The $^{134+137}$Cs pollution map from the airborne monitoring conducted by the MEXT [1] is shown in Fig. 3.2. The results indicate that the catchment area of the Edogawa River was contaminated by 10–100 kBq/m^2 from radioactive cesium discharged in the FDNPP accident. However, the radioactive contamination levels in the catchments of Tokyo Bay's other rivers were lower than those in the catchment of the Edogawa River. Radioactive materials precipitated on the ground surface in the Tokyo metropolitan area are presumably transferred by these rivers until they finally flow into Tokyo Bay, as in the case of artificially discharged environmental contaminants such as heavy metals and pesticides.

Many reports outlining the FDNPP accident have already been released to the public [References 3–8 in Chap. 1]. However, many of these results are analyses of the accident process, whereas few address the environmental radioactive contamination that was caused [1, 7–9]. In particular, after the accident high-concentration radioactive plumes arrived in the greater Tokyo region, and radionuclides washed out with rainfall on March 16 and 22 in 2011 [10]. Nevertheless, the movement through the greater Tokyo region of radioactive contamination produced by the FDNPP accident has been insufficiently analyzed. Clarifying the movement of environmental radioactive contaminants in the densely populated greater Tokyo region is an important issue related to the problem of low-dose exposure to large populations. Evaluation of the migration processes of radioactive cesium from the Tokyo metropolitan area is also necessary from the viewpoint of reduction and removal of radioactive contamination in these areas. In the author's previous paper [11] and in Chaps. 3 and 5, the behaviors of radioactive contaminants in the soil of the Tokyo metropolitan area are discussed. It is estimated that approximately 5–10% of the radioactive cesium that precipitated into the surface soil migrated via rainfall and rivers in the 5 years after the FDNPP accident (Table 5.3). Perhaps these contaminants flowed into Tokyo Bay.

Before the FDNPP accident, the Chernobyl and Three Mile Island (TMI) accidents affected many people of the world. The TMI is located approximately 150 km west of Washington, D.C., but because it avoided the destruction of the pressure vessel, the emissions of radioactive nuclides to the environment included only rare gas and radioactive iodine nuclides, and the release amount of ^{131}I was estimated to be approximately 0.5 TBq [12, 13]. In the case of the Chernobyl accident, Kyiv City is located 130 km south, and four million residents lived in this metropolitan area. It was reported at the Chernobyl Forum by the IAEA in 2006 [14] that a radioactive plume was transported to Kyiv City by the north wind on May 1 1986, immediately after the accident. However, regulations on information disclosure were made by the Soviet government at the time, and the actual state and dynamics of radioactive contamination in Kyiv City are hardly understood even at present. Of course, radioactive contamination by the atomic bombs at Hiroshima and Nagasaki is not forgotten. The results of investigations on the environmental dynamics of radionuclides 60 years after radiation exposure from the atomic bombs have already been reported [15–17]. The greater Tokyo region experienced severe radioactive contamination due to the FDNPP disaster. It was the first time in world history that a densely populated metropolis has undergone massive radioactive contamination.

The study started in August 2011 immediately after the FDNPP accident, with continuous time-series analysis of the distributions and fluctuations of radioactive cesium in sediments and waters in Tokyo Bay and the rivers flowing into Tokyo Bay. Based on the results, the processes of movement from the land and deposition in Tokyo Bay of radioactive cesium that precipitated in the Tokyo metropolitan area after the FDNPP accident were evaluated [2, 11].

Studies on the behavior of radioactive cesium in an urban environment have been performed through simulations [18, 19], but the fluctuation in this radionuclide's spatiotemporal distribution have not been monitored or analyzed over extensive areas for long periods. Numerous studies have reported on the behavior of radioactive materials transported from lakes to lakes and oceans from land contaminated by nuclear disasters such as the Chernobyl and FDNPP accidents. Mixing and depositional processes in the estuary also have significant effects on the dynamics of radionuclides. However, the behavior of cesium as an alkali metal element is often complicated and unknown in estuaries where seawater and river water mix [20–22]. In this chapter, the essential roles that Tokyo Bay and rivers flowing into it play in the movements of radioactive cesium contaminants and the transport and accumulation mechanisms of such contaminants in the greater Tokyo region are clarified.

The NRA (Nuclear Regulation Authority, Japan) has monitored the radioactive contamination derived from the FDNPP accident in the surface sediment of Tokyo Bay since June 2013 [23]. The JCG (Japan Coast Guard) has also been measuring the radioactive contamination of surface sediments in Tokyo Bay since 1981 [24]. Further, survey results of radioactive cesium contamination in the Tokyo Bay area immediately after the accident have been published [25]. However, since their monitoring is spatiotemporally limited, it is insufficient to evaluate the dynamics of radioactive contamination throughout the environments of Tokyo Bay.

6.4 Material and Analytical Methods

6.4.1 Sample Collection

Sediments and waters were sampled in Tokyo Bay and the rivers in its catchment basin. Photograph 6.1 shows a sampling location in Tokyo Bay. Area K is the estuary of the Old Edogawa River very close to the city center. The sampling sites are shown in Fig. 6.1. Sampling was performed at the same sites one to seven times during the study period, which ran from August 20, 2011, to July 12, 2016. Sediment samples were collected at 77 points in Tokyo Bay, ten points in the Edogawa River, and six points in the Sakagawa River. Of the sediment samples collected, 68 were core sediments and 142 were surface sediments. Sediment cores were sampled at

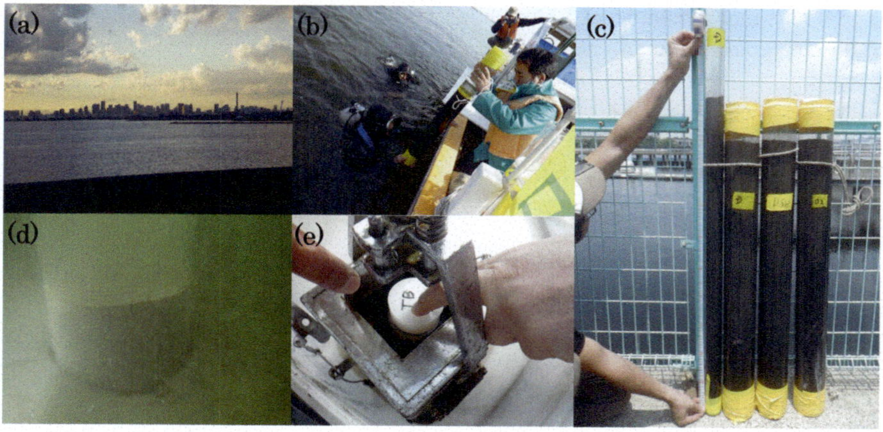

Photo 6.1 Photographs of the landscape in the sampling. (**a**) View of the Tokyo capital from Area X. (**b**) Core sampling by a scuba diver. (**c**) Sediment core was taken from the seafloor. The core sample has a diameter of 10 cm and a maximum length of approximately 90 cm. (**d**) Pull out the acrylic core sampler from the seafloor. (**e**) Surface sediments collected by the Birge-Ekman grab sampler

Site S1 (Fig. 6.1b), where the Sakagawa River flows into the Edogawa River, to evaluate the role of the Sakagawa River in the process of radioactive cesium transport. Soil samples were also collected from the ten sites shown in Fig. 6.1b (K1–K4, S1–S6) for comparison with sediments.

Sediment core sampling was performed using an acrylic pipe with a diameter of 10 cm and a length of 100 cm. The core samples were collected by a diver pushing the pipe into the seabed by hand. Core samples 20–80 cm in length were obtained. Surface sediment specimens were sampled from a boat using a Birge-Ekman bottom grab sampler. Then, on board the boat, after the sediments were recovered, the sediment was inserted into an acrylic pipe with a diameter of 5 cm and a length of 10 cm to obtain samples of the top 5 cm of sediment. Most of the soil and sediment samples consisted of silt and sand with particle sizes of 0.2 mm or less. However, larger pebbles, plant debris, shell fragments, *etc.*, were removed with tweezers. Grain size sorting by sieve was not performed. The sediments were pushed out of the pipes, cut into 1 or 2 cm-thick slices in the depth direction, and then thermally dried to a constant weight in a 60 °C oven to remove the water from the sediments. The dried samples were pulverized in an agate mortar, and the radioactivity of the samples was then measured.

Water samples were obtained from the surface of the water by lowering buckets from boats. Divers obtained bottom water from approximately 1 m above the seabed. Without filtering suspended materials out of the water, the radioactive cesium in 20 L of sample water was concentrated using an ammonium phosphomolybdate (AMP) method [26]. After settling overnight, the AMP precipitate was filtered and collected on a membrane filter (pore size 0.8 μm), and the radioactivity of the dried AMP precipitate was then measured. In this way, it was confirmed in a preliminary experiment that the ionic and suspended radioactive cesium in the sample water could be recovered quantitatively.

6.4.2 Measurement of Radioactivity

The measurement of radioactivity was performed in the same way as described in the other chapters of this book. Radionuclides in the samples were quantified by connecting a 4096-multichannel pulse height analyzer (Lab Equipment, MCA600) to a low-energy HPGe detector (ORTEC, LO-AX/30F) shielded in lead 10 cm thick, sealing the specimens inside a plastic container with a diameter of 5.5 cm and depth of 2.0 cm, and then measuring them via γ-ray spectrometry. The Ge detector calculated the geometric efficiency relative to the sample weight using the American NIST (National Institute of Standards and Technology) Environmental Radioactivity Standards SRM 4350B (river sediment), and SRM 4354 (freshwater lake sediment). The efficiency was corrected to within a range of 2–30 g of the sample weight [27].

The measurement time was set so that the counting error would be less than 5% according to the radioactive intensity of the samples. The nuclides ^{134}Cs (605 keV) and ^{137}Cs (662 keV) were quantified in this study. A ^{134}Cs solution with a known concentration was used to correct the sum peak effect for ^{134}Cs counting. The detection limits of ^{134}Cs and ^{137}Cs under appropriate conditions were 0.6 Bq/kg in sediment samples and 0.3 mBq/L in water samples. Radioactive cesium activity was indicated by the value per sampling day but was corrected for radioactive decay to March 16, 2011, as necessary. In this case, it is denoted as "corrected activity."

6.4.3 Measurements of Heavy Metals and Particle Size Distributions in the Sediments

The heavy metals in the sediments were measured via an XRF method (Rigaku, ZSX-Primus II) using NIST SRM 1646 (estuarine sediment) as the standard sample. The sample measured was prepared from cellulose powder pressed at 0.4 ton/cm^2 in a 4-cm diameter aluminum ring, and 1.2 g of powdered sample was then placed on this disk and repressed at 1.6 ton/cm^2. The correction for the matrix effect was achieved by the X-ray intensity ratio of the peak to the background for each element [28]. Mercury in the sediments was measured by heating vaporization atomic absorption spectrometry (Hiranuma, HG-300).

The particle size distribution of the wet sediment samples was measured using a laser diffractometer (Shimadzu, SALD-3000) with a measurement range of particle sizes from 0.05 to 3000 μm. The dispersion of sedimentary particles was carried out via ultrasonic irradiation (Shimadzu, SUS-200 42 kHz) using sodium hexametaphosphate as a dispersant. In this paper, the particle size obtained is presented as the volume-based average particle diameter.

6.5 Radioactive Cesium Activity in Water Samples Around Tokyo Bay

It was assumed that radioactive cesium flows into Tokyo Bay through the rivers. Thus, the $^{134+137}$Cs activity of river waters and estuary waters of the major rivers and the seawater of Tokyo Bay were analyzed. The river waters were collected from the surface layer (S), and the seawater was collected at the surface layer (S) and 1 m above the seafloor (B). When water samples were taken at the same site on different days, the radioactive cesium concentration at the site was normalized by weighted averaging of the concentrations corrected for radioactive decay. The results are shown in Table 6.1. The $^{134+137}$Cs activity in water standardized for March 16, 2011, ranges from 4.4 to 178 mBq/L, and in the surface water of the Old Edogawa estuary, it is higher than 20 mBq/L. In particular, higher radioactive cesium activity (39–178 mBq/L) is detected in the river water at the confluence of the Sakagawa and Edogawa rivers (Site S1). In the estuaries of the Sumidagawa and Tamagawa rivers, on the other hand, the cesium activity is mostly lower than 10 mBq/L. The water treatment systems in Tokyo discussed in Chap. 5 are located in the catchments of these rivers. Most of the radioactive cesium deposited on the surface and washed away is believed to be transported to water reclamation centers, not rivers. Therefore, the radioactive cesium concentrations of these rivers are thought to be lower than that of the Edogawa River water system. The radioactive cesium concentrations in the Edogawa River system are mostly somewhat lower than that of seawater (74–78 mBq/L) collected 1.5 km offshore from the FDNPP in September 2013.

The radioactive cesium concentration in the Tokyo Bay water system conforms to the radioactive cesium distribution in the sediments of these areas (Fig. 6.2). Surface and bottom seawaters were sampled from the Old Edogawa estuary to the center of the bay. The ratio of the $^{134+137}$Cs activity in the surface water to that in the bottom water is 2.20 ± 0.85 ($n = 14$), excluding Site J in the center of the bay, showing that the activity was is approximately twofold higher in the surface water than in the bottom water. Most of the analyzed samples were collected from spring to autumn, which is the season when Tokyo Bay seawater is stratifying. The vertical distribution of radioactive cesium in Tokyo Bay indicates that the contaminated river water, which has a low density, diffuses as surface water into the bay because the high-density Pacific seawater penetrates deep into the bay. This distribution implies that radioactive cesium flowed into Tokyo Bay via river water.

Moreover, the radioactivity ratio of ^{134}Cs/^{137}Cs converted to the decay-corrected value immediately after the accident is 0.893 ± 0.025 ($n = 52$). This anomaly is considered to be an appropriate value, as Japanese water samples still contain 1–2 mBq/L of ^{137}Cs as background owing to global fallout from atmospheric nuclear tests [29].

Table 6.1 Radioactive cesium concentrations in water samples collected around Tokyo Bay and offshore the FDNPP

Area	Site	Layer	Sampling date	Detected, mBq/L			Corrected[a], mBq/L				Ratio (S/B)	$^{134}Cs/^{137}Cs$ ratio
				^{134}Cs	^{137}Cs	$^{134+137}Cs$	^{134}Cs	^{137}Cs	$^{134+137}Cs$	Weighted average[b]		
Edogawa River (Upstream)	E4	S	2014/3/24	1.7 ± 1.1	6.2 ± 0.9	7.9 ± 1.5	4.6 ± 3.1	6.7 ± 1.0	11.3 ± 3.3	3.5 ± 0.6		0.688 ± 0.478
			2015/11/11	0.4 ± 0.6	2.6 ± 0.6	2.6 ± 0.6	2.0 ± 2.9	2.8 ± 0.7	2.8 ± 0.7			0.688 ± 1.040
			2016/6/24	0.9 ± 0.5	5.2 ± 0.8	6.1 ± 0.9	5.2 ± 3.1	5.9 ± 0.9	11.1 ± 3.2			0.881 ± 0.544
	E5	S	2016/6/24	1.0 ± 0.4	5.8 ± 0.8	6.8 ± 0.9	6.0 ± 2.5	6.6 ± 0.9	12.5 ± 2.7			0.906 ± 0.402
	E2	S	2014/3/24	4.7 ± 0.9	17.5 ± 1.3	22.2 ± 1.6	13.0 ± 2.6	18.8 ± 1.4	31.7 ± 2.9	18.4 ± 1.9		0.692 ± 0.147
			2016/6/24	0.8 ± 0.4	4.1 ± 0.7	4.9 ± 2.4	4.6 ± 2.3	4.7 ± 0.8	9.3 ± 2.4			0.998 ± 0.520
Sakagawa River	S5	S	2014/4/29	4.2 ± 0.8	12.8 ± 1.4	17.0 ± 1.6	12.1 ± 2.2	13.7 ± 1.5	25.8 ± 2.7			0.880 ± 0.187
	S4	S	2014/4/29	12.2 ± 1.2	38.9 ± 1.4	51.1 ± 1.8	34.9 ± 3.3	41.7 ± 1.5	76.6 ± 3.6			0.836 ± 0.085
	S3	S	2016/6/24	1.8 ± 0.8	10.8 ± 1.0	12.6 ± 1.3	10.6 ± 4.5	12.2 ± 1.1	22.7 ± 4.7			0.866 ± 0.380
	S2	S	2016/6/24	2.7 ± 0.6	17.1 ± 1.2	19.8 ± 1.3	16.0 ± 3.3	19.3 ± 1.3	35.3 ± 3.6			0.829 ± 0.182
	S1	S	2014/4/29	31.3 ± 1.9	82.6 ± 2.4	113.9 ± 3.1	89.5 ± 5.6	88.8 ± 2.5	178.3 ± 6.1	67.2 ± 1.9		1.009 ± 0.069
			2014/3/24	19.2 ± 1.5	48.1 ± 1.8	67.4 ± 2.3	53.2 ± 4.2	51.6 ± 1.9	104.8 ± 4.6			1.032 ± 0.090
			2015/11/12	5.3 ± 0.6	20.7 ± 1.1	26.0 ± 1.3	25.4 ± 3.0	23.0 ± 1.2	48.4 ± 3.3			1.101 ± 0.145
			2016/6/24	3.2 ± 0.5	17.5 ± 1.3	20.7 ± 1.3	18.8 ± 2.7	19.8 ± 1.4	38.6 ± 3.1			0.950 ± 0.154
Edogawa River (Downstream)	E6	S	2016/6/24	0.7 ± 0.5	4.9 ± 0.8	5.5 ± 0.9	4.1 ± 3.1	5.5 ± 0.9	9.5 ± 3.2			0.739 ± 0.568
	E7	S	2016/6/24	0.6 ± 0.5	4.3 ± 0.9	5.0 ± 1.0	3.7 ± 3.1	4.9 ± 1.0	8.6 ± 3.3			0.762 ± 0.658
	E1	S	2014/3/24	2.2 ± 0.6	5.5 ± 0.7	7.7 ± 0.9	6.0 ± 1.5	5.9 ± 0.7	11.9 ± 1.7	14.7 ± 1.6		1.010 ± 0.286
			2016/6/24	2.2 ± 0.6	15.4 ± 1.3	17.6 ± 1.5	12.9 ± 3.7	17.4 ± 1.5	30.3 ± 4.0			0.742 ± 0.225
Old Edogawa River	E3	S	2016/7/12	1.9 ± 0.3	13.8 ± 1.0	15.8 ± 1.1	11.7 ± 1.8	15.6 ± 1.2	27.3 ± 2.1			0.747 ± 0.125
	E10	S	2016/3/24	1.4 ± 0.7	8.3 ± 1.0	9.8 ± 1.2	7.7 ± 3.7	9.4 ± 1.2	17.0 ± 3.9	17.3 ± 2.5		0.817 ± 0.405
			2016/7/12	1.2 ± 0.5	8.9 ± 0.9	10.1 ± 1.1	7.5 ± 3.0	10.0 ± 1.1	17.5 ± 3.2			0.747 ± 0.309
	E11	S	2016/3/24	1.5 ± 0.6	9.9 ± 1.0	11.4 ± 1.1	7.9 ± 3.0	11.1 ± 1.1	19.0 ± 3.2			0.708 ± 0.278

Nakagawa River	N1	S	2016/7/12	1.0 ± 0.4	6.7 ± 0.8	7.7 ± 0.9	5.7 ± 2.5	7.6 ± 0.9	13.3 ± 2.7			0.751 ± 0.343
	N2	S	2016/3/24	2.9 ± 0.5	15.2 ± 1.0	18.1 ± 1.1	15.8 ± 2.5	17.0 ± 1.2	32.8 ± 2.8	21.6 ± 1.8		0.924 ± 0.161
		S	2016/7/12	1.2 ± 0.5	7.9 ± 0.9	9.0 ± 1.1	7.1 ± 3.0	8.9 ± 1.1	16.0 ± 3.2		1.4	0.797 ± 0.355
		B		0.9 ± 0.5	5.6 ± 0.9	6.5 ± 1.0	5.3 ± 3.2	6.4 ± 1.0	11.6 ± 3.3			0.826 ± 0.518
Arakawa River (Estuary)	A1	S	2016/7/12	1.4 ± 0.6	8.8 ± 0.9	10.2 ± 1.1	8.4 ± 3.8	9.9 ± 1.1	18.3 ± 3.9			0.852 ± 0.395
	A2	S	2016/3/24	1.8 ± 0.6	9.7 ± 1.1	11.5 ± 1.3	9.9 ± 3.4	10.9 ± 1.3	20.8 ± 3.6	23.1 ± 2.0		0.915 ± 0.327
		S	2016/7/12	2.3 ± 0.5	12.5 ± 1.1	14.7 ± 1.2	13.6 ± 3.3	14.1 ± 1.2	27.7 ± 3.5		1.3	0.967 ± 0.245
		B		1.5 ± 0.5	10.6 ± 1.1	12.1 ± 1.2	9.2 ± 2.9	11.9 ± 1.2	21.1 ± 3.2			0.771 ± 0.258
	AC	S	2016/3/24	0.9 ± 0.5	6.4 ± 1.1	7.2 ± 1.2	4.9 ± 2.7	7.1 ± 1.2	12.0 ± 2.9	15.8 ± 2.2		0.680 ± 0.391
		S	2016/7/12	1.7 ± 0.7	11.7 ± 1.0	13.3 ± 1.2	9.9 ± 4.2	13.2 ± 1.2	23.1 ± 4.3		1.3	0.750 ± 0.323
		B		1.4 ± 0.8	8.2 ± 1.3	9.6 ± 1.5	8.2 ± 5.1	9.3 ± 1.4	17.5 ± 5.3			0.883 ± 0.564
	A3	S	2016/7/12	2.3 ± 0.6	15.4 ± 0.6	17.7 ± 0.9	13.9 ± 3.9	17.4 ± 0.7	31.3 ± 3.9			0.799 ± 0.224
Old Edogawa River (Estuary)	2	S	2014/3/25	6.8 ± 0.7	15.7 ± 1.2	22.5 ± 1.4	18.7 ± 2.1	16.9 ± 1.2	35.6 ± 2.4	20.9 ± 1.1	1.2	1.109 ± 0.147
		B		4.8 ± 0.8	14.1 ± 1.1	19.0 ± 1.3	13.3 ± 2.3	15.2 ± 1.1	28.5 ± 2.5			0.880 ± 0.163
		S	2015/11/13	1.9 ± 0.5	10.6 ± 0.8	12.5 ± 0.9	8.9 ± 2.5	11.8 ± 0.9	20.7 ± 2.6		2.6	0.752 ± 0.218
		B		0.9 ± 0.4	3.2 ± 0.6	4.1 ± 0.8	4.4 ± 2.0	3.5 ± 0.7	7.9 ± 2.2			1.243 ± 0.629
		S	2016/7/12	2.0 ± 0.6	12.3 ± 1.0	14.3 ± 1.2	11.9 ± 3.8	13.9 ± 1.1	25.8 ± 4.0		2.6	0.859 ± 0.284
		B		0.8 ± 0.5	4.7 ± 0.9	5.5 ± 1.0	4.7 ± 2.8	5.3 ± 1.0	9.9 ± 3.0			0.882 ± 0.562
	64	S	2013/10/4	5.1 ± 0.6	15.3 ± 0.8	20.4 ± 1.0	12.0 ± 1.5	16.3 ± 0.8	28.3 ± 1.7	15.2 ± 1.0	3.8	0.738 ± 0.099
		B		1.3 ± 0.5	4.2 ± 0.6	5.5 ± 0.8	3.0 ± 1.2	4.4 ± 0.6	7.5 ± 1.3			0.687 ± 0.280
	D	S	2013/10/4	4.5 ± 0.8	14.5 ± 1.2	19.0 ± 1.4	10.7 ± 1.9	15.4 ± 1.2	26.0 ± 2.3	13.4 ± 0.7	2.2	0.693 ± 0.138
		B		2.4 ± 0.7	5.9 ± 0.9	8.3 ± 1.1	5.7 ± 1.7	6.3 ± 0.9	12.0 ± 1.9			0.901 ± 0.297
		S	2014/3/25	7.0 ± 1.0	16.8 ± 1.2	23.9 ± 1.6	19.5 ± 2.9	18.0 ± 1.3	37.5 ± 3.1		2.3	1.083 ± 0.177
		B		2.7 ± 0.5	8.3 ± 0.9	11.0 ± 1.0	7.4 ± 1.5	8.9 ± 0.9	16.3 ± 1.8			0.839 ± 0.189
		S	2015/11/13	1.7 ± 0.5	9.5 ± 0.9	11.2 ± 1.0	8.0 ± 2.3	10.6 ± 1.0	18.6 ± 2.5		1.4	0.759 ± 0.230

(continued)

Table 6.1 (continued)

Area	Site	Layer	Sampling date	Detected, mBq/L			Corrected[a], mBq/L			Weighted average[b]	Ratio (S/B)	$^{134}Cs/^{137}Cs$ ratio
				^{134}Cs	^{137}Cs	$^{134+137}Cs$	^{134}Cs	^{137}Cs	$^{134+137}Cs$			
		B		1.2 ± 0.5	6.5 ± 0.8	7.8 ± 0.9	5.9 ± 2.3	7.3 ± 0.9	13.2 ± 2.4			0.815 ± 0.326
		S	2016/7/12	1.8 ± 0.6	9.0 ± 0.9	10.8 ± 1.1	11.0 ± 3.9	10.2 ± 1.0	21.2 ± 4.0		3.3	1.080 ± 0.395
		B		nd	5.7 ± 0.9	5.7 ± 0.9	–	6.4 ± 1.0	6.4 ± 1.0			–
	5	S	2013/10/4	3.3 ± 0.4	8.4 ± 0.8	11.7 ± 0.9	7.8 ± 0.9	8.9 ± 0.9	16.7 ± 1.3	13.0 ± 0.8	1.5	0.872 ± 0.135
		B		2.2 ± 0.3	5.6 ± 0.5	7.8 ± 0.6	5.1 ± 0.8	6.0 ± 0.5	11.1 ± 0.9			0.864 ± 0.150
	38	S	2016/7/12	1.2 ± 0.5	7.1 ± 0.9	8.3 ± 1.1	7.0 ± 3.0	8.0 ± 1.0	15.0 ± 3.2	5.4 ± 0.7	3.1	0.874 ± 0.392
		B		nd	4.3 ± 0.7	4.3 ± 0.7	–	4.8 ± 0.7	4.8 ± 0.7			–
Sumidagawa River	M1	S	2016/3/24	1.4 ± 0.6	6.5 ± 0.8	7.9 ± 1.0	7.9 ± 3.4	7.3 ± 0.9	15.1 ± 3.6			1.083 ± 0.495
	M2	S		0.4 ± 0.7	3.2 ± 0.7	3.6 ± 0.9	2.2 ± 3.7	3.6 ± 0.7	5.8 ± 3.7			0.627 ± 1.037
Tamagawa River	T1	S	2016/3/24	0.3 ± 0.4	2.6 ± 0.7	2.9 ± 0.8	1.6 ± 2.0	2.9 ± 0.8	4.5 ± 2.2			0.544 ± 0.702
	T2	S		0.5 ± 0.6	3.3 ± 0.7	3.8 ± 0.9	2.9 ± 3.2	3.7 ± 0.8	6.6 ± 3.3			0.784 ± 0.877
Tsurumigawa River	Z1	S	2016/3/24	0.6 ± 0.6	3.6 ± 0.9	4.2 ± 1.1	3.4 ± 3.4	4.0 ± 1.0	7.4 ± 3.5			0.851 ± 0.871
	Z2	S		0.6 ± 0.4	4.3 ± 0.7	4.9 ± 0.8	3.4 ± 2.1	4.8 ± 0.8	8.2 ± 2.3			0.720 ± 0.461
Tokyo Bay (Center)	68	S	2014/3/25	4.9 ± 0.5	13.1 ± 1.0	18.0 ± 1.1	13.6 ± 1.3	14.0 ± 1.1	27.6 ± 1.7	15.1 ± 0.9	2.7	0.972 ± 0.120
		B		2.0 ± 0.3	4.3 ± 0.6	6.3 ± 0.7	5.6 ± 0.8	4.6 ± 0.7	10.2 ± 1.1			1.210 ± 0.252
	J	S	2015/11/14	0.5 ± 0.4	1.9 ± 0.5	2.4 ± 0.7	2.3 ± 2.2	2.1 ± 0.6	4.4 ± 2.2	5.4 ± 1.7	0.7	1.075 ± 1.061
		B		0.7 ± 0.5	2.8 ± 0.6	3.5 ± 0.8	3.5 ± 2.4	3.1 ± 0.7	6.6 ± 2.5			1.122 ± 0.795
									Weighted average (n = 52)			0.894 ± 0.025
Offshore the FDNPP[c]	Three sites	S	2013/9/5	17.0 ± 9.0	36.9 ± 19.9	53.9 ± 28.8	39.8 ± 20.6	39.1 ± 21.1	78.0 ± 41.6		1.1	1.018 ± 0.761
		B		16.2 ± 3.1	34.4 ± 7.8	50.6 ± 10.9	37.2 ± 7.1	36.4 ± 8.3	73.7 ± 15.4			1.033 ± 0.304

S: Surface water. B: Bottom water 1 m above the seafloor

[a]Decay corrected values to March 16, 2011

[b]Calculated using the data excluding the counting errors of more than 30%

[c]Seawaters were collected at three sites 1.5 km offshore the FDNPP. The water depths are 11–16 m

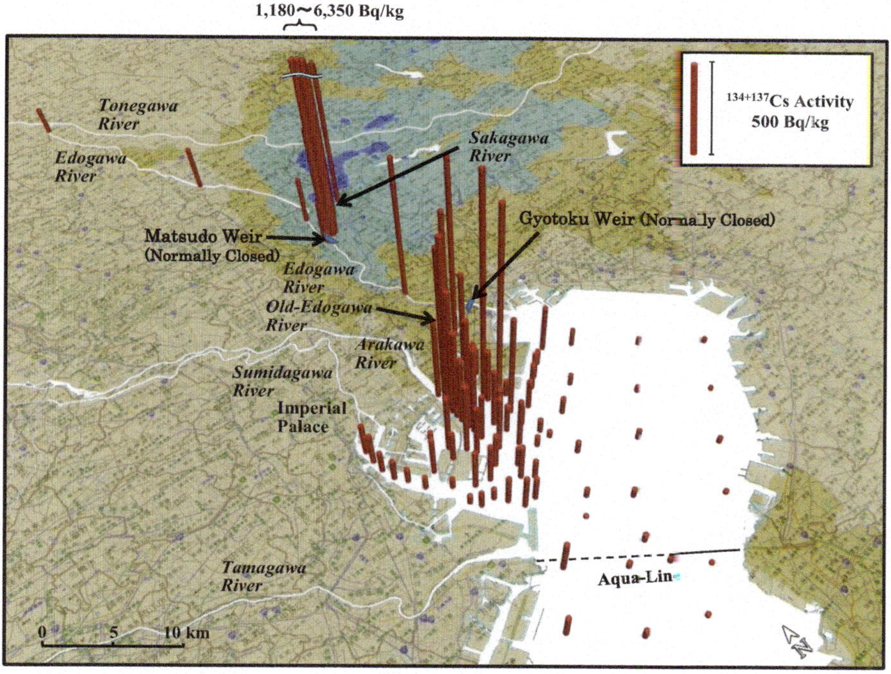

Fig. 6.2 Geographic distributions of $^{134+137}$Cs in the surface sediment in the Tokyo Bay water system. Sediments were collected from August 20, 2011, to July 12, 2016. The $^{134+137}$Cs activities were decay-corrected values to March 16, 2011. The concentrations are shown as an average activity from the surface to 5 cm depth. When there are multiple data at the same site, the concentration expressed as a weighted average value for the counting error

6.6 Spatiotemporal Distribution of Radioactive Cesium in Tokyo Bay Sediments

The measured values obtained from the survey of the Tokyo Bay water system are shown in Table 6.2 (Tokyo Bay) and Table 6.3 (Edogawa Water System). Sampling was performed on different days. Therefore, the radioactive cesium concentrations show values after radioactive decay correction based on March 15, 2011. Multiple measurements collected on different days were subjected to statistical processing. Tables 6.2 and 6.3 show the average activity of $^{134+137}$Cs (sum of ^{134}Cs and ^{137}Cs) in the surface layer of sediments at a depth of 0–5 cm and the inventories of ^{134}Cs, ^{137}Cs, and $^{134+137}$Cs in the sediment cores. The inventory of $^{134+137}$Cs also shows the decay-corrected values based on March 11, 2011. In addition, the average daily flux of $^{134+137}$Cs is shown based on the elapsed time from the accident to sampling. When multiple measurements were taken at the same point, $^{134+137}$Cs radioactivity was evaluated based on the weighted average with counting errors. The deviation of the weighted average was approximated according to the law of uncertainty

Table 6.2 Concentrations, inventories, and fluxes of radioactive cesium in the sediments of Tokyo Bay

Area	Site	Sampling date (n)	Elapsed time[a], day	Contaminated layer[b], cm	Average $^{134+137}Cs$ activity of top 5 cm[c, d], Bq/kg	Inventory, kBq/m^2				Flux[e, c], kBq/m^2 day	$^{134}Cs/^{137}Cs$ ratio[c]
						Detected		Corrected			
						^{134}Cs	^{137}Cs	$^{134+137}Cs$	$^{134+137}Cs$	$^{134+137}Cs$	
X	A3	2011 (3)	–	–	416 ± 4	–	–	–	–	–	–
	01	2011–2015 (6)	–	–	810 ± 6	–	–	–	–	–	–
	33	2011 (1)	–	–	873 ± 7	–	–	–	–	–	–
	02	2011–2016 (7)	–	–	647 ± 5	–	–	–		–	–
		2012/11/1	596	12	762 ± 12	8.5	13.9	22.4	29.2	0.049	1.024
		2014/3/25	1105	58	613 ± 32	47.9	142.0	172.0	265.0	0.240	
		2015/11/13	1703	40	253 ± 34	9.0	38.2	47.2	85.9	0.050	
		2016/7/12	1945	52	352 ± 23	14.0	74.7	88.7	168.4	0.086	0.995
	64	2012–2015 (4)	–	–	470 ± 7	–	–	–	–	–	–
		2013/10/4	933	26	371 ± 12	11.2	24.7	35.9	52.6	0.056	1.009
	65	2012–2015 (5)	–	–	355 ± 6	–	–	–	–	–	–
	04	2011 (2)	–	–	593 ± 8	–	–	–	–	–	–
	D	2011–2016 (7)	–	–	466 ± 5	–	–	–	–	–	–
		2011/8/20	157	26	388 ± 11	9.3	10.8	20.1	21.7	0.138	0.981
		2012/4/2	383	26	509 ± 11	12.1	17.4	29.4	35.0	0.091	0.989
		2012/11/1	596	44	721 ± 17	21.1	34.8	55.8	72.6	0.122	1.014
		2013/10/4	933	44	471 ± 11	17.7	39.2	56.9	83.4	0.089	1.007
		2014/3/25	1105	42	465 ± 13	17.2	44.2	61.5	95.1	0.086	1.008

	2015/11/13	1703	70	366 ± 17	18.0	76.3	94.3	171.0	0.100	1.018
	2016/7/12	1945	78	379 ± 18	16.9	87.1	104.0	200.0	0.103	1.029
G	2016/7/12	1945	70	345 ± 19	15.0	78.5	93.5	178.9	0.092	1.013
50	2012 (1)	–	–	377 ± 11	–	–	–	–	–	–
34	2011–2012 (2)	–	–	271 ± 6	–	–	–	–	–	–
	2012/11/1	596	18	297 ± 9	2.0	3.3	5.3	6.9	0.012	0.969
66	2012–2015 (4)	–	–	245 ± 6	–	–	–	–	–	–
72	2012/11/1	596	16	306 ± 8	2.1	3.4	5.5	7.1	0.012	0.964
03	2011 (2)	–	–	1085 ± 9	–	–	–	–	–	–
	2012/4/2	383	36	207 ± 17[f]	53.0	72.8	126.0	150.1	0.392	1.011
	2012/11/1	596	18	126 ± 7[g]	6.4	10.8	17.2	22.3	0.037	0.968
E	2015 (1)	–	–	398 ± 21	–	–	–	–	–	–
81	2014 (1)	–	–	282 ± 6	–	–	–	–	–	–
F	2012 (2)	–	–	959 ± 18	–	–	–	–	–	–
C	2011–2014 (4)	–	–	140 ± 4	–	–	–	–	–	–
	2011/8/20	157	10	256 ± 11	2.7	3.1	5.8	6.2	0.041	0.972
	2012/4/2	383	16	465 ± 30	4.3	6.1	10.4	12.4	0.032	0.995
	2012/11/1	596	20	115 ± 7	1.4	2.3	3.7	4.8	0.008	1.015
	2014/3/25	1105	10	104 ± 7	0.6	1.5	2.1	3.3	0.003	1.139
06	2011–2012 (3)	–	–	414 ± 9	–	–	–	–	–	–
	2012/11/1	596	8	124 ± 8	0.6	1.1	1.8	2.3	0.004	0.823
82	2014 (1)	–	–	78 ± 7	–	–	–	–	–	–

(continued)

Table 6.2 (continued)

Area	Site	Sampling date (n)	Elapsed time[a], day	Contaminated layer[b], cm	Average $^{134+137}Cs$ activity of top 5 cm[c, d], Bq/kg	Inventory, kBq/m^2				Flux[e, c], kBq/m^2 day	$^{134}Cs/^{137}Cs$ ratio[c]
						Detected		Corrected			
						^{134}Cs	^{137}Cs	$^{134+137}Cs$	$^{134+137}Cs$	$^{134+137}Cs$	
Y	05	2011–2014 (5)	–	–	296 ± 4	–	–	–	–	–	–
		2013/10/4	933	30	278 ± 15	2.4	5.4	7.8	11.4	0.012	0.969
	38	2012–2015 (4)	–	–	147 ± 5	–	–	–	–	–	–
		2012/4/2	383	26	209 ± 11	2.3	3.1	5.3	6.3	0.016	0.982
		2012/11/1	596	16	147 ± 13	0.7	1.1	1.8	2.4	0.004	
		2014/3/25	1105	28	141 ± 11	1.9	4.8	6.7	10.3	0.009	1.025
	54	2012–2014 (2)	–	–	189 ± 7	–	–	–	–	–	–
		2012/11/1	596	28	226 ± 11	1.8	3.0	4.7	6.1	0.010	0.960
	55	2012 (1)	–	–	214 ± 12	–	–	–	–	–	–
	53	2012 (1)	–	–	195 ± 7	–	–	–	–	–	–
	33	2011 (1)	–	–	106 ± 6	–	–	–	–	–	–
	67	2012 (1)	–	–	254 ± 10	–	–	–	–	–	–
	36	2011–2015 (4)	–	–	95 ± 4	–	–	–	–	–	–
		2014/3/25	1105	10	119 ± 9	0.4	0.9	1.3	2.1	0.002	1.059
	76	2012–2014 (2)	–	–	149 ± 6	–	–	–	–	–	–
	32	2011–2015 (4)			136 ± 3	–	–	–	–	–	–
		2012/11/1	596	18	304 ± 9	3.4	5.6	9.0	11.7	0.020	1.006
	79	2012 (2)	–	–	140 ± 8	–	–	–	–	–	–

		2012/11/1	596	18	158 ± 11	1.1	1.9	2.9	3.8	0.006	1.009
	B	2011/8/20	157	16	69 ± 6	2.5	2.8	5.3	5.8	0.037	0.958
	A	2011–2012 (3)	–	–	233 ± 6	–	–	–	–	–	–
		2011/8/20	157	6	71 ± 9	0.3	0.3	0.6	0.7	0.004	0.959
		2012/11/1	596	36	289 ± 12	4.1	6.9	11.1	14.4	0.024	0.992
	16	2011–2015 (3)	–	–	126 ± 3	–	–	–	–	–	–
	62	2012 (1)	–	–	176 ± 7	–	–	–	–	–	–
	61	2012/11/1	596	16	149 ± 7	2.0	3.4	5.3	6.9	0.012	0.949
	60	2012–2014 (2)	–	–	57 ± 3	–	–	–	–	–	–
	59	2012/11/1	596	16	80 ± 5	1.3	2.2	3.5	4.6	0.008	0.984
	56	2012–2014 (2)	–	–	124 ± 11	–	–	–	–	–	–
	58	2012 (1)	–	–	40 ± 3	–	–	–	–	–	–
W	57	2012–2014 (2)	–	–	48 ± 3	–	–	–	–	–	–
		2012/11/1	596	24	56 ± 4	1.0	1.6	2.7	3.5	0.006	1.075
	44	2012 (1)	–	–	62 ± 4	–	–	–	–	–	–
	43	2012/11/1	596	22	72 ± 6	1.5	2.6	4.0	5.2	0.009	1.017
	42	2012 (1)	–	–	107 ± 4	–	–	–	–	–	–
	41	2012/11/1	596	22	113 ± 7	1.9	3.2	5.1	6.6	0.011	0.967
	40	2012 (1)	–	–	165 ± 7	–	–	–	–	–	–
	39	2012 (1)	–	–	167 ± 7	–	–	–	–	–	–
	45	2012/11/1	596	10	205 ± 9	1.9	3.2	5.1	6.6	0.011	0.955
	46	2012–2014 (2)	–	–	174 ± 6	–	–	–	–	–	–

(continued)

Table 6.2 (continued)

Area	Site	Sampling date (n)	Elapsed time[a], day	Contaminated layer[b], cm	Average $^{134+137}$Cs activity of top 5 cm[c, d], Bq/kg	Inventory, kBq/m^2				Flux[e, c], kBq/m^2 day	^{134}Cs/^{137}Cs ratio[c]
						Detected		Corrected			
						^{134}Cs	^{137}Cs	$^{134+137}$Cs	$^{134+137}$Cs	$^{134+137}$Cs	
	47	2012 (1)	–	–	59 ± 4	–	–	–	–	–	–
	M	2012–2014 (2)	–	–	155 ± 5	–	–	–	–	–	–
	49	2012–2015 (3)	–	–	304 ± 8	–	–	–	–	–	–
		2012/11/1	596	26	336 ± 18	5.3	9.0	14.3	18.5	0.031	0.983
		2015/11/13	1703	4	36 ± 8	0.1	0.4	0.5	1.0	0.001	1.176
Z	52	2012/11/1	596	14	140 ± 11	0.7	1.2	1.9	2.5	0.004	0.993
		2014/3/25	1105	5	57 ± 12	0.2	0.5	0.7	1.1	0.001	0.859
		2015/11/13	1703	<5	60 ± 12	0.1	0.3	0.4	0.7	0.000	0.868
	68	2012–2015 (3)	–	–	50 ± 4	–	–	–	–	–	–
		2012/11/1	596	6	35 ± 5	0.1	0.2	0.4	0.5	0.001	0.801
		2014/3/25	1105	16	99 ± 19	0.4	1.0	1.5	2.3	0.002	0.968
		2015/11/13	1703	20	104 ± 18	0.5	2.5	2.8	5.0	0.003	0.984
	69	2012/11/1	596	<5	115 ± 4	0.3	0.6	1.0	1.3	0.002	0.907
		2014/3/25	1105	2.5	19 ± 3	0.0	0.3	0.3	0.4	0.000	0.540
	13	2011/10/1	199	<5	91 ± 4	0.3	0.4	0.7	0.8	0.004	0.987
		2011/12/3	262	<5	11 ± 2	0.1	0.1	0.2	0.3	0.001	0.834
		2014/3/25	1105	2.5	22 ± 4	0.1	0.3	0.4	0.5	0.000	0.791

		2015/11/13	1703	<5	22 ± 6	0.0	0.1	0.1	0.2	0.000	0.914
	J	2015/11/13	1703	8	59 ± 8	0.1	0.4	0.4	0.8	0.000	1.141
	K	2015/11/13	1703	<5	42 ± 8	0.0	0.1	0.2	0.3	0.000	1.258
	H	2015/11/13	1703	<5	81 ± 11	0.1	0.4	0.5	0.9	0.001	1.074
	10	2011/10/1	199	<5	57 ± 3	0.2	0.3	0.5	0.6	0.003	1.005
		2011/12/3	262	<5	62 ± 3	0.5	0.7	1.3	1.4	0.005	0.833
		2014/3/25	1105	2.5	38 ± 8	0.1	0.5	0.6	0.9	0.001	1.062
		2015/11/13	1703	<5	25 ± 5	0.0	0.1	0.2	0.3	0.000	1.811
	07	2011/10/1	199	<5	53 ± 2	0.3	0.5	0.8	0.9	0.005	0.825
		2011/12/3	262	<5	162 ± 3	2.4	3.0	5.4	6.1	0.023	1.003
		2014/3/25	1105	5	56 ± 11	0.3	0.9	1.2	1.7	0.002	0.900
		2015/11/13	1703	<5	32 ± 4	0.1	0.3	0.4	0.7	0.000	1.040
	08	2011/10/1	199	<5	22 ± 2	0.1	0.1	0.2	0.3	0.004	0.890
		2011/12/3	262	<5	45 ± 3	0.5	0.6	1.1	1.2	0.005	0.999
		2014/3/25	1105	5	57 ± 11	0.2	0.7	0.9	1.4	0.001	0.930
	11	2011/10/1	199	<5	26 ± 3	0.0	0.2	0.2	0.2	0.001	0.218
		2011/12/3	262	<5	11 ± 2	0.1	0.1	0.2	0.3	0.001	1.385
		2014/3/25	1105	2.5	56 ± 9	0.2	0.5	0.7	1.0	0.001	0.991
		2015/11/13	1703	<5	46 ± 10	0.0	0.2	0.2	0.4	0.000	0.957
	14	2011/10/1	199	<5	52 ± 4	0.2	0.2	0.4	0.5	0.002	1.002
		2011/12/3	262	<5	52 ± 3	0.4	0.4	0.8	0.9	0.003	1.309
		2014/3/25	1105	2.5	19 ± 9	0.0	0.2	0.3	0.4	0.000	1.187
	17	2011/10/1	199	<5	50 ± 3	0.2	0.3	0.5	0.6	0.003	0.989
		2011/12/3	262	<5	46 ± 3	0.3	0.5	0.9	1.0	0.004	0.790
		2014/3/25	1105	2.5	34 ± 9	0.1	0.4	0.5	0.7	0.001	0.874
		2015/11/13	1703	<5	26 ± 5	0.0	0.1	0.1	0.2	0.000	1.283

(continued)

Table 6.2 (continued)

Area	Site	Sampling date (n)	Elapsed time[a], day	Contaminated layer[b], cm	Average $^{134+137}Cs$ activity of top 5 cm[c, d], Bq/kg	Inventory, kBq/m^2 Detected		Corrected		Flux[e, c], kBq/m^2 day	$^{134}Cs/^{137}Cs$ ratio[c]
						^{134}Cs	^{137}Cs	$^{134+137}Cs$	$^{134+137}Cs$	$^{134+137}Cs$	
	09	2011/10/1	199	<5	38 ± 3	0.2	0.3	0.5	0.6	0.003	0.920
		2014/3/25	1105	–	7 ± 2	0.0	2.1	0.2	0.2	0.000	–
	12	2011/10/1	199	–	2 ± 1	0.0	0.0	0.0	0.0	0.000	–
		2014/3/25	1105	2.5	31 ± 7	0.1	0.2	0.3	0.5	0.001	0.941
	15	2011/10/1	199	<5	22 ± 2	0.1	0.2	0.3	0.3	0.002	0.910
		2014/3/25	1105	5	37 ± 11	0.1	0.4	0.5	0.8	0.001	0.960
	18	2011/10/1	199	<5	11 ± 1	0.4	0.4	0.8	0.9	0.004	1.211
		2014/3/25	1105	–	2 ± 1	0.0	0.1	0.1	0.1	0.000	–
	L	2015/11/13	1703	<5	36 ± 5	0.0	0.2	0.3	0.5	0.000	1.045
	30	2011/10/1	199	<5	22 ± 2	0.2	0.2	0.5	0.5	0.003	1.071
	21	2011/10/1	199	<5	10 ± 1	0.2	0.4	0.6	0.6	0.003	0.712
		2014/3/25	1105	–	1 ± 1	0.0	0.0	0.0	0.0	0.000	–
	24	2011/10/1	199	–	10 ± 1	0.0	0.1	0.1	0.2	0.001	–
V	19	2011/10/1	199	<5	136 ± 5	1.0	1.2	2.2	2.4	0.012	1.024
		2011/12/3	262	<5	234 ± 6	4.2	4.4	9.7	11.0	0.042	0.962
	22	2011/10/1	199	<5	118 ± 5	0.5	0.6	1.1	1.2	0.006	1.003
	23	2011/10/1	199	<5	42 ± 3	0.2	0.3	0.5	0.5	0.003	0.754

	20	2011/10/1	199	<5	48 ± 3	0.1	0.3	0.5	0.5	0.003	0.517
		2011/12/3	262	<5	75 ± 5	0.6	0.8	1.4	1.6	0.006	0.995
		2014/3/25	1105	–	3 ± 1	0.0	0.2	0.2	0.2	0.000	–
	80	2014/3/25	1105	–	nd	0.0	0.0	0.0	0.0	0.000	–
								Weighted average[h] (n = 117)			1.006 ± 0.003

Values in parentheses are the number of sampling
[a]Elapsed time is from March 16, 2011, to the sampling date
[b]Contaminated layer means the thickness from the surface to the deepest layer where ^{134}Cs is detected
[c]Calculated using the decay-corrected value to March 16, 2011
[d]Weighted average value for multiple samples
[e]Calculated using inventory and elapsed time
[f]The concentration maximum appeared under the 5 cm depth
[g]The peak of radioactive cesium disappeared
[h]Value for the counting error within 5%. Data not listed in this table also included in the calculation

Table 6.3 Concentrations, inventories, and fluxes of radioactive cesium in the sediment of the Edogawa and Sakagawa rivers

Area	Site	Sampling date	Elapsed time[a], day	Contaminated layer[b], cm	$^{134+137}Cs$ activity of top 5 cm[c, d], Bq/kg	Inventory[c], kBq/m^2			Flux[c, e], kBq/m^2 day	$^{134}Cs/^{137}Cs$ ratio[c]
						^{134}Cs	^{137}Cs	$^{134+137}Cs$	$^{134+137}Cs$	
Edogawa River (Upstream)	E1	2011/12/3	262	<5	120 ± 2	4.4	4.3	8.7	0.033	1.039
		2015/11/11	1701	25		107	107	214	0.126	1.005
	E2	2011/12/3	262	<5	203 ± 3	7.8	7.7	15.5	0.060	1.031
		2015/11/12	1702	15		12.0	11.7	23.7	0.014	1.012
	E3	2011/12/3	262	<5	222 ± 4	19.5	18.8	38.3	0.146	1.036
		2015/11/12	1702	5		1.4	1.6	3.0	0.002	1.027
	E4	2011/12/3	262	<5	199 ± 4	15.4	15.5	30.9	0.118	0.998
		2015/11/12	1702	24		8.1	7.9	16.0	0.010	1.026
Sakagawa River (A tributary to Edogawa River)	S5	2014/4/29	1140	13	4160 ± 40	339	336	675	0.592	1.009
	S4	2014/4/29	1140	14	1180 ± 11	98.0	98	196	0.172	0.996
	S6	2014/4/29	1140	41	1310 ± 22	460	449	909	0.797	1.023
	S3	2014/4/29	1140	5	1810 ± 14	70.0	70.4	140	0.123	0.994
	S2	2014/4/29	1140	14	5230 ± 46	152	152	303	0.440	1.001
	S1	2014/4/29	1140	50	6350 ± 64	571	565	1140	0.997	1.011
		2015/11/13	63 (1703)[f]	12	3020 ± 44	33.1	32.8	65.9	1.030	1.012
		2016/6/25	288 (1928)[f]	32	3630 ± 67	145	141	285	0.993	1.029

Edogawa River (Downstream)	E5	2011/12/3	262	<5	581 ± 8	13.6	13.4	27.0	0.103	1.023
		2014/3/24	1104	<5		44.0	43.3	87.3	0.079	1.036
		2015/11/12	1702	20		7.7	8.1	15.8	0.009	0.959
	E6	2011/12/3	262	<5	697 ± 6	35.0	34.6	69.6	0.266	1.000
		2015/11/12	1702	31.5		141	141	282	0.166	1.003
	E7	2011/12/3	262	<5	881 ± 10	30.3	32.0	62.3	0.238	0.947
		2014/3/24	1104	<5		6.9	6.8	13.7	0.012	1.027
	E8	2012/4/3	384	<5	706 ± 17	11.8	11.4	23.2	0.060	1.038
	E9	2012/4/3	384	<5	464 ± 10	11.8	11.9	23.7	0.062	1.000
	E10	2012/4/3	384	<5	133 ± 5	4.4	4.7	9.1	0.024	0.949
									Average[d]	1.009 ± 0.025

[a]Elapsed time is from March 16, 2011, to the sampling date
[b]Contaminated layer means the thickness from the surface to the deepest layer where ^{134}Cs is detected
[c]Calculated using the decay-corrected value to March 16, 2011
[d]Weighted average value for multiple samples
[e]Calculated using inventory and elapsed time
[f]The sediments deposited after the flood on September 11, 2015

propagation. The radioactivity ratio of ^{134}Cs and ^{137}Cs based on March 16, 2011, was also calculated to ensure the reliability of the analysis. The ^{134}Cs/^{137}Cs activity ratio immediately after the accident is believed to have been almost 1.0 [30–32]. Figure 6.2 shows the geographic distribution of the average concentration of $^{134+137}$Cs at each sediment sampling site in the Tokyo Bay water system.

6.6.1 Distribution of Radioactive Cesium in the Surface Sediments of Tokyo Bay

The highest $^{134+137}$Cs concentration of all measurements in Tokyo Bay was 1340 ± 13 Bq/kg found in surface sediments (0–5 cm) sampled on November 1, 2012, at Site 01 in the Old Edogawa estuary. Six samplings were performed at the site between October 1, 2011, and November 13, 2015. These $^{134+137}$Cs concentrations tended to decrease over time, but the water depth was as shallow as approximately 5 m, and the river water from the Old Edogawa flowed in, so their concentrations and vertical distributions were greatly affected. The weighted average concentration of $^{134+137}$Cs during the whole monitoring period was 810 ± 6 Bq/kg. Since an anoxic water mass is dominant in the inner part of Tokyo Bay [4, 5] the sediment is not affected much by the activity of benthic organisms. However, this estuary not only shallow is but also hosts navigation of many vessels, and the sediment is highly likely to undergo physical mixing. Therefore, divers visually inspected the seafloor and collected sediments that appeared not to be disturbed. However, it was challenging to collect a stable sedimentary layer from the same place each time.

The $^{134+137}$Cs concentrations in the surface sediments of Tokyo Bay were high in the estuary where the Arakawa and Old Edogawa rivers flow in but low in the central part of the bay. As shown in Fig. 6.1, depending on the $^{134+137}$Cs concentration and the characteristics of the water mass, the contaminated sediments in the Tokyo Bay sea areas can be classified into V (the Tamagawa River estuary), W (the Sumidagawa estuary), X (the Arakawa and Old Edogawa estuaries), Y (off the coast of Area X), and Z (central part of Tokyo Bay). Throughout the monitoring, the $^{134+137}$Cs concentrations of surface sediments were highest in Area X and decreased significantly toward Area Z. The radioactivities of $^{134+137}$Cs are expressed as weighted averages; that in Area X is 424 Bq/kg (78–1340 Bq/kg, n = 55); Area Y, 131 Bq/kg (40–371 Bq/kg, n = 40); and Area Z, 17 Bq/kg (1–162 Bq/kg, n = 50). At the estuary of the Tamagawa River (Area V), which flows through the suburbs of western Tokyo, the weighted average is 57 Bq/kg (nd–234 Bq/kg, n = 8). At the estuary of the Sumidagawa River (Area W), which flows through central Tokyo the weighted average is 103 Bq/kg (36–336 Bq/kg, n = 16). Both concentrations indicate lower contamination than in the Old Edogawa estuary (Area X) This distribution of radioactive cesium in Tokyo Bay sediments is in good agreement with the pollution levels of radioactive cesium in the catchments of these incoming rivers.

From Fig. 6.2, radioactive cesium deposited in the sediments of Tokyo Bay is presumed to originate from the Old Edogawa and Edogawa rivers, which will are discussed in Sects. 6.7 and 6.8.

6.6.2 *Inventory and Flux of Radioactive Cesium in the Tokyo Bay Sediment*

The inventory and flux of radioactive cesium deposition in Tokyo Bay sediments from the basin based on the 5-year investigation of the FDNPP accident were estimated. Tables 6.2 and 6.3 show the inventory, flux, and $^{134}Cs/^{137}Cs$ activity ratio of radioactive cesium in sediments collected from Tokyo Bay and the Edogawa water system. The inventory was estimated for the samples in which the radioactive cesium contamination was confirmed to be due to the FDNPP accident by the vertical distribution of ^{134}Cs in the sediment core. The thickness of this contaminated sedimentary layer is also shown in the tables. The $^{134}Cs/^{137}Cs$ activity ratio of 117 sediment samples with a counting error of the activity readings within 5% is 1.006 ± 0.003 (weighted average), which is consistent with the radioactive cesium released by the FDNPP accident [30–32]. This agreement ensures the reliability of the measured data.

The inventory values of $^{134+137}Cs$ are 2.28–265 kBq/m^2 for Area X, 0.67–14.4 kBq/m^2 for Area Y, 0.04–6.10 kBq/m^2 for Area Z, 3.48–6.63 kBq/m^2 for Area W, and 0.00–11.0 kBq/m^2 for Area V. Area X in the Old Edogawa estuary shows an overwhelmingly large inventory compared to other marine areas of Tokyo Bay. This inventory reveals approximately the same pollution level as the radioactive cesium inventory of soils in highly contaminated areas in the middle of the Edogawa watershed. Additionally, since the concentration at Site A3 in the Arakawa River upstream of the Old Edogawa estuary decreases significantly, radioactive cesium that accumulated in Area X is assumed to have been supplied from the Old Edogawa and Edogawa water systems. Nevertheless, even if core samples were collected at the same sampling site, such as Site 02 or Site C, the inventory changed significantly on different sampling days. As described in Chap. 3, the soils contaminated by radioactive cesium show large local deviations. A similar trend is observed in the Tokyo Bay sediments. However, due to the physical mixing of seawater, it is unlikely that local segregation occurs during the process of radioactive cesium settling in the sea. Since the water depths at the sites where the cores were collected are as shallow as 5–10 m, radioactive cesium deposited on the sediment surface is probably disturbed by tidal currents. Nevertheless, the spatiotemporal distribution of radioactive cesium obtained at Site D suggests that the core at this point recorded a steady depositional process throughout the monitoring period. Therefore, the core analysis taken from this site can be used for quantitative analysis of radioactive cesium contamination in Tokyo Bay.

Dividing the $^{134+137}$Cs inventory by the elapsed time (days) yields the average daily flux for sediment. The fluxes (arithmetic means) from the accident to the date of sediment collection are 0.084 ± 0.088 (n = 22) for Area X, 0.013 ± 0.010 (n = 13) for Area Y, 0.002 ± 0.003 (n = 50) for Area Z, 0.011 ± 0.010 (n = 6) for Area W, and 0.009 ± 0.014 (n = 8) kBq/m^2 day for Area V. The $^{134+137}$Cs flux in central Tokyo Bay is approximately 1/40 that in the Old Edogawa estuary. This large difference in flux in a narrow sea area such as Tokyo Bay suggests that radioactive cesium flowing into Tokyo Bay from the Old Edogawa was deposited in the estuary without diffusing into the bay. At Site D, where a steady depositional process is recorded, the average flux of the seven monitoring runs through July 2016 is 0.104 ± 0.019 kBq/m^2 day. The average flux for each collection period of the core sediments shown in Table 6.7 is 0.103 ± 0.051 kBq/m^2 day, which is evaluated as a typical representative value of the $^{134+137}$Cs flux in the Old Edogawa estuary due to the FDNPP accident.

6.6.3 *Spatiotemporal Fluctuations in Radioactive Cesium in the Tokyo Bay Sediment*

The vertical distributions of radioactive cesium in the sediments of Tokyo Bay and their temporal changes were investigated using core samples. Sediment cores were collected between August 2011 and July 2016 using a sampler with an inner diameter of 10 cm and a length of 100 cm. Figure 6.3 shows the vertical distribution of $^{134+137}$Cs radioactivity from the Old Edogawa estuary (Site 02) to the center of the bay (Site J) in 2014 and 2015. The highest activity of radioactive cesium was 1070 Bq/kg, which was detected in a layer at depths of 34–36 cm near the estuary, where the Old Edogawa flows. In this core, ^{134}Cs was detected until in the layer at depths of 54–56 cm. The concentration of radioactive cesium decreased rapidly in the layers shallower than 34–36 cm were $^{134+137}$Cs radioactivity peaked and measured 250–400 Bq/kg in the surface sediment layer. The vertical distribution of $^{134+137}$Cs in the Site 02 core has not been significantly affected by physical mixing despite its estuary location. This distribution suggests that contaminated suspended particles flowing from the Old Edogawa River steadily accumulated in this area. The vertical distribution of radioactive cesium showed a sharp decrease from the estuary (Site 02) to the inner part of Tokyo Bay (Site J). In the case of Site J, the peak $^{134+137}$Cs of 44 Bq/kg is detected in a 5 cm-deep layer. This distribution suggests that most of the radioactive cesium from the Old Edogawa was deposited rapidly in the estuary and scarcely diffused into the center of the bay. This behavior of radioactive cesium is fortunate in that the entire Tokyo Bay has not been exposed to massive radioactive contamination due to the FDNPP accident.

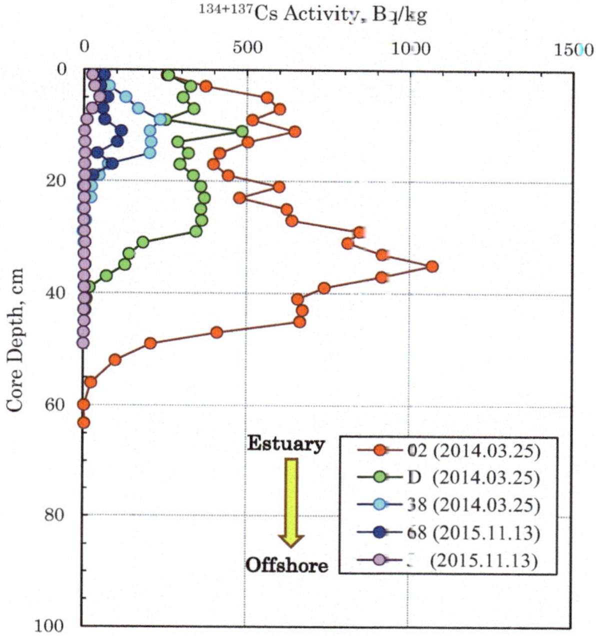

Fig. 6.3 Spatiotemporal changes in the vertical distribution of $^{134+137}$Cs in the sediment cores collected from the Old Edogawa estuary toward central Tokyo Bay. The date in the parentheses is the core collection date

Since core samples collected at Site D, 2.5 km south of the estuary of the Old Edogawa River, reveal the steadiest deposition, the temporal changes in the vertical distribution of radioactive cesium at this site are investigated. Site D is located 1.5 km south of Site 02 and is not significantly affected by the flow of river water from the Old Edogawa and Arakawa rivers. The results are shown in Fig. 6.4. In August 2011, in the sediment core that was quickly collected immediately after the FDNPP accident, the radioactive cesium concentration in the sediment was 284 Bq/kg in the 0–1 cm layer at the surface, but the highest level was 433 Bq/kg in the layer at depths of 4–6 cm. The peak in $^{134+137}$Cs moved to the layer at depths of 12–14 cm layer by October 2013, and its concentration was 547 Bq/kg. In July 2016, 5 years and 4 months after the accident, the peak of $^{134+137}$Cs was buried to a layer at depths of 64–66 cm. Although the peak concentration was 329 Bq/kg, the value corrected for radioactive decay since the time of the accident was 632 Bq/kg, indicating that the concentration peak of radioactive cesium more than 5 years after the accident had been buried but was retained in the sediment.

However, this peak depth does not necessarily indicate the history of the steady pollutant influx from the Old Edogawa River. Many rivers flooded in the Tokyo metropolitan area due to the Kanto-Tohoku heavy rain from September 9–11, 2015. As discussed in Sect. 6.6, the Edogawa water system also flooded. The flood carried material down the river and into Tokyo Bay. As shown in Fig. 6.5, traces of flood sediments were recorded as anomalies in water content and porosity (deposits of low-density sediments) in the layers at depths of 8–22 cm collected in November

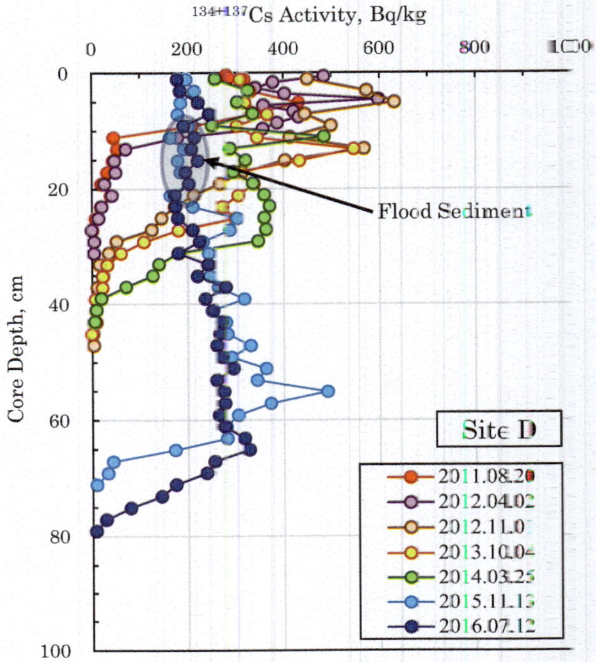

Fig. 6.4 Temporal changes in the vertical distribution of $^{134+137}$Cs in the sediment core collected at Site D. The black circle in the core collected on Nov. 13, 2015, shows the flood sedimentary layer that flowed in due to the Kanto-Tohoku heavy rainfall from September 9–11, 2015

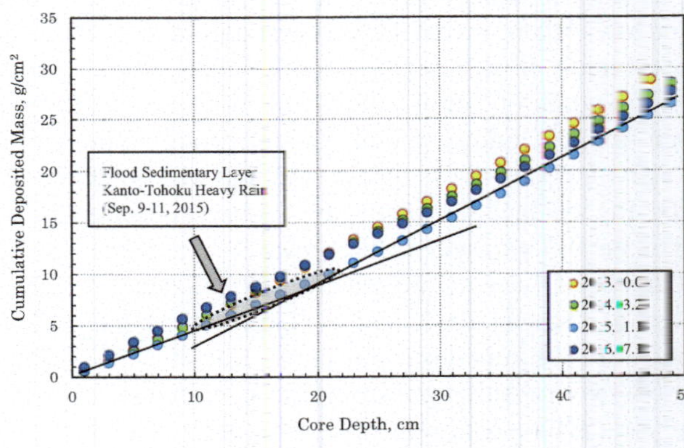

Fig. 6.5 Evidence for the flood sediment deposited at Site D. Coastal flood sediments have high water content and small particle size, so their apparent density will be smaller [33]. The cumulative mass of 8–22 cm layer in the core collected on November 13, 2015, is lower than that of the other cores

2015. It is known that sediments with high water content and low apparent density accumulate in estuaries during floods [33]. In other words, the cores collected in November 2015 and July 2016, shown in Fig. 6.4, had flood sediment layers with thicknesses of 10 cm or more. A quantitative analysis of the distribution of radioactive cesium in the Site D core is discussed in Sect. 6.8.

6.7 Spatiotemporal Distributions of Radioactive Cesium in the Sediments of the Old Edogawa, Edogawa, and Sakagawa Rivers

6.7.1 *Concentration of Radioactive Cesium in the Sediment of the Edogawa Water System*

As Sects. 6.5 and 6.6 clearly indicate, the source of radioactive contamination in Tokyo Bay is presumed to be the Edogawa River system. The downstream reach of the Edogawa River from Gyotoku Weir has been renamed the Old Edogawa River, as shown in Fig. 6.2. This weir is closed except during a major flood, when the water of the Edogawa River flows into the Old Edogawa River. Most of the river water flows into Tokyo Bay near Site 33 in Fig. 6.1, but some water also flows near Site 03. The middle watershed of the Edogawa River is a highly contaminated area due to the FDNPP accident. This contaminated area is the catchment area of the Sakagawa River, which joins the Edogawa River. Table 6.4 shows the concentrations of radioactive cesium in the soils of the Sakagawa River catchment area collected in 2011 and those of the Sakagawa riverbank collected in 2014. The radioactive cesium concentrations are higher in the catchment area than in the riverbank.

Table 6.3 shows the radioactive decay-corrected $^{134+137}$Cs concentrations of surface sediments at depths of 0–5 cm in the Edogawa and Nakagawa rivers. The weighted average $^{134+137}$Cs concentration in the Sakagawa surface sediment is 1760 ± 8 Bq/kg (n = 8, 0–5 cm). This concentration indicates that the soil contaminated with radioactive cesium is transported from the catchment areas, and deposited in the Sakagawa sediment. The Edogawa River branches off from the Tonegawa River, but according to the results of airborne monitoring [1] in Fig. 3.2, the forested area in Gunma Prefecture upstream of the Tonegawa River is contaminated with 60–300 kBq/m^2 of radioactive cesium. However, upstream of the Edogawa River branching off from the Tonegawa River, the concentration of $^{134+137}$Cs in the sediment is 170 ± 2 Bq/kg (n = 20, Sites E1–E4), and that at the downstream sites in the Sakagawa River is 473 ± 3 Bq/kg (n = 14, Sites E5–E10). Therefore, the contaminated area upstream of the Tonegawa River is not considered an essential source of radioactive cesium to supply Tokyo Bay. On the other hand, radioactive cesium deposited in the Sakagawa catchment in the northern part of Chiba Prefecture is estimated to be a significant source of pollution in Tokyo Bay.

6.7.2 *Inventories and Fluxes of Radioactive Cesium in the Sediments of the Edogawa and Sakagawa Rivers*

The Matsudo Weir is located about 30 m downstream from Site S1 and is usually closed to regulate the water flow in the Sakagawa River. Therefore, the flow velocity

Table 6.4 Concentrations and inventories of radioactive cesium in soil samples collected around Sakagawa River

Site	Sampling date	Detected, Bq/kg			Corrected[a], Bq/kg			Inventory, kBq/m^2	^{134}Cs/^{137}Cs ratio
		^{134}Cs	^{137}Cs	$^{134+137}$Cs	^{134}Cs	^{137}Cs	$^{134+137}$Cs		
K1	2011/5/19	3010 ± 22	3170 ± 24	6180 ± 33	3210 ± 24	3180 ± 24	6390 ± 34	83	1.009 ± 0.011
K2	2011/6/11	31,200 ± 97	33,700 ± 101	64,900 ± 140	33,900 ± 105	33,900 ± 102	67,800 ± 146	881	1.000 ± 0.004
K3	2011/6/11	16,300 ± 60	17,600 ± 66	33,900 ± 89	17,700 ± 65	17,700 ± 66	35,400 ± 93	460	1.000 ± 0.005
K4	2011/5/19	970 ± 12	1070 ± 13	2040 ± 18	1060 ± 13	1080 ± 13	2140 ± 18	57	0.982 ± 0.017
S5	2014/4/29	168 ± 4	452 ± 7	620 ± 8	480 ± 13	485 ± 7	965 ± 15	25	0.990 ± 0.030
S4	2014/4/29	904 ± 22	2480 ± 34	3380 ± 41	2580 ± 62	2660 ± 37	5250 ± 72	137	0.970 ± 0.027
S6	2014/4/29	445 ± 12	1140 ± 18	1590 ± 22	1270 ± 33	1230 ± 20	2500 ± 39	65	1.033 ± 0.032
S3	2014/4/29	391 ± 10	1020 ± 16	1410 ± 19	1120 ± 29	1100 ± 17	2220 ± 34	58	1.018 ± 0.031
S2	2014/4/29	271 ± 6	700 ± 10	971 ± 12	775 ± 18	752 ± 11	1530 ± 21	40	1.031 ± 0.028
S1	2014/4/29	551 ± 15	1480 ± 23	2030 ± 28	1580 ± 42	1590 ± 25	3170 ± 49	82	0.994 ± 0.031

The K series samples are the soil of the residential areas, and the S series samples are the soil of the Sakagawa riverbank

[a]Decay-corrected value to March 16, 2011

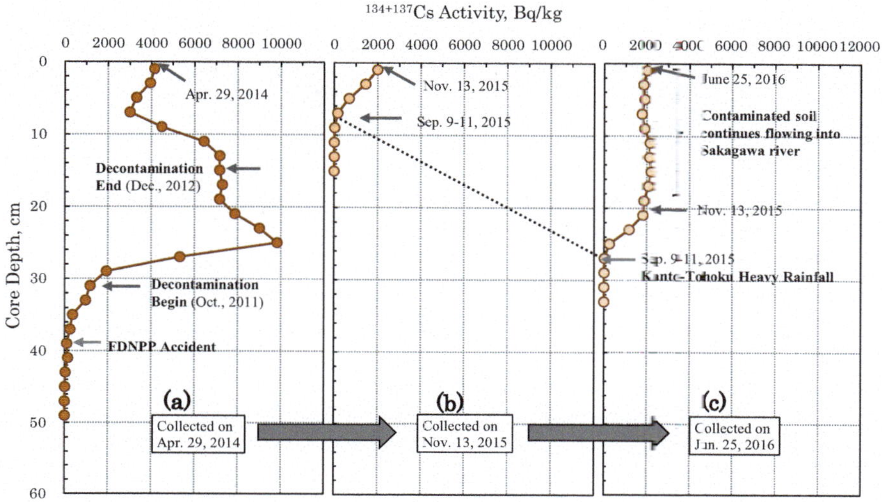

Fig. 6.6 (**a–c**) Temporal changes in the vertical distribution of $^{134+137}$Cs in the sediment core at Site S1 in Sakagawa River. When the Matsudo weir was opened because of the Kanto-Tohoku heavy rainfall, large proportions of the contaminated sediment deposited at Site S1 flowed out. Decontamination works in Kashiwa City were carried out between "Begin" and "End" in figure (**a**) [34]

of water in the Sakagawa River decreases rapidly above this weir, meaning that suspended solids are deposited at this point. Table 6.3 shows the radioactive cesium concentrations in the surface sediments (0–5 cm depths) of the Sakagawa and Edogawa rivers from upstream to downstream. In the Edogawa River, the radioactive cesium concentration increases approximately 2.8 times from Sites E1–E4 upstream of the confluence with the Sakagawa River to Sites E5–E10 downstream. Since the Sakagawa is a narrow river with a complicated flow path, the concentrations of radioactive cesium fluctuate greatly depending on the sediment sampling point. Nevertheless, the concentration of radioactive cesium in Sakagawa sediments is approximately 10–40 times higher than that in the upper Edogawa. These values indicate that the contamination of Edogawa sediment by radioactive cesium supplied from the Sakagawa River. The soil in the catchment area of the Sakagawa is highly radioactively contaminated and provides a credible origin for the radioactive cesium in the region. The results suggest that the behavior of radioactive cesium in this water system can be estimated from its inventory and flux.

Figure 6.6 shows the vertical distribution of radioactive cesium in the sediment core at Site S1. The vertical distribution of radioactive cesium before and after the FDNPP accident was recorded in the S1(a) core sampled on April 29, 2014. If it is presumed that the 38–40 cm layer in which ^{134}Cs is detected was deposited immediately after the accident, the rate of deposition of this sediment is approximately 1.0 cm/month. Thus, as shown in Fig. 6.6a, the sediment layer deposited in October 2011 is the layer from 32 to 34 cm; this time is when decontamination work started in Kashiwa City [34], which is in the catchment area of the Sakagawa River. The decontamination work was completed in December 2012, and the sedimentary layer corresponds to a depth of 14–16 cm. For this reason, as shown in Fig. 6.6a, the vertical distribution of $^{134+137}$Cs in core S1(a) records the radioactive sludge discharge history due to decontamination work in Kashiwa City. Presumably, the radioactive cesium in the highly contaminated layer represents contaminated soil from the high-contamination zone accompanied by a discharge of contaminated sludge caused by the decontamination work in Kashiwa City. Some of the decontamination wastewater that was not properly treated flowed into the Sakagawa River, and it is highly likely that this contaminated sludge was deposited at Site S1.

However, eastern Japan, including the Tokyo metropolitan area, experienced heavy rain called the Kanto-Tohoku heavy rainfall between September 9 and 11, 2015. The total precipitation in the watershed of the Tokyo metropolitan area was more than 500 mm, and many rivers flooded. More than 200 mm of rain also fell on the Sakagawa catchment area [35–37]. A new sediment core S1(b) was collected at the same location as the S1(a) core on November 13, 2015, approximately 3 months after the heavy rain event. Figure 6.6b shows the analytical results for this core. The highly contaminated sediment layer observed in core S1 (a) disappeared, but new contaminated sediment was deposited after the flood. A newer core S1(c) was also collected on June 25, 2016, approximately 10 months after the flood event. The mass fluxes of these sediment cores are shown in Table 6.5. The mass flux of core S1(b) is considerable, indicating that the sedimentation process at this point was strongly affected by flooding. Throughout the entire sedimentation period, the concentrations of $^{134+137}$Cs in the sediment are 3000–10,000 Bq/kg for S1(a) and approximately 2000 Bq/kg for S1(b) and S1(c). After the flood event, no highly contaminated sediment flowed in, but approximately 2000 Bq/kg of contaminants continued to flow into the Sakagawa River. Perhaps a large influx of low-contamination soil could also have diluted the radioactive cesium concentration. The temporal changes in inventory and flux at Site S1 site were calculated using these measurement results. The results are shown in Table 6.5.

From the mass fluxes and radioactive cesium concentrations shown in Table 6.5, the inventory and flux for each sampling period can be calculated. The $^{134+137}$Cs fluxes of the S1(a), (b), and (c) cores were 0.997 kBq/m^2 day, 1.03 kBq/m^2 day, and 0.993 kBq/m^2 day, respectively, after decay correction to the values on March 16, 2011. From immediately after the FDNPP accident until June 2016, the radioactive cesium flux at the Sakagawa Site S1 did not change at all. Furthermore, as shown in parentheses in Table 6.5, this flux decreased with the radioactive decay of $^{134+137}$Cs (Fig. 6.7). The value was 0.517 kBq/m^2 day in June 2016. These values suggest that the amount of radioactive pollutants flowing into the Sakagawa River from its catchment areas remained constant for 5 years after the accident. The $^{134+137}$Cs flux at Site D offshore from the Old Edogawa estuary, which is downstream of the Sakagawa and Edogawa rivers, is approximately 0.05 kBq/m^2 day (Fig. 6.8), so the $^{134+137}$Cs flux of Site S1 is approximately ten times higher than that at Site D.

In any case, the amount of radioactive cesium supplied from the contaminated land to Tokyo Bay, excluding the effects of decreases due to radioactive decay, changed little during the 5 years after the FDNPP accident. A more detailed analysis of where this radioactive cesium on land originated is required.

Table 6.5 Concentrations, inventories, and fluxes of radioactive cesium in the sediment at Site S1 in Sakagawa River

Sampling date	Core depth, cm	Sampling interval	Elapsed time[a], day	Total mass, kg/m^2	Mass flux, kg/m^2 day	Activity[b], kBq/kg			Inventory[b], kBq/m^2			Flux[b], kBq/m^2 day
						^{134}Cs	^{137}Cs	$^{134+137}Cs$	^{134}Cs	^{137}Cs	$^{134+137}Cs$	$^{134+137}Cs$
2014/4/29 Figure 6.6a	0–44	2011/3/16 2014/4/29	1140 (1140)	163	0.143	3.50 (1.22)	3.46 (3.22)	6.96 (4.45)	571 (200)	565 (526)	1140 (726)	0.997 (0.637)
2015/11/13 Figure 6.6b	0–10[c]	2015/9/11[d] 2015/11/13	63 (1703)	50.5	0.802	0.66 (0.14)	0.65 (0.58)	1.30 (0.72)	33 (7)	33 (29)	66 (36)	1.030 (0.575)
2016/6/25 Figure 6.6c	0–26	2015/9/11[d] 2016/6/25	288 (1928)	89.0	0.309	1.62 (0.28)	1.58 (1.40)	3.20 (1.67)	145 (24)	141 (124)	285 (149)	0.993 (0.517)
	20–26[c]	2015/9/11[d] 2015/11/13	63 (1928)	30.7	0.487	1.01 (0.17)	1.03 (0.92)	2.05 (1.09)	31 (5)	32 (28)	63 (33)	0.996 (0.529)
	0–20	2015/11/13 2016/6/25	225 (1928)	58.3	0.259	1.94 (0.33)	1.87 (1.65)	3.81 (1.98)	113 (19)	109 (96)	222 (116)	0.987 (0.513)

[a]The value in parentheses is days from March 16, 2011

[b]Decay-corrected value to March 16, 2011. The value in parentheses is detected value for the interval examined

[c]These sedimentary layers deposited at the same time

[d]The contaminated sediment layer was flowed out due to the Kanto-Tohoku heavy rainfall on September 9–11, 2015

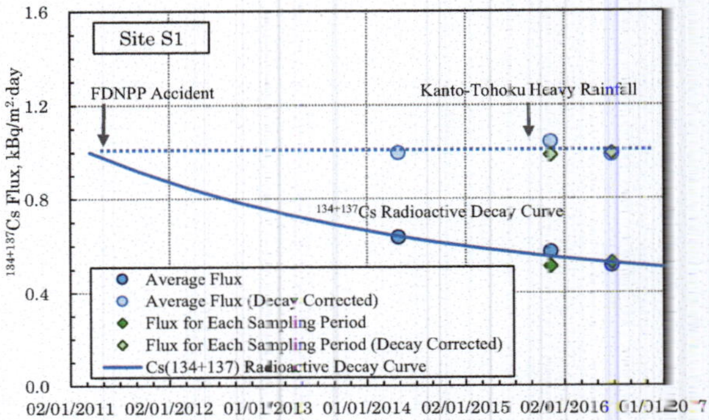

Fig. 6.7 Temporal changes in fluxes of $^{134+137}$Cs to the sediment at Site S1 in Sakagawa River

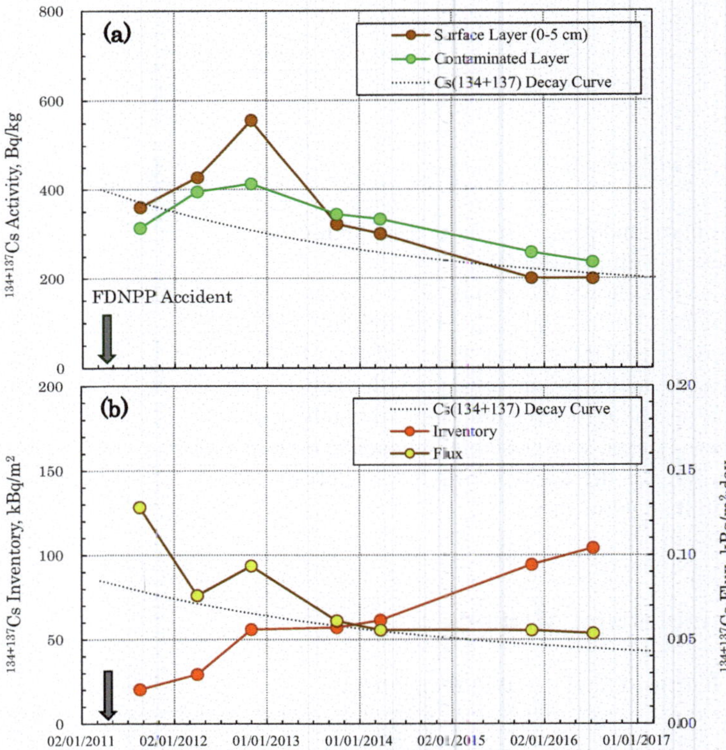

Fig. 6.8 (**a**, **b**) Temporal changes in concentrations, inventories, and fluxes of $^{134+137}$Cs at Site D in the estuary sediment of Old Edogawa River. The contaminated layer was defined as the layer including the detectable ^{134}Cs. Flux was calculated by dividing the inventory by elapsed time since the FDNPP accident

6.8 Source, Migration, and Deposition of Radioactive Cesium in the Tokyo Bay Water System

6.8.1 *Inventory and Flux of Radioactive Cesium Due to the FDNPP Accident in the Sediment at Site D in the Old Edogawa Estuary*

As shown in Fig. 6.2 and Table 6.3, the radioactive cesium concentrations in sediments in the Edogawa River system increase in the order of Sakagawa-lower Edogawa-upper Edogawa. This order suggests that radioactively contaminated soil that has flowed into the Sakagawa from highly contaminated areas around Kashiwa and Matsudo cities is transported to Tokyo Bay via the Edogawa and Old Edogawa rivers. In other words, the source of radioactive cesium contamination in Tokyo Bay is contaminated soil in highly polluted areas, and the Edogawa River system is considered to play an essential role in its transport. Based on the vertical distribution of radioactive cesium in the core sediment deposits at Site D, as shown in Fig. 6.4, the changes over time in the activity, inventory, and flux of the $^{134+137}$Cs in the sediment were analyzed.

Table 6.6 shows the temporal changes in the concentration of radioactive cesium in the sediments and inventory at Site D, offshore from the estuary where the Old Edogawa River flows into Tokyo Bay. The $^{134+137}$Cs concentrations of the surface 0–5 cm layer and the total radioactively contaminated layers were the highest in the sediment collected in November 2012. As shown in Table 6.7, the daily flux of $^{134+137}$Cs for each sampling period also peaked in November 2012 when the concentration maximum. These values mean that it took approximately one and a half years for the contaminated soil in the Sakagawa catchment area to be transported to Tokyo Bay. The inventory of $^{134+137}$Cs increased from 20 to 104 kBq/m^2 during the 4 years and 11 months from August 2011 to July 2016. Theoretically, by July 2016, the $^{134+137}$Cs activity should have decayed to 53% of the value immediately after the accident. The inventory increased by five times despite the ^{134}Cs decrease due to radioactive decay. This increase suggests that even now, radioactive cesium continues to flow into Tokyo Bay constantly. Table 6.7 shows the average flux of $^{134+137}$Cs at Site D and the flux for each sampling period calculated from the values in Table 6.6. The estimated average flux immediately after the accident was 0.128 kBq/m^2 day, but by July 2016, it had decreased to 0.054 kBq/m^2 day. The flux estimated for each sampling period fluctuates significantly from 0.003 to 0.128 kBq/m^2 year (0.059–0.177 kBq/m^2 day after decay correction).

However, both average fluxes are approximately 0.103–0.104 kBq/m^2 day after 5 years and 4 months had passed since the FDNPP accident, but the material that flows and accumulates in Tokyo Bay has not changed much over time. This flux is approximately 10% of that at Sakagawa Site S1, which is the contamination source. These results are shown in Fig. 6.8. The radioactive decay curve of the $^{134+137}$Cs activity is based on the hypothesis that the ^{134}Cs/^{137}Cs activity ratio emitted during the accident was 1.0 [30–32]. In the post-2013 period, when the dynamics of

Table 6.6 Concentrations and inventories of radioactive cesium in the sediments collected at Site D in the Old Edogawa estuary

Sampling date	Elapsed time[a], day	$^{134+137}$Cs activity, Bq/kg					Inventory[c], kBq/m^2						Ratio ^{134}Cs/^{137}Cs[d]
		Surface layer, 0–5 cm		Contaminated Layer[b]			^{134}Cs		^{137}Cs		$^{134+137}$Cs		
		Detected	Corrected	Depth, cm	Detected	Corrected	Detected	Corrected	Detected	Corrected	Detected	Corrected	
2011/8/20	157	360	388	0–26	314	338	9.3	10.7	10.8	10.9	20.1	21.7	0.991 ± 0.021
2012/4/2	383	427	509	0–26	395	471	12.1	17.2	17.0	17.4	29.4	35.0	0.989 ± 0.009
2012/11/1	596	556	721	0–44	413	538	21.1	36.5	34.7	36.0	55.8	72.6	1.014 ± 0.015
2013/10/4	933	322	471	0–44	344	512	17.7	41.9	39.2	41.6	56.9	83.4	1.007 ± 0.012
2014/3/25	1105	300	465	0–42	334	518	17.2	47.7	44.1	47.3	61.5	95.1	1.008 ± 0.013
2015/11/13	1703	200	366	0–70[e]	260	473	18.0	86.4	76.3	84.9	94.3	171	1.018 ± 0.012
2016/7/12	1945	200	379	0–78[e]	238	456	16.9	101	87.1	98.5	104	200	1.025 ± 0.015
											Weighted average[d]		1.007 ± 0.005

Detected: Value on the sampling date. Corrected: Decay-corrected value to March 16, 2011

[a]Days from March 16, 2011

[b]Contaminated layer is the thickness from the surface to the deepest layer where ^{134}Cs is detected

[c]Cumulative value from March 16, 2011, to the sample collection

[d]Weighted average value in each layer of the core

[e]Sedimentary layer included the layer of 14 cm thickness deposited by the flood from August 9–11, 2015

Table 6.7 Fluxes of radioactive cesium in the sediments collected at Site D in the Old Edogawa estuary

Sampling date	Elapsed time, day	Sampling interval, day	Average flux[a], kBq/m^2 day						Each sampling interval			
									Inventory, kBq/m^2		Flux, kBq/m^2 day	
			^{134}Cs		^{137}Cs		$^{134+137}$Cs		$^{134+137}$Cs		$^{134+137}$Cs	
			Detected	Corrected	Detected	Corrected	Detected	Corrected	Detected	Corrected	Detected	Corrected
2011/8/20	157	157	0.059	0.068	0.069	0.070	0.128	0.138	20.1	21.7	0.128	0.138
2012/4/2	383	226	0.032	0.045	0.044	0.046	0.076	0.091	9.3	13.3	0.041	0.059
2012/11/1	596	213	0.035	0.061	0.058	0.060	0.094	0.122	26.4	37.6	0.124	0.177
2013/10/4	933	337	0.019	0.045	0.042	0.045	0.061	0.089	1.1	10.8	0.003	0.032
2014/3/25	1105	172	0.016	0.043	0.040	0.043	0.056	0.086	4.6	11.7	0.027	0.068
2015/11/13	1703	598	0.011	0.051	0.045	0.050	0.055	0.100	32.8	75.9	0.055	0.127
2016/7/12	1945	242	0.009	0.052	0.045	0.051	0.054	0.103	9.7	29.0	0.040	0.120
						Average		0.104 ± 0.019	Average			0.103 ± 0.051

Detected: Value on the sampling date. Corrected: Decay-corrected value to March 16, 2011

[a]Average value from March 16, 2011, to the date of sample collection

contamination seem to have reached a steady state, the fluctuations in radioactive cesium concentration and inventory in Site D sediments are almost consistent with the decrease due to radioactive decay of $^{134+137}Cs$. This fact means that the constant inflow of ^{137}Cs with a long half-life into Tokyo Bay continues even after ^{134}Cs has disappeared. The annual supply of radioactive cesium in Area X is assumed to be approximately 0.24 TBq (see Sect. 6.9).

6.8.2 *Contamination via Global Fallout of ^{137}Cs in Tokyo Bay Sediment Before the FDNPP Accident*

Information on the radioactive cesium concentration in Tokyo Bay sediment before the FDNPP accident was relatively deficient compared to that for Japanese water samples [38]. However, clarifying the background concentration and flux of radioactive cesium flowing into Tokyo Bay is very important for evaluating radioactive cesium contamination in Tokyo Bay caused by the FDNPP accident. The global fallout from atmospheric nuclear tests started in the 1950s and was an essential source of ^{137}Cs pollution in the environment before the Chernobyl and FDNPP accidents. In Japan, global fallout measurements began in 1954. Figure 6.9 shows the annual precipitation measured by NIRS in Chiba City on the east coast of Tokyo Bay [39]. The global fallout reached its maximum value of 1.585 kBq/m^2 in 1963 but has since decreased sharply and was less than 0.001 kBq/m^2 in 1984. However, it increased to 0.161 kBq/m^2 in 1986 when the Chernobyl nuclear accident occurred and decreased to 0.002 kBq/m^2 in 1987. In other words, the background level of

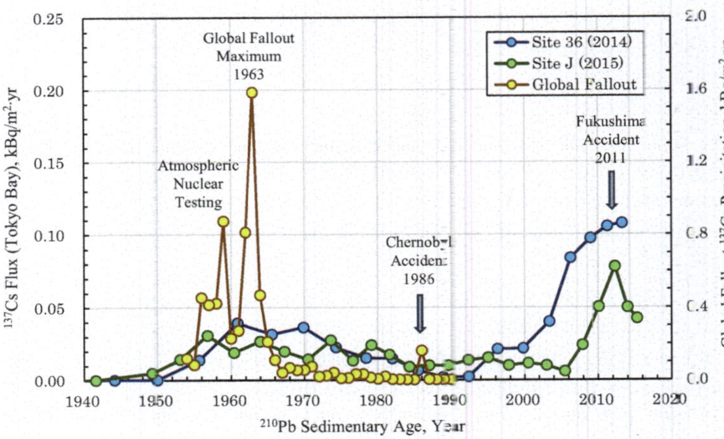

Fig. 6.9 Nuclear events prior to the FDNPP accident recorded in the sediment cores in central Tokyo Bay. Annual precipitations of ^{137}Cs by atmospheric global fallout were measured by NIRS in Chiba City in the east of Tokyo Bay [39]

^{137}Cs in Tokyo Bay is less than 0.001 kBq/m^2 at present, and it is not necessary to consider the effects of the Chernobyl accident.

The inventory of ^{137}Cs in the Tokyo Bay sediment has already been reported as 0.37–0.51 kBq/m^2, but this value might be underestimated due to the short core used in the analysis [38]. The cores from Site 36 (taken in 2014) and Site J (taken in 2015) shown in Fig. 6.9 were dated using the ^{210}Pb method. The deepest layers where ^{137}Cs was detected were 32–34 cm (1956) at Site 36 and 48–50 cm (1948) at Site J. The ^{134}Cs derived from the FDNPP accident was found in upper layers of 8–10 cm (2003) at Site 36 and 12–14 cm (2003) at Site J. This distribution of ^{134}Cs is reasonable, given the effect of physical mixing of the sediments. However, the sedimentation rate increased at Site J, which is far from the mouths of the Arakawa and Old Edogawa rivers. The average concentrations of ^{137}Cs in these cores from 1955 to 1970, when global fallout was remarkable, are 16.8 ± 6.8 Bq/kg for Site 36 and 10.6 ± 3.1 Bq/kg for Site J. These results may indicate that Site J is affected by the dilution effect of suspended particle flow from the Tamagawa and Tsurumigawa rivers.

In addition, in the 1954–1985 sedimentary deposits before the Chernobyl accident, the inventories of ^{137}Cs are 0.796 kBq/m^2 at Site 36 and 0.713 kBq/m^2 at Site J. These inventories are slightly larger than previously reported [38]. Since the global fallout inventory at this time was 6.56 kBq/m^2, the ^{137}Cs inventory in central Tokyo Bay is approximately 11–12% of the global fallout. However, between 1954 and 1985, the ^{137}Cs global fallout must have been precipitated onto the surface of the seawater, so the inventories recorded in the sediment cores are quite small. Presumably, ^{137}Cs, which moved directly from the atmosphere into seawater due to global fallout, was likely to have been dissolved in seawater and not deposited on the seafloor.

On the other hand, the average fluxes of ^{137}Cs in Tokyo Bay sediment resulting from atmospheric nuclear tests were 0.027 kBq/m^2 year at Site 36 and 0.024 kBq/m^2 year at Site J. As shown in Fig. 6.8, the flux of $^{134+137}$Cs at Site D offshore of the Old Edogawa estuary was approximately 19.7 kBq/m^2 year (0.054 kBq/m^2 day) in 2016. In other words, radioactive cesium due to the FDNPP accident continues to flow into Tokyo Bay at a rate that is more than 700 times faster than that in the era of atmospheric nuclear tests.

6.8.3 *Estimation of the Sedimentation Process for Radioactive Cesium in the Edogawa Water System and Tokyo Bay*

A model calculation was conducted, and the movement process of radioactive cesium from the land to the aquatic system was simulated based on values observed in various environments [40–43]. The results in Figs. 6.2 and 6.3 show that radioactive cesium deposited on the ground surface in the northeastern part of the Tokyo metropolitan accumulated in the estuaries in the inner part of Tokyo Bay but did not diffuse to the center of the bay. The fact that radioactive cesium was selectively deposited in the estuaries suggests that radioactive cesium adsorbed onto suspended particles large enough to settle quickly when they enter Tokyo Bay. Therefore, the particle size distribution of sedimentary materials was measured in the Tokyo Bay and Edogawa water systems. Figure 6.10 shows the analytical results for the particle size distribution calculated on a volume basis.

The average particle size of the park soil in Kashiwa City, which is a highly contaminated area, is approximately 80 μm but is divided into three sizes: 5–10, 20–80, and 200–800 μm at Site S5 in the Sakagawa headwaters. The large particles are not washed away by the river water but are likely to remain in place. Site S1 is the point where the Sakagawa merges with the Edogawa. Since the Matsudo Weir blocks the flow, the flow velocity of the water decreases. The average particle size of the sediment is approximately 10 μm, as fine particles less than 0.5 μm in size are also deposited (red circle in Fig. 6.10). The average grain size of sediments decreases from the Old Edogawa estuary to offshore. Site 02 has 30% of 50–200 μm particles and Site D has approximately 10% of 50–100 μm particles. At Site 52, approximately 6 km offshore from Site D, these large particles disappear, and the fraction of particles with sizes of 10 μm or smaller increases. These results mean that the contaminated particles flowing from the Old Edogawa River are spreading in this estuary area, with the large particles settling there first. This size distribution suggests that the coagulated deposition of colloidal particles by salting out did not occur in the estuary. Because the average particle size of the sediment decreases from the estuary toward the offshore area, large particles among sedimentary substances flowing from the river accumulate first as they diffuse through the seawater. If Stokes' law is applied in a simplified manner, the settling velocity of suspended particles would be 0.07 m/day at 1 μm, 1.7 m/day at 5 μm, 7 m/day at 10 μm, and 170 m/day at 50 μm. These results suggest that radioactive cesium deposited in the Old Edogawa estuary had adsorbed onto relatively large particles with particle sizes of approximately 5–50 μm.

Cesium cations are well known to penetrate between layers of 2:1 clay minerals, such as vermiculite and mica, and are strongly immobilized by ionic adsorption [44–49]. Thus, radioactive cesium precipitated on the surface from the atmosphere is retained by adsorbing onto clay minerals in the soil. Recently, it was directly confirmed that radioactive cesium released from the FDNPP was adsorbed onto clay minerals using new technologies such as EXAFS and NMR [50–52]. In addition, fine particles of radioactive cesium with high specific activity, so-called cesium balls or hot particles, have been found in eastern Japan [53–57]. However, when radioactive cesium is present in the air in gaseous form or as ultrafine dust, the effect of a washout due to rainfall is thought to play an essential role in the deposition of radioactive cesium on the ground. Although the physicochemical species of mobile radioactive cesium observed in the early days of the FDNPP accident are not precise, the presence of fine particles from radioactive cesium solid solutions such as cesium balls can also be noted. However, in Tokyo Bay, radioactive cesium deposited on the ground surface can be considered to have been adsorbed onto fine clay minerals in the soil and transported to the estuaries around Tokyo Bay. In this case, it is assumed

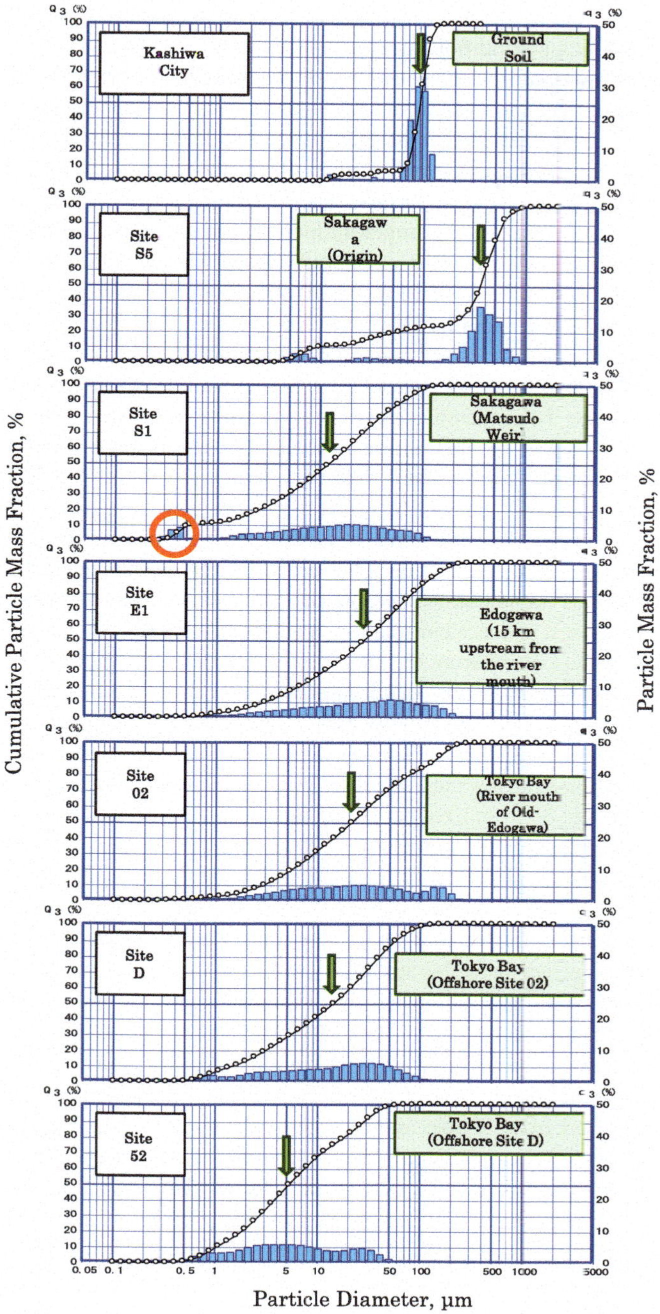

Fig. 6.10 Geographic changes in the particle size distributions of surface sediments in the Tokyo Bay water system. The vertical arrow indicates the average particle diameter

that when river water mixes with seawater, the contaminated fine particles settle to the seabed due to the salting-out effect associated with seawater [44, 58, 59]. Moreover, the coagulum accumulates in estuary sediments. However, such a depositional mechanism cannot explain the distribution of radioactive cesium in Tokyo Bay. Anthropogenic heavy metal elements also flow into Tokyo Bay through a process similar to that for radioactive cesium. Therefore, the distributions of radioactive cesium and heavy metals in Tokyo Bay were compared.

Anthropogenic heavy metal concentrations in the surface sediments of Tokyo Bay were estimated, assuming that concentrations in the age before the Meiji Restoration (1868), when industrial modernization began in Japan, were the natural background values. The background concentrations were 100 mg/kg for zinc, 40 μg/kg for mercury, and 10 mg/kg for lead. As shown in Fig. 6.11, an apparent particle size effect between the concentration of heavy metals and the average sediment particle size in the Tokyo Bay water system was confirmed. If the sedimentary particles are assumed to be spherical, the concentrations of radioactive cesium and heavy metals in the sediment must follow the inverse square law for the size of the sedimentary particles [58–60]. This figure illustrates that heavy metals selectively accumulate in the sediments in central Tokyo Bay, where the average particle size of the heavy metals is relatively small. This phenomenon occurs when heavy metals artificially flow into the aquatic environment and are adsorbed onto the surfaces of suspended particles in water. Nevertheless, the slope is less than 2. Since many rivers flow into Tokyo Bay, there may be a dilution effect due to suspended fine particles with low anthropogenic heavy metal contamination. Additionally, larger particle sizes deviate slightly from the inverse square law. This bias is probably due to the relatively high background concentration of sediment components, as the effect of heavy metal leaching during the advection process is smaller for larger particles.

As shown in Fig. 6.12, unlike the case of heavy metals, no negative correlation is found between radioactive cesium concentration and particle size. The average particle size of the soil in the highly contaminated zone is generally 50–100 μm, and the radioactive cesium activity ranges from 1000 to 35,000 Bq/kg. The source of radioactive cesium contamination in Tokyo Bay is believed to be flows from the highly contaminated areas of the Sakagawa catchment area. When flowing into the Sakagawa, sediments are classified into average particle sizes of 8–20 μm (4000–35,000 Bq/kg) and 250–350 μm (1500–5000 Bq/kg). Large particles settle in the Sakagawa River, and small particles move to the junction with the Edogawa River. From the Sakagawa River to Tokyo Bay, there appears to be a positive correlation between particle size and radioactive cesium concentration, in contrast to the particle size effects observed for heavy metals. This distribution suggests that radioactive cesium is not adsorbed onto suspended particles in the water during transport from the Sakagawa River to Tokyo Bay. In other words, the polluted particles deposited in Area X of Tokyo Bay are soil particles with a particle size of several micrometer to several tens of micrometer that have adsorbed large amounts of radioactive cesium in the Sakagawa catchment area. The low radioactivity of radioactive cesium in Area Z in central Tokyo Bay, where the sediment particles are small, is thought to be due to the dilution effect of particles with low contamination

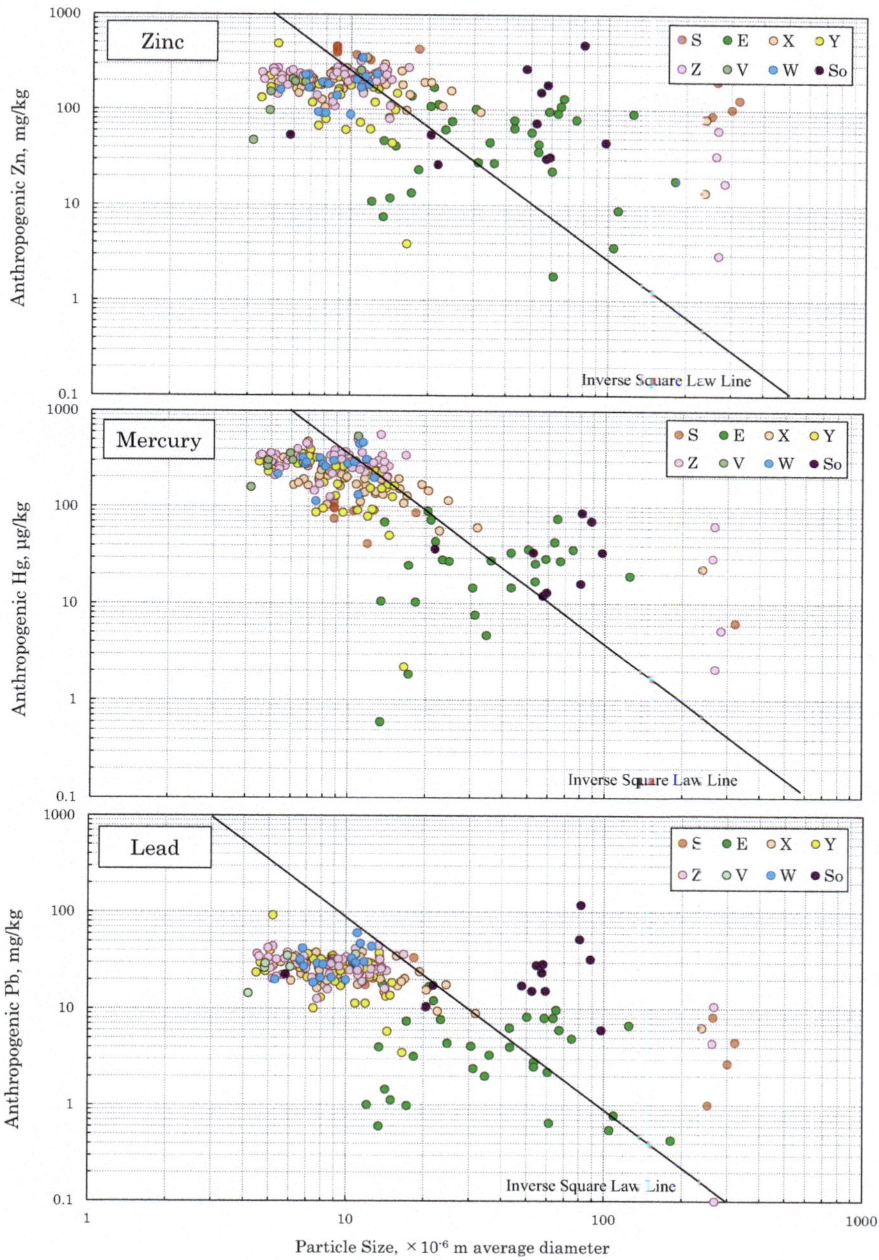

Fig. 6.11 Relationship between the anthropogenic heavy metal concentrations and the particle size of surface sediments in the Tokyo Bay water system. The anthropogenic concentrations of heavy metals were estimated assuming that the concentrations of sediments collected in Tokyo Bay deposited older than 1900 are the background concentration [2]

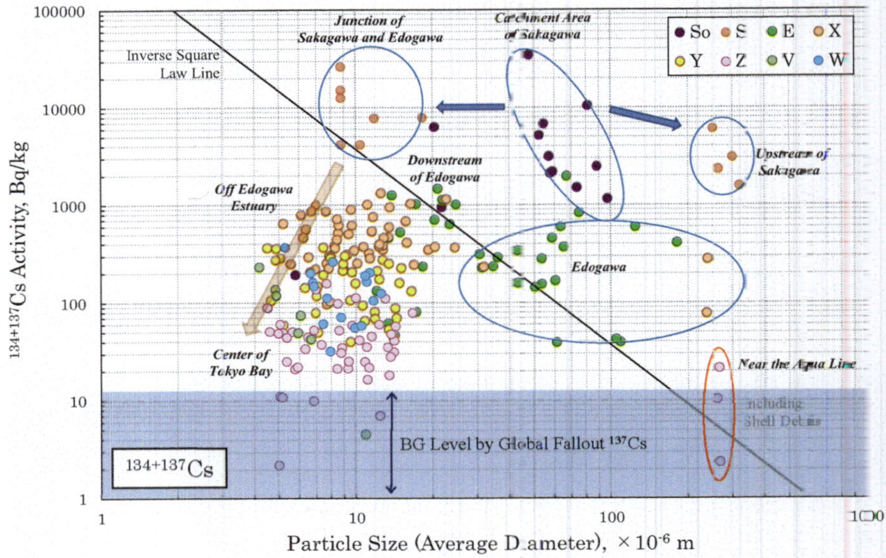

Fig. 6.12 Relationship between the $^{134+137}$Cs activities and the grain size of surface sediments in the Tokyo Bay water system. S_o: Soil in the Sakagawa catchment area, S: Sakagawa River, E: Edogawa and Old Edogawa Rivers, X: Old Edogawa estuary, Y: Offshore the Old Edogawa estuary, Z: Center of Tokyo Bay, V: Tamagawa estuary, W: Sumidagawa estuary. The vertical arrow indicates the range of background ^{137}Cs referred from the JCG report [24]

levels that flowed in from rivers other than the Edogawa water system. The distributions in Area W (estuary of the Sumidagawa River) and Area V (estuary of the Tamagawa River) also support the possibility of dilution effects. This dilution effect is also observed with heavy metals.

Nevertheless, the relationship between cesium radioactivity and particle size in Tokyo Bay is very different from that observed in the estuary of the Abukuma River, which flows through areas most polluted by the FDNPP accident [61–63]. This difference is probably because the Abukuma River estuary faces the Pacific Ocean, where the particle sizes of polluted suspended particles are mixed and separated due to tides and waves. On the other hand, Tokyo Bay is closed, so it is not affected by ocean waves. As a result, large particles contaminated with radioactive cesium preferentially accumulate in the river estuaries as the flow velocity decreases, and small particles are carried to the center of the bay where they settle. Such a mechanism can explain the distribution in Fig. 6.12. The size range of sediments in Area X is 5–30 μm, where the highest concentrations and large amounts of radioactive cesium were deposited in the Old Edogawa estuary. The settling velocity of particles of these sizes is approximately 2–60 m/day, and in a water depth of 5–10 m, contaminated particles can quickly settle to the bottom of the sea well within their residence time [64]. The behavior of radioactive cesium in the Tokyo Bay water system means that the fate of radioactive cesium was determined when it precipitated on the ground surface.

6.9 Balance of Radioactive Cesium Flowing into Tokyo Bay from the Edogawa Watershed

The inventory of $^{134+137}$Cs standardized to March 16, 2011, was estimated for the source and the sink of each area in Fig. 6.13. Table 6.8 shows the balance of radioactive cesium between the Edogawa water system and the Tokyo Bay sediment based on the analytical results. The amount of $^{134+137}$Cs deposited in the Edogawa River basin is estimated to be 8.33 TBq from the results of the MEXT airborne monitoring [1]. Judging from the analytical results of the core samples, the average inventory of $^{134+137}$Cs in Area X, which is 10 km^2 and located approximately 8 km southeast of Chiyoda Ward in central Tokyo, is 131 kBq/m^2 ($n = 10$). Therefore, the total inventory of $^{134+137}$Cs in this area is 1.31 TBq. Similarly, the average inventory in Area Y is 5.52 kBq/m^2 ($n = 11$), and the total inventory is 0.22 TBq.

By July 2016, approximately 70% (1.31/1.88 TBq) of radioactive cesium deposited in the study areas has accumulated in Area X. If the amount of deposition in Area Y is included, the total would account for more than 80%. Assuming that the deposited radioactive cesium was all supplied to Area X from the Edogawa River

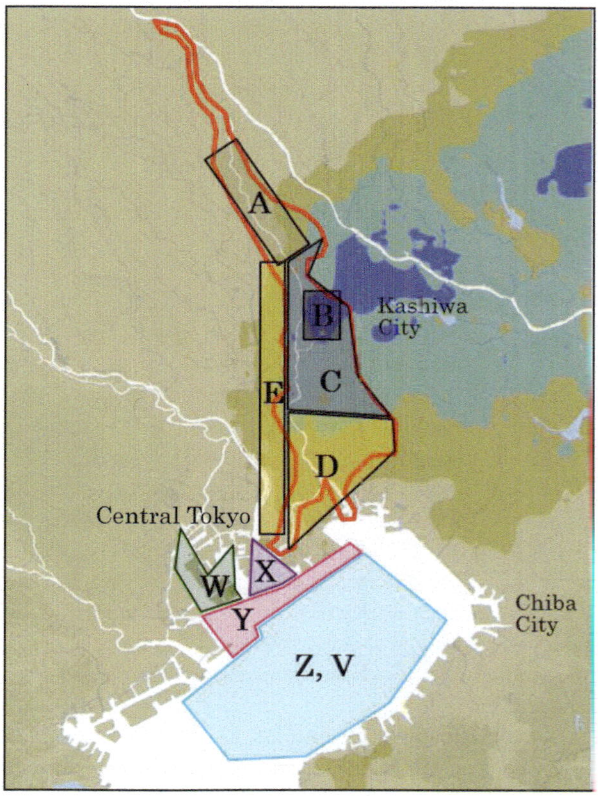

Fig. 6.13 Classification region map for the estimation of radioactive cesium balance in Tokyo Bay and the Edogawa river watershed. The area surrounded by the red line shows the Edogawa river catchment. Area E indicates the Nakagawa river catchment of a branch of the Edogawa river

Table 6.8 Balance of the radioactive cesium precipitated in Tokyo Bay and the Edogawa watershed

	Region[a]		Area, km^2	$^{134+137}Cs$ activity[b] Inventory, kBq/m^2	Each amount, TBq	Total amount, TBq	Ratio to total amount, %
Source[c]	Edogawa Upstream	A	50	10	0.50	**8.33**	
	Kashiwa and Matsudo Cities	B	20	80	1.60		
	Edogawa Midstream	C	65	45	2.93		
	Edogawa Downstream	D	80	20	1.60		
	Nakagawa Watershed	E	85	20	1.70		
Sink[d]	Old-Edogawa Estuary	X	10	131	1.31	**1.53**	**15.7**
	Offshore Area X	Y[e]	40	5.52	0.22		**2.6**
	Central Bay	Z[e], V[e]	330	0.73	0.24	0.24	2.9
	Sumidagawa Estuary	W[e]	20	5.55	0.11	0.11	1.3

[a]Region shows in Fig. 6.13
[b]Decay-corrected value to March 16, 2011
[c]Estimated by the MEXT airborne monitoring [1]
[d]Calculated by the analytical values of the core sediments
[e]Radioactive cesium flowing from rivers other than the Edogawa river system also deposited in these areas

system, approximately 16% (1.31/8.33 TBq) of radioactive cesium precipitated in the catchment area during the 5 years was deposited in Area X after the accident.

On the other hand, it is estimated that critical sources of radioactive cesium for Areas V, W, and Z are other than the Edogawa River system. In particular, in addition to the inflow from the watershed, radioactive plumes from the FDNPP must have precipitated directly onto the sea surface. The inventory for Areas Z and Y in central Tokyo Bay is 0.71–0.80 kBq/m^2, and the cumulative global fallout measured from 1954 to 1995 in Chiba City on the east coast of Tokyo Bay was 6.56 kBq/m^2 [39]. These numbers yield a ratio is equivalent to 11–12%. As shown in Table 5.2, it is estimated that the wastewater treatment system in Tokyo discharged 4.5–12% of the radioactive cesium deposited due to the FDNPP accident. The agreement between these two values for the discharge rate of radioactive cesium from the land is reasonable considering the behavior of the precipitated radioactive cesium.

Since radioactive cesium of 131 kBq/m^2 accumulated in Area X during the 5.4 years after the FDNPP accident, this value represents 20 times the global fallout of ^{137}Cs. Most of the radioactive cesium flowing into Tokyo Bay was not only contaminated soil from the catchment areas but also contaminated particles that were discharged by decontamination work in highly contaminated zones. As shown in Table 6.7, 0.054 kBq/m^2 day (19.7 kBq/m^2 year) of radioactive cesium is currently flowing into Tokyo Bay via the Old Edogawa River (2016). In other words, most of this radioactive cesium is ^{137}Cs, which has a long half-life of 30.1 years, so for now, the inflow of radioactive cesium of approximately 0.2 TBq/year will continue.

At the southern end of Area W is the unfinished Central Breakwater landfill that is still being reclaimed. Here, sludge incineration ash discharged from the wastewater treatment systems in Tokyo is buried. As described in Table 5.1, the cumulative amount of $^{134+137}Cs$ contained in this incinerated ash by the end of 2015 was 0.6 TBq. It must be remembered that radioactive cesium precipitated in urban areas of Tokyo is also transported to Tokyo Bay via such routes and is accumulating.

6.10 Conclusions

Changes in radioactive cesium contamination of Tokyo Bay sediments surrounded by the Tokyo metropolitan area during 5 years after the FDNPP accident are discussed. From August 2011 to July 2016, monitoring surveys were conducted on the spatiotemporal distribution of radioactive cesium released by the FDNPP accident across a 400 km^2 area in the northern part of Tokyo Bay and its incoming rivers. Most of the radioactive cesium flowed from highly contaminated areas in the northeastern Tokyo metropolitan area through the Old Edogawa River into Tokyo Bay and accumulated off the estuary without spreading to the center of the bay. The $^{134+137}Cs$ concentration increased immediately after the accident, peaked at the end of 2012 at 556 Bq/kg, and then decreased according to radioactive decay. However, the inventory of $^{134+137}Cs$ offshore from the Old Edogawa estuary increased to 104 kBq/m^2 in July 2016. At this time, approximately 70% of the radioactive cesium accumulated in the Tokyo Bay sediment was deposited in sediments of Area X in the Old Edogawa estuary.

On the other hand, the concentration of $^{134+137}Cs$ in sediments of the central part (Area Z) of Tokyo Bay was 17 $\pm$ 0.3 kBq/kg (weighted average, $n = 50$). Most likely, radioactive cesium was transported to Tokyo Bay as contaminated soil particles from highly polluted areas in the northeastern Tokyo metropolitan area through rivers such as the Sakagawa, Edogawa, and Old Edogawa rivers. Observation results suggest that rivers play an essential role in transporting radioactive cesium from land sources. Additionally, radioactive cesium precipitated from the atmosphere and adsorbed on soil particles enters the rivers as polluted particles but selectively settles and accumulates at the estuaries around Tokyo Bay, where the flow velocity of the river water decreases. Therefore, it is estimated that radioactive cesium has not spread to the center of the bay. The flux of $^{134+137}Cs$ offshore of the Old Edogawa estuary decreased sharply immediately after the accident, but afterward, it mostly agreed with the value predicted based on radioactive decay.

Conversely, the inventory in sediments is increasing. The average flux from 2014 to 2016 was approximately 0.055 kBq/m^2. It is calculated that 8.33 TBq of $^{134-137}$Cs was deposited in the basin of the Edogawa water system and that 1.53 TBq of these radionuclides migrated through the river and was deposited in the estuary of the Old Edogawa river by July 2016. Currently, approximately 0.2 TBq of radioactive cesium per year continues to flow into the inner part of Tokyo Bay. The radioactive cesium that has flowed into Tokyo Bay is retained firmly in its sediments, so Tokyo Bay plays an essential role as a reservoir of radioactive cesium discharged from the Tokyo metropolitan area.

References

1. MEXT (Ministry of Education, Culture, Sports, Science and Technology of Japan) (2011) MEXT's airborne monitoring results in Iwate, Shizuoka, Nagano, Yamanashi, Gifu, and Toyama prefectures and revisions to airborne monitoring that take into account the effects of natural nuclides [in Japanese]. https://radioactivity.nsr.go.jp/ja/contents/5000/4899/24/1910_111112nd.pdf
2. Yamazaki H, Ishida M, Hinokio R, Yamashiki YA, Azuma R (2018) Spatiotemporal distribution and fluctuation of radiocesium in Tokyo Bay in the five years following the Fukushima Daiichi Nuclear Power Plant (FDNPP) accident. PLoS One 13:e0193414. https://doi.org/10.1371/journal.pone.0193414
3. MOE (Ministry of Environment, Japan) (2009) Enclosed coastal seas in Japan (88 areas). Environmental Guidebook [in Japanese]. https://www.env.go.jp/water/heisa/heisa_net/downloadData/Guidebook/guidebook00.pdf. International EMECS Center [in English]. http://www.emecs.or.jp/en/encsea
4. MLIT (Ministry of Land, Infrastructure, Transport and Tourism of Japan) (2014) Water environment in Tokyo Bay [in Japanese]. https://www1.kaiho.milt.go.jp/KANKYO/TB_Renaissane/RenaissanceProject/Handouts/7th/H03.pdf
5. MLIT (2014) Overview of Tokyo Bay and the catchment area [in Japanese]. https://www.ktr.mlit.go.jp/ktr_content/000621812.pdf
6. Guo X, Yanagi T (1994) Three dimensional structure of tidal currents in Tokyo Bay, Japan. La mer 32:173–185
7. Kakehi S, Kaeriyama H, Ambe D, Ono T, Ito S et al (2016) Radioactive cesium dynamics derived from hydrographic observations in the Abukuma River Estuary, Japan. Environ Radioact 153:1–9. https://doi.org/10.1016/j.jenvrad.2015.11.015
8. Yamasaki S, Imoto J, Furuki G, Ochiai A, Ohnuki T et al (2016) Radioactive Cs in the estuary sediments near Fukushima Daiichi Nuclear Power Plant. Sci Total Environ 551-552:155–162. https://doi.org/10.1016/j.scitotenv.2016.01.155
9. Saegusa H, Ohyama T, Iijima K, Onoe H, Takeuchi R et al (2016) Deposition of radiocesium on the river flood plains around Fukushima. J Environ Radioact 164:36–46. https://doi.org/10.1016/j.jenvrad.2016.04.020
10. Tsuruta H, Nakajima T (2012) Radioactive materials in the atmosphere released by the accident of the Fukushima Daiichi Nuclear Power Plant. Chikyukagaku (Geochemistry) 46:99–111. [in Japanese with English abstract]. https://doi.org/10.14934/chikyukagaku.46.99
11. Ishida M, Yamazaki H (2017) Radioactive contamination in the Tokyo metropolitan area in the early stage of the Fukushima Daiichi Nuclear Power Plant (FDNPP) accident and its fluctuation over five years. PLoS One 12:e0187687. https://doi.org/10.1371/journal.pone.0187687
12. USNRC (US Nuclear Regulatory Commission) (1989) Programmatic Environmental Impact Statement related to decontamination and disposal of radioactive wastes resulting from March

28, 1979 accident Three Mile Island Nuclear Station, Unit 2. https://ntrl.ntis.gov/NTRL/dashboard/searchResults/titleDetail/NUREG0683SUPN3.xhtml

13. NSAC (Nuclear Science Advisory Committee) (1980) Analysis of Three Mile Island-Unit 2 accident. NSAC-80-1. http://www.iaea.org/inis/collection/NCLCollectionStore/_Public/13/677/13677904.pdf
14. IAEA (International Atomic Energy Agency) (2006) Environmental consequences of the Chernobyl accident and their remediation: twenty years of experience. Report of Chernobyl Forum Expert Group 'Environment'. IAEA Radiological Assessment Reports Series. https://www-pub.iaea.org/MTCD/publications/PDF/Pub1239_web.pdf
15. Saito-Kokubu Y, Yasuda K, Magara M, Miyamoto Y, Sakurai S et al (2007) Geographical distribution of plutonium derived from the atomic bomb in the eastern area of Nagasaki. J Radioanal Nucl Chem 273:183–186. https://doi.org/10.1007/s10967-007-0733-9
16. Saito-Kokubu Y, Esaka F, Yasuda K, Magara M, Miyamoto Y et al (2007) Plutonium isotopes derived from Nagasaki atomic bomb in the sediment of Nishiyama reservoir at Nagasaki, Japan. Int J Appl Radiat Isotopes 65:465–468. https://doi.org/10.1016/j.apradiso.2006.10.010
17. Saito-Kokubu Y, Yasuda K, Magara M, Miyamoto Y, Sakurai S et al (2008) Depositional records of plutonium and ^{137}Cs released from Nagasaki atomic bomb in sediment of Nishiyama reservoir at Nagasaki. J Environ Radioact 99:211–217. https://doi.org/10.1016/j.jenvrad.2007.11.010
18. Gallego E (2006) MUD; a model to investigate the migration of ^{137}Cs in the urban environment and drainage and sewage treatment systems. J Environ Radioact 85:247–264. https://doi.org/10.1016/j.jenvrad.2004.10.017
19. Pratama MA, Yoneda M, Yamashiki Y, Shimada Y, Matsui Y (2014) Modeling migration of Cs-137 in sewer system of Fukushima City using urban environment and drainage system (MUD). Int J Eng Tech 6:402–408. https://doi.org/10.7763/IJET.2014V6.732
20. USGS (US Geological Survey) (1973) Distribution of radionuclides in bottom sediments of the Columbia River estuary. Geological Survey Professional Paper 433-L. https://pubs.usgs.gov/pp/0433l/report.pdf
21. USCS (1975) Distribution of radionuclides in the Columbia River streambed, Hanford reservation to Longview, Washington. Geological Survey Professional Paper 433-O. https://pubs.usgs.gov/pp/0433l/report.pdf
22. Bera G, Yeager KM, Shim M, Shiller AM (2015) Anthropogenic stable cesium in water and sediment of a shallow estuary, St. Louis Bay, Mississippi. Estuar Coast Shelf Sci 157:32–41. https://doi.org/10.1016/j.ecss.2015.02.004
23. NRA (Nuclear Regulation Authority, Japan) (2013) Readings of Sea Water at Tokyo Bay. http://radioactivity.nsr.go.jp/en/list/290/list-1.html
24. JCG (Japan Coast Guard) (2016) Report of Radioactivity Surveys. Results of Surveys in 2016 [in Japanese]. http://www1.kaiho.mlit.go.jp/KANKYO/OSEN/housha.html
25. Koibuchi Y (2013) Tracing fine-grained sediment transport around Tokyo Bay using cesium-134 and cesium-137 originating from Fukushima Daiichi Power Plant. In: Rodriguez GR (ed) Coastal processes III. WIT Press, Southampton, pp 191–201. https://doi.org/10.2495/CP130171
26. Aoyama M, Hirose K (2008) Analysis of environmental radionuclides. In: Povinec PP (ed) Radiometric determination of anthropogenic radionuclides in seawater. Elsevier, London. pp 137–162. https://doi.org/10.1016/S1569-4860(07)11004-4
27. Yasuhara M, Yamazaki H (2005) The impact of 150 years of anthropogenic pollution on the shallow marine ostracode fauna, Osaka Bay, Japan. Mar Micropaleontol 55:63–74. https://doi.org/10.1016/j.marmicro.2005.02.005
28. Gohda S, Yamazaki H, Hirata M, Nagasawa T (1987) XRF analysis of rock and sediment using standard rock samples. BUNSEKI KAGAKU (Analytical Chemistry) 36:199–203. [in Japanese with English abstract]
29. Oikawa S, Watabe T, Takata H, Suzuki C, Nakahara M et al (2013) Long term temporal changes of ^{90}Sr and ^{137}Cs in sea water, bottom sediment and marine organism samples – from

the Chernobyl accident to immediately after the Fukushima accident. BUNSEKI KAGAKU (Analytical Chemistry) 62:455–474. [in Japanese with English abstract]. https://doi.org/10.2116/bunsekikagaku.62.455

30. Chaisan K, Smith JT, Bossew P, Kirchner G, Loptev GV (2013) Worldwide isotope ratios of the Fukushima release and early-phase external dose reconstruction. Sci Rep 3:2520. https://doi.org/10.1038/srep02520
31. Komori M, Shozugawa K, Nogawa N, Matsuo M (2013) Evaluation of radioactive contamination caused by each plant of Fukushima Daiichi Nuclear Power Station using $^{134}Cs/^{137}Cs$ activity ratio as an index. BUNSEKI KAGAKU (Analytical Chemistry) 62:475–483. [in Japanese with English abstract]. https://doi.org/10.2116/bunsekikagaku.62.475
32. Nishizawa Y, Yoshida M, Sanada Y, Torii T (2016) Distribution of the $^{134}Cs/^{137}Cs$ ratio around the Fukushima Daiichi nuclear power plant using an unmanned helicopter radiation monitoring system. J Nucl Sci Technol 53:468–474. https://doi.org/10.1080/00223131.2015.1071721
33. Sekiguchi H, Yamazaki H, Nakagawa R, Ishida M, Azuma R et al (2013) Implications of flood event layers in littoral sedimentary environments. J JSCE, Ser B2. Coast Eng 69:591–595. [in Japanese with English abstract]
34. Kashiwa City Government (2013) Decontamination project in Kashiwa City, 2nd edn. [in Japanese]. http://www.city.kashiwa.lg.jp/houshasenkanren/3331/3332/p032036_d/fil/jissikeikaku250613.pdf
35. Japan Meteorological Agency (2020) Search past weather data [in Japanese]. https://www.data.jma.go.jp/obd/stats/etrn/select/prefecture00
36. MLIT. Edogawa River Office (2015) River Management Report 2015 [in Japanese]. http://www.ktr.mlit.go.jp/ktr_content/content/000642546.pdf
37. MLIT (2015) Overview of the Kanto-Tohoku heavy rainfall disaster [in Japanese]. http://www.bousai.go.jp/fusuigai/suigaiworking/pdf/dai1kai/siryo1.pdf
38. Yamada M, Nagaya Y (2000) $^{239+240}Pu$ and ^{137}Cs in sediments from Tokyo Bay: distribution and inventory. J Radioanal Nucl Chem 245:273–279. https://doi.org/10.1023/A:1006741917854
39. JCAC (Japan Chemical Analysis Center). Radioactivity survey data in Japan. Environmental and dietary materials. http://www.kankyo-hoshano.go.jp/en/07/07.html
40. Håkanson L (2004) Modelling the transport of radionuclides from land to water. J Environ Radioact 73:267–287. https://doi.org/10.1016/j.jenvrad.2003.10.003
41. Håkanson L (2004) A new generic sub-model for radionuclide fixation in large catchments from continuous and single-pulse fallouts, as used in a river model. J Environ Radioact 77:247–273. https://doi.org/10.1016/j.jenvrad.2004.03.010
42. Smith JT, Wright SM, Cross MA, Monte L, Kudelsky AV et al (2004) Global analysis of the riverine transport of ^{90}Sr and ^{137}Cs. Environ Sci Technol 38:850–857. https://doi.org/10.1021/es0300463
43. Monte L, Boyer P, Brittain JE, Håkanson L, Lepicard S et al (2005) Review and assessment of models for predicting the migration of radionuclides through rivers. J Environ Radioact 79:273–296. https://doi.org/10.1016/j.jenvrad.2004.08.002
44. Förstner U (1983) Metal transfer between solid and aqueous phases. In: Förstner U, Wittmann GTW (eds) Metal pollution in the aquatic environment. Springer, New York, NY, pp 247–270
45. Stumm W, Morgan JJ (1996) The solid-solution interface. In: Stumm W, Morgan JJ (eds) Aquatic chemistry, 3rd edn. John Wiley & Sons, New York, NY, pp 516–608
46. Sawhney BL (1972) Selective sorption and fixation of cations by clay minerals: a review. Clay Clay Miner 20:93–100. https://doi.org/10.1346/CCMN.19720200208
47. Evans DW, Alberts JJ, Clark RA III (1983) Reversible ion-exchange fixation of cesium-137 leading to mobilization from reservoir sediments. Geochim Cosmochim Acta 47:1041–1049. https://doi.org/10.1016/0016-7037(83)90234-X
48. Bostick BC, Vairavamurthy MA, Karthikeyan KG, Chorover J (2002) Cesium adsorption on clay minerals: an EXAFS spectroscopic investigation. Environ Sci Technol 36:2670–2676. https://doi.org/10.1021/es0156892

49. Zachara JM, Smith SC, Liu C, McKinley JP, Serne RJ et al (2002) Sorption of Cs^+ to micaceous subsurface sediments from the Hanford site, USA. Geochim Cosmochim Acta 66:193–211. https://doi.org/10.1016/S0016-7037(01)00759-1
50. Qin H, Yokoyama Y, Fan Q, Iwatani H, Tanaka K, Sakaguchi A et al (2012) Investigation of cesium adsorption on soil and sediment samples from Fukushima Prefecture by sequential extraction and EXAFS technique. Geochem J 46:297–302. https://doi.org/10.2343/geochemj.2.0214
51. Motokawa R, Endo H, Yokoyama S, Nishitsuji S, Kodayashi T et al (2014) Collective structural changes in vermiculite clay suspensions induced by cesium ions. Sci Rep 4:6585. https://doi.org/10.1038/srep06585
52. Tokuda Y, Norikawa Y, Masai H, Ueda Y, Nihei N et al (2016) Nuclear magnetic resonance study of Cs adsorption onto clay minerals. Radio Issu Fukushima's Revital Fut 1:3–11. https://doi.org/10.1007/978-4-431-55848-4_1
53. Adachi K, Kajino M, Zaizen Y, Igarashi Y (2013) Emission of spherical cesium-bearing particles from an early stage of the Fukushima nuclear accident. Sci Rep 3:2554. https://doi.org/10.1038/srep02554
54. Itoh S, Eguchi T, Kato N, Takahashi S (2014) Radioactive particles in soil, plant, and dust samples after the Fukushima nuclear accident. Soil Sci Plant Nutr 60:440–550. https://doi.org/10.1080/00380768.2014.907735
55. Abe Y, Iizawa Y, Terada Y, Adachi K, Igarashi Y, el al. (2014) Detection of uranium and chemical state analysis of individual radioactive microparticles emitted from the Fukushima nuclear accident using multiple synchrotron radiation X-ray analysis. Anal Chem 86:8521–8525. https://doi.org/10.1021/ac501998d
56. Yamaguchi N, Mitome M, Kotone H, Asano M, Adachi K et al (2016) Internal structure of cesium-bearing radioactive microparticles released from Fukushima nuclear power plant. Sci Rep 6:20548. https://doi.org/10.1038/srep20548
57. Furuki G, Imoto J, Ochiai A, Yamasaki S, Nanba K et al (2017) Caesium-rich micro-particles: a window into the meltdown events at the Fukushima Daiichi Nuclear Power Plant. Sci Rep 7:42731. https://doi.org/10.1038/srep42731
58. Stumm W, Morgan JJ (1996) Particle-particle interaction: colloids, coagulation, and filtration. In: Stumm W, Morgan JJ (eds) Aquatic chemistry, 3rd edn. John Wiley & Sons, New York, NY, pp 818–871
59. Thill A, Mousteir S, Garnier JM, Estournel C, Naudin JJ et al (2001) Evolution of particle size and concentration in Rhône river mixing zone: influence of salt flocculation. Cont Shelf Res 21:2127–2140. https://doi.org/10.1016/S0278-4343(01)00047-4
60. Förstner U (1983) Grain-size effects. In: Förstner U, Wittmann GTW (eds) Metal pollution in the aquatic environment. Springer, New York, NY, pp 121–132
61. Yamashiki Y, Onda Y, Smith HG, Blake WH, Wakahara T et al (2013) Initial flux of sediment-associated radiocesium to the ocean from the largest river impacted by Fukushima Daiichi Nuclear Power Plant. Sci Rep 4:3714. https://doi.org/10.1038/srep03714
62. Tanaka K, Iwatani H, Sakaguchi A, Fan Q, Takahashi Y (2015) Size-dependent distribution of radiocesium in riverbed sediments and its relevance to the migration of radiocesium in river systems after the Fukushima Daiichi Nuclear Power Plant accident. J Environ Radioact 139:390–397. https://doi.org/10.1016/j.jenvrad.2014.05.002
63. Pratama M, Yoneda M, Shimada Y, Matsui Y, Yamashiki Y (2015) Future projection of radiocesium flux to the ocean from the largest river impacted by Fukushima Daiichi Nuclear Power Plant. Sci Rep 5:8408. https://doi.org/10.1038/srep08408
64. Wada A, Takano T, Hoizumi T (1994) Estimation of bay-water retention time. Ann J Hydr Eng, B1, JSCE 34:331–336. [in Japanese with English Abstract]

Chapter 7
The Behavior of Radioactive Cesium Precipitated in Forests

Abstract A characteristic of the natural environment in the Tokyo metropolitan area is that forests and mountainous areas are close to urban areas. Such a geographic environment is an essential factor in analyzing the dynamics of radioactive contamination in the capital Tokyo. After the Fukushima disaster, the hinterland was exposed to high levels of radioactive contamination. Radioactive nuclides released from the FDNPP were transported to the forested areas through the air as radioactive plumes and deposited there. In such forest areas, the radionuclides deposited from radioactive plumes are strongly adsorbed not only onto the soil but also onto trees, deciduous leaves, humus, etc., causing environmental radioactive contamination. According to the MEXT airborne monitoring, radioactive cesium deposition may have reached 600 kBq/m^2 or more in some places. These radioactivity levels are approximately the same as those in the contaminated area of Fukushima Prefecture, where the FDNPP is located. Since the decontamination of radioactively contaminated areas destroys the forest ecosystem, the SCJ (Science Council of Japan) has noted problems regarding the effects of radioactive contamination in forest areas on forestry and agriculture. Radioactive nuclides that are strongly adsorbed onto trees and litter are very likely to migrate from forest to downstream city as water-soluble humic acid and fulvic acid complexes along with the decomposition of litter. The decomposition and solubilization of these substances have progressed for more than 9 years after the accident, so it is necessary to discuss the redistribution of radionuclides in the environment. The analysis of radioactive contamination in forests cannot be avoided in resolving the radioactive contamination of the Tokyo metropolitan area.

Keywords Forest · Radioactive cesium · Soil · Litter · Bark · Spring water · Spatiotemporal distribution

---

Electronic Supplementary Material The online version of this chapter (https //doi.org/10.1007/978-981-15-7368-2_7) contains supplementary material, which is available to authorized users.

H. Yamazaki, *Radioactive Contamination of the Tokyo Metropolitan Area*,
https://doi.org/10.1007/978-981-15-7368-2_7

7.1 Introduction

The Metropolitan Area Improvement Law legally defines the Tokyo metropolitan area. As shown in supplementary figures and tables in Chap. 2, it comprises the administrative areas of Tokyo, Kanagawa, Chiba, Saitama, Ibaraki, Tochigi, Gunma, and Yamanashi. The territory has a total area of 36,900 km^2, of which the flat land is 14,030 km^2, and the urban area is 3930 km^2, according to the statistics of the MLIT (Ministry of Land, Infrastructure, Transport, and Tourism, Japan). The area of greenery and forests is approximately 22,000 km^2, which corresponds to approximately 59% of the Tokyo metropolitan area. Most of the Tokyo metropolitan area consists of the Kanto Plain with an area of 17,000 km^2, and the capital, Tokyo, is located on the southeastern side facing Tokyo Bay. Most of the Kanto Plain is less than 100 m above sea level, but forests and mountainous areas with elevations more than 1000 m spread to the west, northwest, and north. The main areas that have been radioactively contaminated by the FDNPP accident are Fukushima Prefecture and the Tokyo metropolitan area. According to statistics from the MAFF (Ministry of Agriculture, Forestry, and Fisheries, Japan) [1], Fukushima Prefecture has approximately 9740 km^2 of forest, which is 71% of the area. The forest area of the Tokyo metropolitan area, including Tokyo, is approximately 17,600 km^2, accounting for 48% of the total area. The total forest area in Ibaraki, Tochigi, and Gunma prefectures where radioactive contamination was severe is 9590 km^2, which is almost equal to the forest area in Fukushima Prefecture. Incidentally, the forest area in Tokyo is approximately 790 km^2, which is 36% of the total area. Most of the forests are located in the western part of Tokyo. Although few people live in the forested areas of the Tokyo metropolitan area, most of the densely populated major cities are located on plains bordering forest areas. The total population is 44.0 million, of which 38.1 million live in the city area.

A characteristic of the natural environment in the Tokyo metropolitan area is that forests and mountainous areas are close to urban areas. Such a geographic environment is an essential factor in analyzing the dynamics of radioactive contamination in the capital Tokyo. After the Fukushima disaster, the hinterland was exposed to high levels of radioactive contamination. According to the MEXT airborne monitoring, radioactive cesium deposition may have reached 600 kBq/m^2 or more in some places. These radioactivity levels are approximately the same as those in the contaminated area of Fukushima Prefecture, where the FDNPP is located (see Fig. 3.2).

Radioactive nuclides released from the FDNPP were transported to the forested areas through the air as radioactive plumes and deposited there. Furthermore, radioactive nuclides are detected at high concentrations in mountainous areas with elevations of more than 1000 m (see Fig. 2.5). The presence of radioactive nuclides in the mountainous areas indicates that the radioactive nuclides released from the FDNPP diffused to high altitudes during transport through the atmosphere. When the radioactive plume was advected over the flat lands and reached the mountainous areas, it was carried by updrafts to high levels, cooled by adiabatic expansion, became fog and rain, and was deposited on the surface. The FDNPP is located in a coastal area, and the height of the fumes was reported to be several hundred meters when the reactor buildings collapsed due to hydrogen explosions. However, the mountainous areas at high-elevations, which form the hinterland of the Tokyo metropolitan area, are receiving radioactive contamination. In such highland forest areas, the radionuclides deposited from radioactive plumes are strongly adsorbed not only onto the soil but also onto trees, deciduous leaves, humus, *etc.*, causing environmental radioactive contamination. Nevertheless, it is unlikely that radioactive cesium, an alkali metal element ordinarily present as water-soluble cations, is firmly retained in the forest environment.

In the early stages of the FDNPP accident, due to the difficulty of forest decontamination work, the Japanese government substantially abandoned radioactivity decontamination father than 20 m from the forest edge in the guidelines for "Appropriate methods for forest decontamination 2011" [2, 3]. Radioactive materials deposited in forests are recycled in forest ecosystems. The risk of their moving out of the forest to adjacent urban areas over time and becoming a new source of radioactive contamination in urban areas is extremely high. Additionally, there have been many reports that wild animals living in these contaminated areas are highly contaminated by radioactivity. Although wildlife and birds living in contaminated areas play crucial roles as carriers of radioactive material, they mainly roam in forested areas, so their ecological behavior is challenging to determine.

Since the decontamination of radioactively contaminated areas destroys the forest ecosystem, the SCJ (Science Council of Japan) has noted problems regarding the effects of radioactive contamination in forest areas on forestry and agriculture [4]. However, the Japanese government still resists deforestation. Wildlife contamination, enrichment through the food chain, and recycling in polluted forest ecosystems are important issues but are not addressed in this book. Radioactive nuclides that are strongly adsorbed onto trees and litter are very likely to migrate from forest to downstream city as water-soluble humic acid and fulvic acid complexes along with the decomposition of litter. The decomposition and solubilization of these substances has progressed for more than 9 years after the accident, so it is necessary to discuss the redistribution of radionuclides in the environment. The analysis of radioactive contamination in forests cannot be avoided in resolving the radioactive contamination of the Tokyo metropolitan area.

7.2 Samples and Methods

The behavior of radioactive cesium precipitated in the forest was monitored at Site D in Gunma Prefecture, as shown in Fig. 3.1. According to the MEXT's airborne monitoring (Fig. 3.2), $^{134+137}$Cs precipitation is 10–30 kBq/m^2 in this area. The site is a copse forest at an elevation of 997 m. When the first sampling was performed on April 29, 2011, deciduous leaves were just developing. On the ground, leaves from the previous fall were deposited, and beneath them were humified fallen leaves. These humified leaves are called litter. As shown in Photo 7.1, the soil layer was found below the litter layer and 10–20 cm below the surface. The soil samples were

Photo 7.1 Sedimentary layer of forest in Kita-Karuizawa, Gunma Prefecture. Site D in Fig. 3.1. In winter, the surface soil freezes to a depth of approximately 10 cm

Photo 7.2 Leached water from the litter layer in the Kita-Karuizawa forest. This place is approximately 10 m away from the place in Photo 7.1. Light brown, transparent water is constantly springing. This water contains soluble humic substances. From January to March, snow accumulates, and the water freezes

taken from a point approximately 5 m from a deciduous tree. This position is under the leaves of the tree and scarcely exposed to direct sunlight in the summertime.

A well with a depth of 1 m and a diameter of 0.5 m was dug at a distance of approximately 10 m from the soil sampling point, and the water that pooled there was sampled. In addition, leaching water from the litter layer, which collected in a depression approximately 20 m from the soil sampling point, was sampled (Photo 7.2). This leachate was transparent but slightly brown due to soluble humic substances. Radioactive cesium in water samples was measured after concentration by the AMP coprecipitation method (Reference 26 in Chap. 6). Moreover, zeolites with particle sizes of 2–5 mm were immersed in the site water for approximately half a year, and the adsorbed radioactive cesium was measured.

Photo 7.3 Cherry Tree. This place is approximately 10 m away from Photo 7.1. At a height of 150 cm above the ground, the diameter of the trunk is 45 cm

Barks samples were collected from locations approximately 150 cm above the ground on the cherry and oak trunks shown in Photos 7.3 and 7.4. A bark area of approximately 300 cm^2 was peeled off, and the front and back surfaces were scraped to a thickness of approximately 0.5 mm using a knife to obtain a sample for measurement. Multiple bark samples were collected from the same tree.

The radioactivity of each sample was measured according to the method described in Sect. 3.2. A Ge semiconductor detector was shielded with 10 cm of lead and connected to a 4096-channel pulse height analyzer. The sample was sealed in a plastic container with a diameter of 5.5 cm and a thickness of 2.0 cm and measured in close contact with the Ge detector. The measurement time was set so that the counting error was 5% or less. The detection limit of radioactive cesium by this gamma-ray measurement system was less than 0.6 Bq/kg

Chapters 3, 5, and 6 discuss the dynamics of radioactive cesium in urban parts of the Tokyo metropolitan area. The process of radioactive cesium enrichment in urban ecosystems is elucidated [5, 6]. The behavior of radioactive cesium in forests has also been gradually understood. The forest functions as a reservoir for radioactive materials [7–20]. Japan suffered from radioactive cesium contamination of its staple food, rice, by atmospheric nuclear tests in the 1950s and 1960s. Studies have also been conducted on the incorporation of radionuclides into crops and the separation of radionuclides from contaminated food [21–25]. However, radioactive materials

Photo 7.4 Oak Tree. This place is approximately 5 m away from the cherry tree in Photo 7.3. At a height of 150 cm above the ground, the diameter of the trunk is 50 cm

accumulated in the forest ecosystem due to accidents may leave the forest over time, and there is a high risk of such materials becoming a new source of radioactive contamination in urban and agricultural areas. In the Tokyo metropolitan area, most of the forests in the catchments have high concentrations of radioactive pollution, so it is necessary to continue to monitor them reliably.

7.3 Temporal Changes in the Vertical Distribution of Radioactive Cesium in the Forest Surface

Soil samples were collected from a forest on a plateau in Gunma Prefecture, approximately 230 km southwest of the FDNPP. These sampling points compose Site D in Fig. 3.1. The first soil sample was collected on April 29, 2011. Tables 3.3 and 3.4 show the initial analytical results. A sedimentary layer with a depth of 20 cm from the surface was collected, which contained fallen leaves and litter. The soil layer was deeper than approximately 5 cm. In this section, the analytical results for radioactive cesium are shown as bar graphs in Fig. 7.1. The inventories at this point,

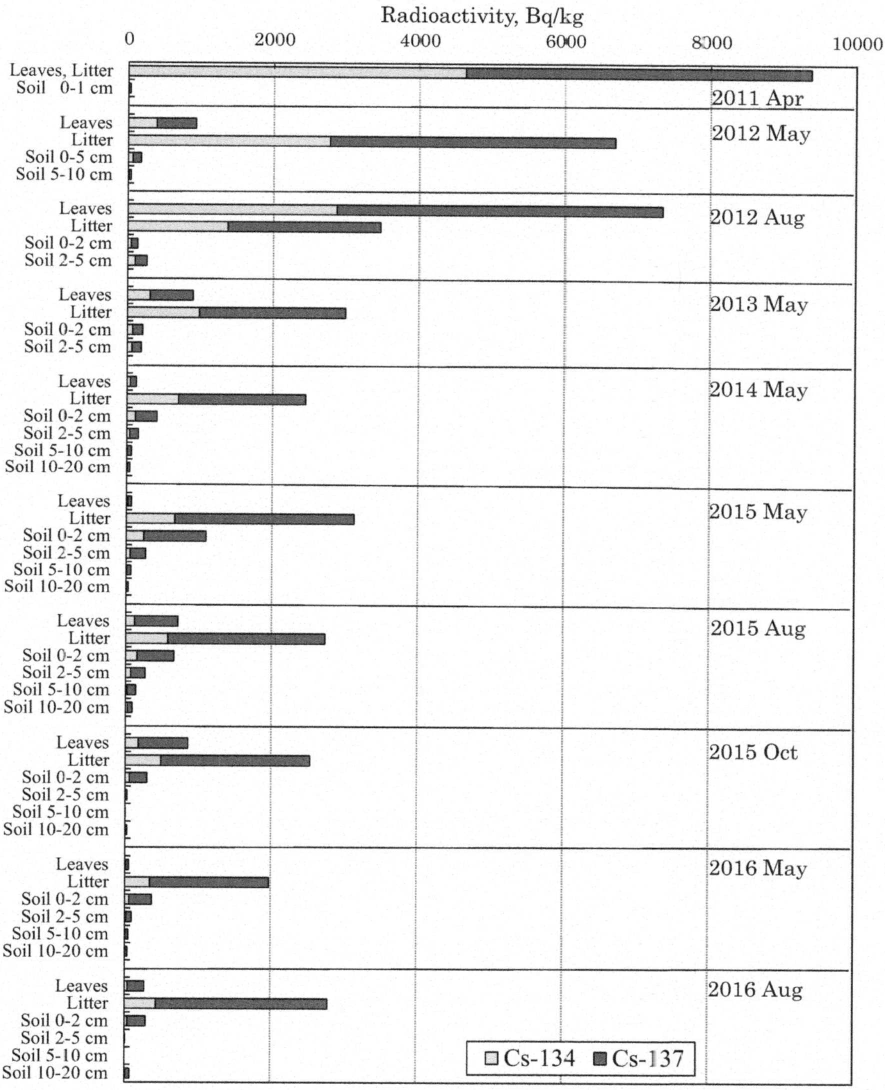

Fig. 7.1 Temporal changes in the vertical distributions of radioactive cesium in the forest sediments in the Kita-Karuizawa. Core samples were collected from the area of 3 m^2 in the background of Photo 7.3

which were radioactive decay-corrected to March 16, 2011, were 1630 kBq/m^2 for ^{131}I, 183 kBq/m^2 for ^{134}Cs, and 189 kBq/m^2 for ^{137}Cs. Since almost no ^{131}I and radioactive cesium were detected in the soil under litter, the inventory of the litter was calculated, assuming a density of 800 kg/m^3 for litter. In the surface soil (0–1 cm depth) collected from the nearby point, which was approximately 20 m from this sampling point and not covered with leaves or litter, ^{131}I was 21.2 kBq/m^2, and $^{134+137}Cs$ was 4.8 kBq/m^2. Although these samples were taken at very close distances, their concentrations and inventories varied quite widely. These differences may suggest that the radioactive plume, which was advected in the atmosphere, came into contact with the tree branches, at which time the radionuclides were attached to the bark, washed away by rainfall and adsorbed on fallen leaves and litter under the trees. By the time of the accident, the tree leaves in the area had not yet grown.

Figure 7.1 shows the results of monitoring the vertical distribution of radioactive cesium in forest soil from April 2011 to August 2016. Until one and a half years after the FDNPP accident, the fallen leaves were contaminated with high concentrations of radioactive cesium. However, in the spring of 2013, the concentration of radioactive cesium in fallen leaves decreased significantly. This change was probably due to the dilution effect of fallen leaves from autumn 2012, which are considered less contaminated. Since then, except for 2015, the concentration of radioactive cesium in fallen leaves has remained low. However, the fallen leaves were contaminated with $^{134+137}Cs$ to approximately 1000 Bq/kg in August and October 2015. The cause is not well understood, but as shown in Fig. 5.2, the concentration of radioactive cesium in the sludge incineration ash, which was treated with rainwater, also increased slightly in 2015 in Tokyo. Since decommissioning work was being carried out by the TEPCO at the FDNPP site, dust contaminated with radioactive materials could undeniably have been scattered and transferred to Tokyo and Gunma prefectures. Monitoring shows that radioactive cesium adsorbs to tree leaves effectively, so the fallen leaves in 2015 may have functioned as adsorbents for the leaked radioactive cesium.

The secular change in radioactive cesium contamination was examined in the lower litter layer because new leaves are added to the forest surface each year, and new litter layers are added. In 2011, when the accident occurred, fallen leaves and litter were not separately measured, but their concentrations were 4640 Bq/kg for ^{134}Cs and 4740 Bq/kg for ^{137}Cs. The following year, in August 2012, the fallen leaf layer contained 2880 Bq/kg for ^{134}Cs and 4470 Bq/kg for ^{137}Cs. However, the concentration of radioactive cesium in the litter collected in May 2013 decreased sharply, and after this time, no significant fluctuations were observed until August 2016.

The average concentrations of radioactive cesium in the litter layer from March 2013 to 2016 were 600 ± 220 Bq/kg for ^{134}Cs and 2060 ± 300 Bq/kg for ^{137}Cs. Since ^{134}Cs in August 2012 was 1380 Bq/kg and ^{137}Cs was 2100 Bq/kg, in August 2016, the concentration should have decreased by radioactive decay to 25% of initial value for ^{134}Cs and 90% for ^{137}Cs. However, the actual measured values were approximately 30% for ^{134}Cs and 110% for ^{137}Cs. Radioactive cesium contamination increased slightly over time. In other words, radioactive cesium that was precipitated in this forest immediately after the FDNPP accident in 2011 released half of the deposited amount somewhere in the first year and a half, after which the remaining radioactive cesium was strongly absorbed by litter and retained in the forest. Such a rapid decrease in concentration immediately after the accident was also measured in the sludge incineration ash in Tokyo, as shown in Fig. 5.3. These phenomena mean that radioactive cesium contained in the radioactive plume released from the FDNPP reactor may have contained at least two physical-chemical species with different adsorption characteristics. In general, cesium cations strongly adsorb to humic substances such as clay minerals and litter, so cesium species that rapidly flowed out in the early stage of the accident are likely to have been ultrafine particles of nonionic species (References 40–47 in Chap. 3).

During the 5 years of monitoring, the concentration of radioactive cesium in the soil below the litter layer was deficient. This vertical distribution suggests that radioactive cesium adsorbed on the litter separates from the litter and penetrates into the ground with difficulty. It is well known that radioactive cesium is strongly adsorbed to aluminum silicate zeolite and clay minerals such as vermiculite, illite, and mica and retained in the soil (References 44–52 in Chap. 6). Nevertheless, radioactive cesium was strongly adsorbed to litter. The litter eventually decomposed and transformed into organic acids such as humic and fulvic acids. Elucidation of the behavior of radioactive cesium adsorbed on litter is vital for understanding the dynamics of radioactive cesium in the forest environment.

Radioactive cesium has long been known to adsorb well on litter and humus [26]. However, although these humic substances have many functional groups, their structure is complicated and not constant [27–31]. Therefore, little is known about the adsorption mechanism of radioactive cesium on litter and humus [32–37].

7.4 Temporal Changes in Radioactive Cesium Adsorbed on Bark

The behavior of radioactive cesium on bark was monitored using a cherry tree and an oak tree. Figure 7.2 shows the time-dependent changes in radioactive cesium concentration on the surface and back of the barks from these trees. Due to the complex nature of the bark surface, collecting samples for measurement was not effortless. The samples were carefully stripped using a small scalpel from the surface and the back to a thickness of approximately 0.5 mm. As shown in Photos 7.3 and 7.4, the condition of the bark differs significantly between cherry and oak. The bark of the cherry tree has small horizontal grooves, while those of oak are vertical. The bark samples were collected approximately 150 cm from the ground surface.

Five months after the FDNPP accident, the concentration of $^{134+137}$Cs on the bark surface of the cherry tree was 6300 Bq/kg, which was almost identical to that of fallen leaves and litter at the same time. However, no radioactive cesium was detected on the back of the bark. The radioactive cesium transported by the radioactive plume that spread on March 15, 2011 (see Fig. 2.1), was trapped on tree trunks and branches. At this time, there were no leaves yet. In the cherry tree, the concentration of radioactive cesium was slightly lower in the 2015 and 2016 samples, but in other cases, it was more than 2000 Bq/kg. Despite the decrease due to the radioactive decay of ^{134}Cs, the radioactive cesium attached to the surface of the cherry tree remained in place for 5 years after the accident. In the case of the oak tree, the concentration of radioactive cesium was slightly lower than that of the

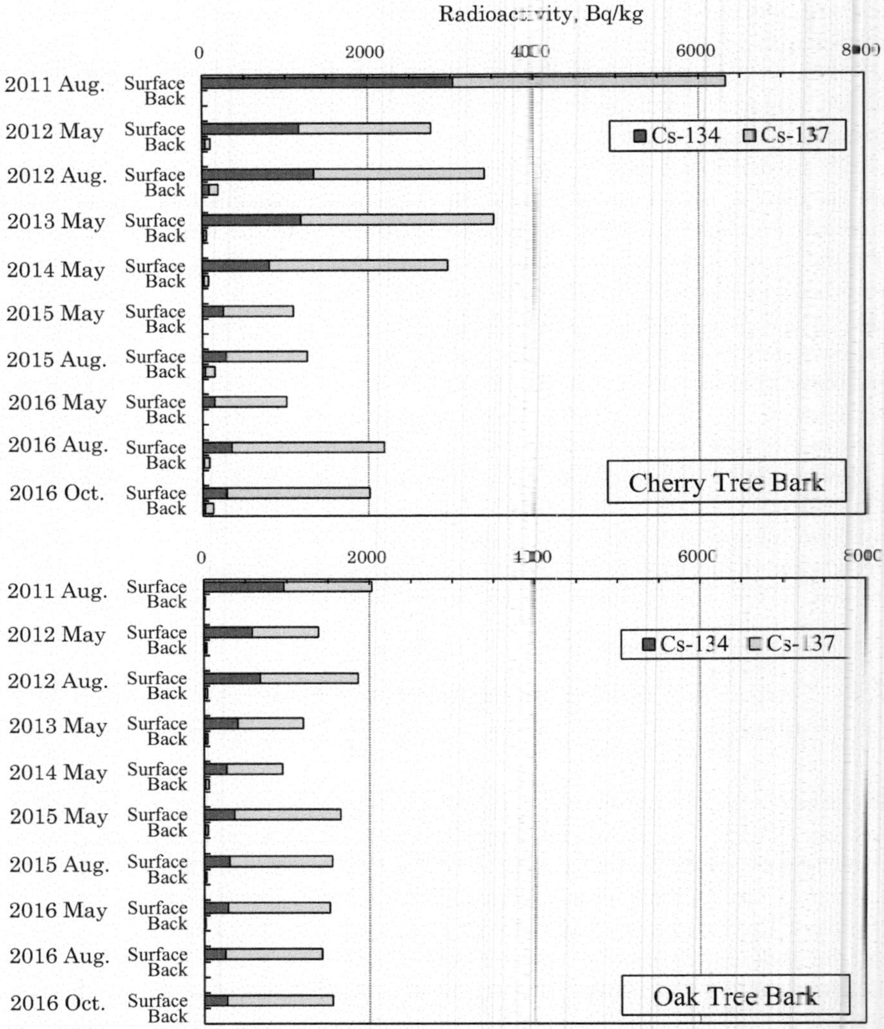

Fig. 7.2 Temporal changes in the radioactive cesium activities on the cherry and oak barks in the Kita-Karuizawa forest. The tree of cherry and oak is 5 m apart

cherry tree, but there was almost no change in the concentration over time. The lower concentration of radioactive cesium in the oak tree than the cherry tree may reflect differences in the structure of the furrows on the bark surfaces.

Throughout the entire period, no radioactive cesium was detected on the back side of the bark of either the cherry tree or the oak tree, indicating that radioactive cesium adsorbed on the bark surface did not penetrate into the tree. The sample on the back of the bark contains the cambium of the tree. Therefore, the absence of radioactive cesium in this sample means that the intake of radioactive cesium by a tree is not involved in the cell division and xylem formation processes. Of course, considering long-term forest radioactive contamination, it is necessary to discuss the route by which radioactive cesium deposited on the surface is absorbed by roots and distributed to trees [38–40]. However, the forests in the Tokyo metropolitan area were heavily polluted by fallout radioactive cesium in the early days of the accident, but this material penetrated only a few tens of centimeters into the surface of the forest. Therefore, radioactive cesium absorbed by the roots is negligibly small compared to the contamination of bark, fallen leaves, and litter.

7.5 Behavior of Radioactive Cesium in Forest Water

Radioactive cesium deposited in forests, once adsorbed on trees, may move and be redistributed into the atmosphere by wind or washed off by rainfall and discharged into water systems. Migration through the food chain by wildlife also needs to be considered. The monitoring of wildlife was not performed in this study because the continued capture of wild animals is stressful. While the release of radioactive cesium from trees to the atmosphere is likely to be the source of the recontamination by radioactive material in urban areas, it is challenging to collect atmospheric components in forest areas. Therefore, the behavior of radioactive cesium was monitored only in aquatic systems.

Figure 7.3 shows the results of collecting radioactive cesium in water using a zeolite well known to adsorb radioactive cesium effectively. A well water was dug with a depth of 1 m to prevent rainwater and surface water from directly flowing in. Zeolite debris was pack in a plastic net and immersed to a depth of approximately 30 cm. As shown in Photo 7.2, the water that springs naturally from the litter layer accumulates in depressions near the well. A zeolite bag was also immersed there. The well water was clear and colorless but inhabited by many aquatic insects except in winter. In winter, approximately 10 cm of the surface layer froze. The litter spring was colored brown by soluble humic substances but was transparent. Therefore, the zeolites after immersion were also colored brown.

Monitoring began in August 2011. In the first 2 years, the zeolites were soaked for 1 year and then collected. However, after August 2013, the zeolite bags were replaced in May and August. Figure 7.3 shows the analytical results. The vertical arrow in this figure indicates that when the zeolite bag was recovered, its water was also collected. Figure 7.4 shows the analytical results for the water. During the 5-year monitoring period, radioactive cesium adsorbed on the zeolites in the well water was almost undetectable. This result means that no radioactive cesium was dissolved as cations in the well water. In winter, when rainfall decreases, the water level in the well drops to approximately 50 cm, and in summer, it rises to 1 m with abundant water, so this increase or decrease in water level indicates a change in the groundwater level. This variation means that the groundwater, which is the source of the well water, is continually replaced, but that the groundwater was not exposed to radioactive cesium contamination for 5 years.

On the other hand, no radioactive cesium contamination was observed in the litter spring water from August 2011 to August 2012. However, since August 2012,

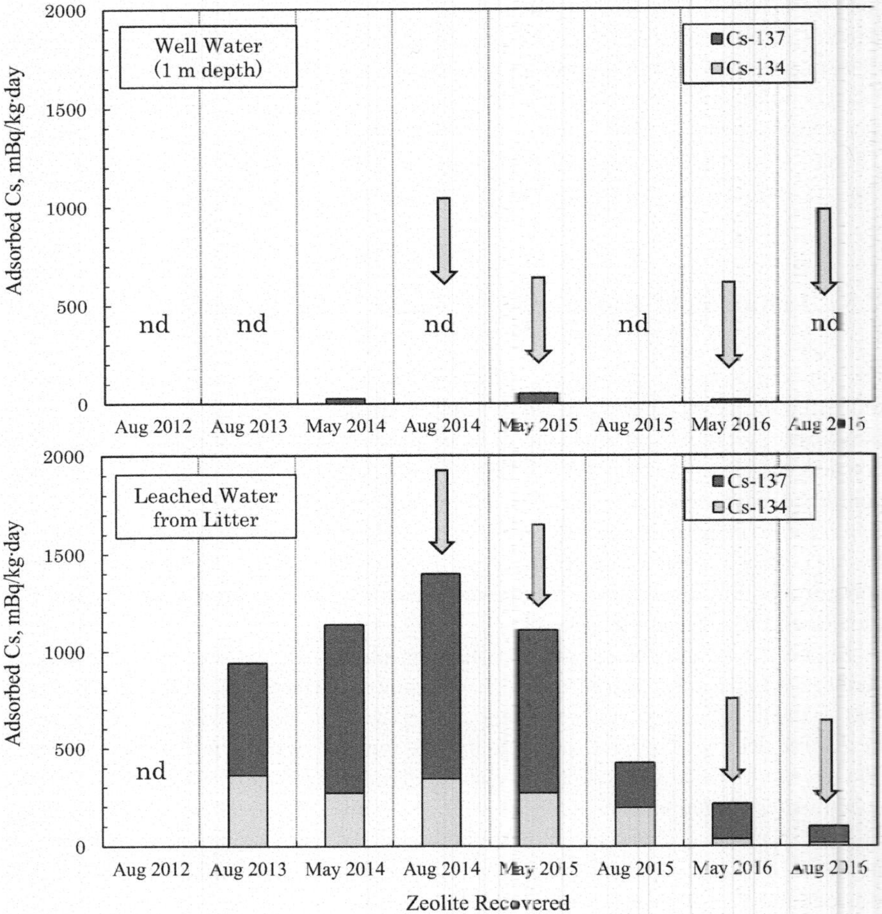

Fig. 7.3 Temporal changes in the radioactive cesium activities adsorbed on zeolite soaked in the waters in the Kita-Karuizawa forest. nd: not detected (nd <30 mBq/kg zeolite). The vertical arrows are the same as the water sampling days shown in Fig. 7.4

radioactive cesium contamination has been recognized. The highest contamination was observed in the 3 months from May to August 2014, and the amount of $^{134+137}$Cs adsorbed on the zeolite material reached 1400 mBq/kg/day. The amount of adsorption after this time decreased significantly and was 90 mBq/kg/day in August 2016. This fluctuation indicates that radioactive cesium deposited on leaves and litter immediately after the FDNPP accident had a component that desorbed from the litter layer and leached into the water. The remaining radioactive cesium was adsorbed on the litter and retained for 6 years in the litter layer. As shown in Fig. 7.1, the radioactive cesium concentration in the litter layer hardly changed during this monitoring period. Because the physical-chemical species of radioactive cesium dissolved in the litter spring water adsorbed on the zeolite material, it was assumed

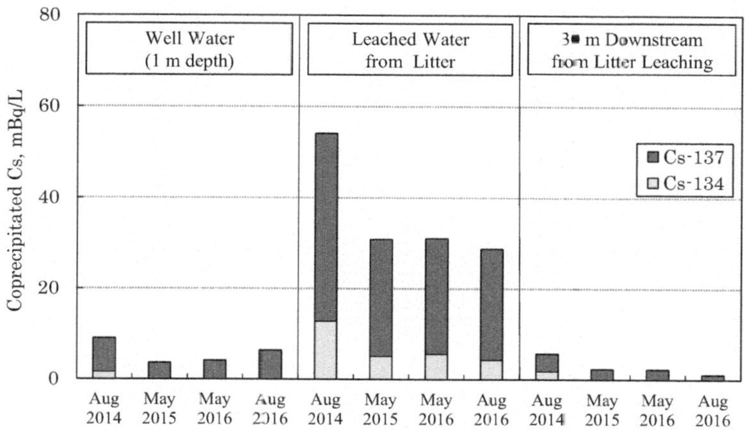

Fig. 7.4 Temporal changes in the radioactive cesium activities in the waters in the Kita-Karuizawa forest. Radioactive cesium in the water samples was coprecipitated by the AMP method [Reference 26 in Chap. 6], and the radioactivity of the ammonium phosphomolybdate precipitate measured

to be a cation or a low molecular weight humic acid complex. As the litter decomposes, what is adsorbed there is expected to become ionized and desorbed. However, it is strange that such a dissolution process ended 4 years after the accident, despite the continued decomposition of litter. A large amount of radioactive cesium remains in the litter layer.

Figure 7.4 shows the results for the concentration of radioactive cesium in water by the AMP coprecipitation method. Since this sample was not filtered in advance, radioactive cesium in the suspended particles was detected as the same ionic cesium species. Samples could not be measured until May 2014. Sample waters were collected when the zeolite bag with the vertical arrow shown in Fig. 7.3 was recovered.

In the well water, almost no radioactive cesium adsorbed on zeolite was detected, but with the AMP method, radioactive cesium of approximately 5–10 mBq/L was coprecipitated and collected. This concentration can be considered mostly in suspension in water and collected on filter paper with ammonium phosphomolybdate precipitate. No evident temporal change in concentration was observed. In the litter spring, the maximum concentration of radioactive cesium adsorbed on zeolite showed the highest value in August 2014. The AMP method also showed a maximum concentration of 54 mBq/L in this sample. This value decreased to 29 mBq/L 2 years later in August 2016, but the rate of decrease was slower than that of the zeolite-adsorbed species. This result also suggests that radioactive cesium in the form of particles or soluble organic polymers bound to humic substances gradually desorbed from the litter. However, the concentration of radioactive cesium collected by the AMP method approximately 30 m downstream from the sampling point was deficient, and the mobility of the radioactive cesium species released from the litter is likely to be reduced.

7.6 Leaching of Radioactive Cesium Adsorbed on Litter and Soil

To estimate the state of radioactive cesium adsorbed on the litter, the leaching behavior of radioactive cesium was investigated using litter collected in August 2011 near a well. If radioactive cesium is weakly bound to soil or litter, it can be leached with a weakly acidic solution of low ionic strength such as rainwater. Elution of radioactive cesium bound to insoluble polymeric organic acids, such as humic acid, requires a relatively strong alkaline solution. If radioactive cesium is adsorbed between layers of clay minerals by ion exchange, it should be leachable with ammonium ions [41, 42].

One hundred milliliters of solvent was added to 50 g of wet litter and left for 1 week with occasional stirring. The solvent was filtered using a membrane filter with a pore size of 0.22 μm, and the elution rate of radioactive cesium was measured. The remaining solvent was discarded by decantation, a new 50 mL solvent was added, and the mixture was left for 1 week with occasional stirring. The solvents used were rainwater, seawater, and 0.1 M—NH_4NO_3 solution. The extraction procedure was performed continuously for the same sample by changing the solvents in the order of rainwater, seawater, and NH_4NO_3 solution. In the natural environment, radioactive cesium adsorbed on litter first encounters rainwater, which has a low ionic strength but is a weakly acidic solvent. Since seawater has an ionic strength of 0.7, many substances that adsorb ions are leached out by seawater. Ammonium ions are well known to displace cesium adsorbed on clay minerals [41, 42]. The extraction procedure was repeated once for rainwater, twice for seawater, and three times for the NH_4NO_3 solution. Similar dissolution experiments were performed on various soil samples to compare with litter. Table 7.1 shows the results.

Although the extracted amount of radioactive cesium adsorbed on the litter was very small in rainwater, 0.23% was eluted. The accuracy of the extraction experiment is minimal, and this extraction rate is significantly higher than that of the soil sample. The extraction rate of radioactive cesium in soil contaminated with relatively high concentrations was below the detection limit except for schoolyard soil in Tokyo. Radioactive cesium cannot be exchanged with rainwater having low ionic strength because the cesium is strongly adsorbed onto clay minerals in the soil. In any case, it is shown that radioactive cesium is hardly eluted from forest litter or urban soils by rainwater.

The radioactive cesium adsorbed on the litter was extracted twice with seawater and three times with the NH_4NO_3 solution. The extraction rates were 1.06% and 0.62%, respectively. The total extracted amount of the six extraction treatments was a small value of 1.91% for the whole sample. The lowest extraction rate was 0.55% of non-decontaminated garden soil in Fukushima Prefecture. On the other hand, the elution of radioactive cesium was 2.41% with seawater and 34.8% with the NH_4NO_3 solution, in the schoolyard soil treated with a dustproof agent. Inorganic salts such as magnesium chloride, sodium chloride, and potassium chloride and high molecular weight organic compounds such as acrylic polymers and natural resins were blended

Table 7.1 Results of the sequential elution experiments of radioactive cesium adsorbed on the litter and soils by rainwater, seawater, and NH_4NO_3 solution

	Activity, Bq/kg[a]			$^{134+137}$Cs extracted[b], %								
				Solvent								Total extracted (%)
				Rainwater (pH 5.84)	Seawater (pH 7.98)			0.1 M—NH_4NO_3 solution				
Sample	^{134}Cs	^{137}Cs	$^{134+137}$Cs	First	First	Second	Sum	First	Second	Third	Sum	
Litter, Site D in Fig. 3.1 (Kita Karuizawa, Gunma Prefecture)	4020	4070	8090	**0.23**	0.71	0.35	**1.06**	0.26	0.17	0.18	**0.62**	*1.91*
Park Soil (Kashiwa City, Chiba Prefecture)	5300	5330	10,630	**0.00**	0.00	0.11	**0.11**	0.73	0.51	0.60	**1.83**	*1.94*
Park Soil (Chiyoda Ward, Tokyo)	1170	1180	2350	**0.00**	0.47	0.39	**0.86**	3.82	3.51	0.00	**7.33**	*8.19*
Schoolyard Soil treated to prevent the soil dust scattering (Koto Ward, Tokyo)	1700	1740	3440	**0.15**	1.51	0.89	**2.41**	17.53	11.47	5.75	**34.8**	*37.3*
Garden Soil before decontamination[c] (Date City, Fukushima Prefecture)	127,000	124,000	251,000	**0.00**	0.01	0.00	**0.01**	0.24	0.19	0.11	**0.54**	*0.55*
Garden Soil after decontamination[c] (Date City, Fukushima Prefecture)	12,800	12,700	25,500	**0.00**	0.19	0.30	**0.49**	10.02	3.11	2.09	**15.2**	*15.7*
Estuary Sediment (Abukuma River, Fukushima Prefecture)	4800	4790	9590	**0.00**	0.00	0.00	**0.00**	0.23	0.08	0.00	**0.31**	*0.31*

[a]Decay-corrected value to March 16, 2011

[b]In the leaching experiment, elutions were carried out sequentially by changing the solvent in the order of rainwater, seawater, and 0.1 M—NH_4NO_3 solution. One-stage leaching was for a week and was stirred twice a day for 5 min each

[c]Residents received no explanation about the contents of the decontamination work. The inhabitants did not know why the decontamination was carried out or how. Probably, the contaminated surface soil removed, and the house subjected to high-pressure washing with water

in the dustproof agent. Since such compounds had been sprayed on the schoolyard soil before the FDNPP accident occurred, the adsorption of radioactive cesium to soil was probably inhibited by these agents. Therefore, radioactive cesium, which had not adequately adsorbed onto the clay minerals, was eluted together with the dustproof agent by extraction with seawater or NH_4NO_3. Similar release behavior of radioactive cesium was observed in the soil of decontaminated gardens in Tokyo parks and Fukushima Prefecture. The elution rate with ammonium cations was consistent with an elution rate of approximately 20% when lake sediment was treated with 100 mM—NH_4^+ solvent [42]. This result confirmed that radioactive cesium was adsorbed onto clay minerals in such garden soil. However, in some soils, radioactive cesium was hardly eluted even with ammonium cations. This difference in elution rate reflects the difference in soil composition.

From the results of the sequential elution experiments, the radioactive cesium adsorbed on the litter was not easily leached by ammonium cations. Given this outcome, it is unlikely that radioactive cesium is bound to the organic ligands in the litter by purely ionic bonds. The main component of litter is a high polymer organic substance, such as soluble fulvic acid or insoluble humic acid. The basic structure of these humic substances is an aromatic or aliphatic chain, which contains conjugated double bonds and various functional groups, such as carboxyl, phenol, and carbonyl groups, and is highly reactive. Thus, there is a high possibility that radioactive cesium is chemically bound. On the other hand, since fulvic acid and humic acid are polymers having molecular weights in the hundreds of thousands, it cannot be ruled out that radioactive cesium is physically included in the structure by agglutination. From the data obtained here, it is not possible to determine which mechanism was predominant in the adsorption of radioactive cesium.

Presumably, the high molecular weight organic acids that retain radioactive cesium in the forest eventually decompose to release radioactive cesium as a soluble low molecular weight compound. This species of radioactive cesium is taken up by the ecosystem and accumulated elsewhere, causing secondary radioactive contamination.

References

1. MAFF (Ministry of Agriculture, Forestry, and Fisheries, Japan) (2017) Forest and artificial forest rates by prefecture [in Japanese]. https://www.rinya.maff.go.jp/j/keikaku/genkyou/h29/1.html
2. Prime Minister's Office of Japan. Nuclear Emergency Response Headquarters (2011) About publication of appropriate method of decontamination of forest [in Japanese]. https://www.kantei.go.jp/jp/singi/genshiryoku/dai21/21_09_gensai.pdf
3. MOE (Ministry of the Environment, Japan) (2016) Decontamination information site. About decontamination of forests [in Japanese]. https://www.josen.env.go.jp/about/efforts/forest.html
4. SCJ (Science Council of Japan) (2014) Radiological contamination and its impact on forests, forestry and timber-related industries due to the Fukushima nuclear accident - current situation and problems [in Japanese]. https://www.scj.go.jp/ja/info/kohyo/pdf/kohyo-22-h140901.pdf

5. Seleznev AA, Yaemoshenko IV, Ekidin AA (2010) Accumulation of ^{137}Cs in puddle sediments within urban ecosystem. J Environ Radioact 101:643–646. https://doi.org/10.1016/j.jenvrad.2010.03.015
6. Ozaki H, Takahashi H, Watai C, Inamochi R, Nagashima T et al (2015) High level of cesium-137 concentration in road dust samples collected in the eastern area of Japan. J Environ Chem 26:131–140. [in Japanese with English Abstract]
7. Tsukada H, Hasegawa H, Hisamatsu S, Yamasaki S (2002) Transfer of ^{137}Cs and stable Cs from paddy soil to polished rice in Aomori, Japan. J Environ Radioact 59:351–363. S0265-931X(01)00083-2 [pii]
8. Yoshida S, Muramatsu Y, Dvornik AM, Zhuchenko TA, Linkov I (2004) Equilibrium of radiocesium with stable cesium within the biological cycle of contaminated forest ecosystems. J Environ Radioact 75:301–313. https://doi.org/10.1016/j.jenvrad.2003.12.008
9. Shaw G, Venter A, Avila R, Bergman R, Bulgakov A et al (2005) Radionuclide migration in forest ecosystems - results of a model validation study. J Environ Radioact 84:285–296. https://doi.org/10.1016/j.jenvrad.2003.09.006
10. Tsukada H, Takeda A, Hisamatsu S, Inaba J (2008) Concentration and specific activity of fallout ^{137}Cs in extracted and particle-size fractions of cultivated soils. J Environ Radioact 99:875–881. https://doi.org/10.1016/j.jenvrad.2007.11.014
11. Yamaguchi N, Takata Y, Hayashi K, Ishikawa S, Kuramata M et al (2012) Behavior of radiocaesium in soil-plant systems and its controlling factor: a review. Bull Natl Inst Agro Environ Sci 31:75–129. [in Japanese with English Abstract]
12. Koarashi J, Moriya K, Atarashi-Andoh M, Matsunaga T, Fujita H et al (2012) Retention of potentially mobile radiocesium in forest surface soils affected by the Fukushima nuclear accident. Sci Rep 2:1005. https://doi.org/10.1038/sreo01005
13. Tsuiki M, Maeda T (2012) Spatial distribution of radioactive cesium fallout on grasslands from the Fukushima Daiichi Nuclear Power Plant in 2011. Grassl Sci 58:153–160. https://doi.org/10.1111/j.1744-697X.2012.00257.x
14. Hashimoto S, Ugawa S, Nanko K, Shichi K (2012) The total amounts of radioactively contaminated materials in forests in Fukushima, Japan. Sci Rep 2:416. https://doi.org/10.1038/srep00416
15. Hashimoto S, Matsuura T, Nanko K, Linkov I, Shaw G et al (2013) Predicted spatio-temporal dynamics of radiocesium deposited onto forests following the Fukushima nuclear accident. Sci Rep 3:2564. https://doi.org/10.1038/srep02564
16. Fujii K, Ikeda S, Akama A, Komatsu M, Takahashi M et al (2014) Vertical migration of radiocesium and clay mineral composition in five forest soils contaminated by the Fukushima nuclear accident. Soil Sci Plant Nutr 60:751–764. https://doi.org/10.1080/00380768.2014.926781
17. Murakami M, Ohte N, Suzuki T, Ishii N, Igarashi Y et al (2014) Biological proliferation of cesium-137 through the detrital food chain in a forest ecosystem in Japan. Sci Rep 4:3599. https://doi.org/10.1038/srep03599
18. Ishi Y, Hayashi S, Takamura N (2017) Radiocesium transfer in forest insect communities after the Fukushima Dai-ichi Nuclear Power Plant accident. PLoS One 12:e0171133. https://doi.org/10.1371/journal.pone.0171133
19. Imamura N, Komatsu M, Ohashi S, Hashimoto S, Kajimoto T et al (2017) Temporal changes in the radiocesium distribution in forests over the five years after the Fukushima Daiichi Nuclear Power Plant accident. Sci Rep 7:1. https://doi.org/10.1038/s41598-017-08261-x
20. Teramage MT, Carasco L, Coppin F (2019) Impact of drying and wetting cycles on ^{137}Cs ageing in forest soils contaminated with different input forms. J Environ Radioact 203:93–97. https://doi.org/10.1016/j.jenvrad.2019.03.007
21. Yamagata N, Yamagata T (1958) Separation of radioactive cesium in Biological materials. Bull Chem Soc Jpn 31:1063–1068. https://doi.org/10.1246/bcsj.31.1063
22. Tensho K, Yeh K, Mitsui S (1961) Mechanism of specific cesium absorption by paddy rice. J Sci Soil Manure Jpn 32:139–144. [in Japanese]
23. NARO (National Agriculture and Food Research Organization) (2012) Potassium application to reduce radioactive cesium in brown rice [in Japanese]. http://www.naro.affrc.go.jp/publicity_report/press/laboratory/narc/027913.html

24. Yoshitaka T (2012) Behavior of radioactive substances in soil and effects on agricultural crops: arrangement of measures and issues. NOUGYOTOKEIZAI (Agric Econ) 78:101–112. [in Japanese]. http://www.t-yasu.net/Main/Welcome.html
25. Ohse K, Kihou N, Kurishima K, Inoue T, Taniyama I (2015) Changes in concentration of ^{131}I, ^{134}Cs and ^{137}Cs in leafy vegetable, soil and precipitation in Tsukuba city, Ibaraki, Japan, on the first 4 months after the Fukushima Daiichi nuclear power plant accident. Soil Sci Plant Nutr 61:225–229. https://doi.org/10.1080/00380768.2014.989541
26. Ritchie JC (1962) Distribution of fallout cesium-137 in litter, humus, and surface soil layers under natural vegetation in the Great Smoky Mountains. Dissertation, The University of Tennessee. http://trace.tennessee.edu/utk_gradthes/1427
27. Kumada K, Sato O (1962) Chromatographic separation of green humic acid from podzol humus. Soil Sci Plant Nutr 8:31–33
28. Kumada K (1965) Studies on the color of humic acid. Part 1 On the concepts of humic substances and humification. Soil Sci Plant Nutr 11:151–156
29. Kumada K, Sato O, Ohsumi Y, Ohta S (1967) Humus composition of mountain soils in central Japan with special reference to the distribution of P type humic acid. Soil Sci Plant Nutr 13:151–168
30. Kawada H (1972) Studies on the humus form of forest soil. Part 1 On the examination of analytical method for the determination of humus form and a proposal improved method. Bull Gov For Exp Sta 248:15–68
31. Kawada H (1975) Studies on the humus form of forest soil. Part 2 Humus form of the brown forest soil. Bull Gov For Exp Sta. 270:89–110
32. Takahasi S, Suchara I, Suchariva J, Okamoto K, Umegaki K et al (2018) Downward migration of Cs-137 within the humus layer under temperate coniferous stands in the Czech Republic. J Soils Sediments 18:276–286. https://doi.org/10.1007/s11368-017-1802-5
33. Kruyts Z, Titeux H, Delvaux B (2004) Mobility of radiocesium in three distinct forest floors. Sci Total Environ 319:241–252. https://doi.org/10.1016/S0048-9697(03)00369-3
34. Generalova VA, Onoshko MP (2006) Effect of humus acids on migration of radiostrontium and radiocesium in valley deposits of the Sozh river. Radiochemistry 48:99–104. https://doi.org/10.1134/S1066362206010206
35. Niedrée B (2013) Fate and effects of radionuclides in soil, state of knowledge. In: Effects of ^{137}Cs and ^{90}Sr on structure and functional aspects of the microflora in agricultural used soils. Forschungszentrum Jülich, Jülich, pp 4–18. ISSN: 1866-1793 ISBN: 978-3-89336-843-3
36. Söderlind M, Hakanen M, Lehto J (2016) Sorption of cesium on boreal forest soil I: the effect of grain size, organic matter and mineralogy. J Radioanal Nucl Chem 309:637–645. https://doi.org/10.1007/s10967-015-4612-5
37. Suchara I (2017) The distribution of ^{137}Cs in selected compartments of coniferous forests in the Czech Republic. In: Gupta DK, Walther C (eds) Impact of cesium on plants and the environment. Springer, New York, NY, pp 71–100. https://doi.org/10.1007/978-3-319-41525-3
38. Korobova E, Ermakov A, Linnik V (1998) ^{137}Cs and ^{90}Sr mobility in soils and transfer in soil-plant systems in the Novozybkov district affected by the Chernobyl accident. Appl Geochem 13:803–814. S0883-2927(98)00021-3 [pii]
39. Kato H, Onda Y, Gomi T (2012) Interception of the Fukushima reactor accident-derived ^{137}Cs, ^{134}Cs and ^{131}I by coniferous forest canopies. Geophys Res Lett 39:20403. https://doi.org/10.1029/2012GL052928
40. Mimura T, Mimura M, Kobayashi D, Komiyama C, Sekimoto H et al (2014) Radioactive pollution and accumulation of radionuclides in wild plants in Fukushima. J Plant Res 127:5–10. https://doi.org/10.1007/s10265-013-0599-6
41. Sholkovitz ER (1985) Redox-related geochemistry in Lakes: Alkali metals, alkaline-earth element, and ^{137}Cs. In: Stumm W (ed) Chemical processes in lakes. John Willey & Sons, New York, NY
42. Evans DW, Alberts JJ, Clark RA III (1983) Reversible ion-exchange fixation of cesium-137 leading to mobilization from reservoir sediments. Geochim Cosmochim Acta 47:1041–1049. https://doi.org/10.1016/0016-7037(83)90234-X

Chapter 8
Radioactive Contamination of Fishes in Aquatic Ecosystems

Abstract This chapter shows the results of a 5-year monitoring survey of radioactive contamination of fishes and shellfishes in lightly radioactively contaminated aquatic ecosystems. One of these ecosystems is Tokyo Bay and the Edogawa water system in the Tokyo metropolitan area, and the other is a small closed pond called Kejonuma in Miyagi Prefecture, approximately 150 km north of the FDNPP. In these water systems, water-soluble radioactive cesium deposited in the hydrosphere immediately after the accident was taken into the hydrosphere and is still recycled in the aquatic ecosystem. In general, radioactive cesium fixed to clay minerals plays an essential role in the transport process of radioactive cesium on the ground surface. However, radioactive cesium adsorbed to clay is thought to be excreted almost undigested even if organisms ingest it. When radioactive cesium in the ionized state enters the hydrosphere, it may continue to recycle in the aquatic ecosystem through the food chain. This recycling suggests that there is a high risk of long-term radiation contamination in aquatic ecosystems. The other form of cesium is ultrafine particles emitted from the reactor as a cesium solid-solution. However, little is known about the behavior of these microparticles in the hydrosphere. In this chapter, the effective half-life, biological half-life, and concentration factor were evaluated from the temporal changes of radioactive cesium contamination of seafood inhabiting the Tokyo Bay water system and Kejonuma pond. These ecological parameters seem to be strongly influenced by fish species and age.

Keywords Fish · Radioactive cesium · Freshwater · Effective half-life · Biological half-life · Concentration factor

8.1 Introduction

Radioactive materials released into the environment from a nuclear accident can be classified into nuclear fuel materials such as uranium and plutonium and fission products produced by the combustion of nuclear fuel. The reactor explosions in the Chernobyl accident were accompanied by fire, so significant amounts of nuclear fuel material such as uranium and plutonium were scattered into the environment. On the

H. Yamazaki, *Radioactive Contamination of the Tokyo Metropolitan Area*,
https://doi.org/10.1007/978-981-15-7368-2_8

other hand, although the melting of nuclear fuel occurred in the Fukushima nuclear accident, the explosion of the pressure vessel fortunately did not occur, so the release of uranium and plutonium to the environment was small compared to that in the Chernobyl accident. However, the Unit 1 reactor (460 MWe) and the Unit 2 and 3 reactors (784 MWe) operating at their rated powers released more fission nuclides than the Chernobyl accident (References 2–8 in Chap. 1). The Unit 3 reactor used MOX fuel (mixed oxide fuel) in which plutonium is mixed with uranium, but approximately 90 and 140 mass fraction fission nuclides were mainly released. Fortunately, it has been estimated that the FDNPP accident released much less plutonium into the environment than the Chernobyl accident [1–3]. Among the materials released, ^{90}Sr, ^{131}I, ^{134}Cs, and ^{137}Cs are nuclides important for health physics. In addition, radioactive noble gases derived from nuclear fission are released continuously into the atmosphere from nuclear power plants and nuclear fuel reprocessing plants regardless of accidents. After the FDNPP accident, the ^{85}Kr and ^{133}Xe concentrations in the Japanese atmosphere exceeded 1 Bq/m^3 and 1000 Bq/m^3, respectively [4]. The author could not measure ^{90}Sr, which is a nuclide that emits only beta radiation, in monitoring the Fukushima nuclear accident because it involves great expense and considerable. The measurable radioactive nuclides are ^{131}I, ^{134}Cs, and ^{137}Cs, which are usually measured in environmental science except for the period immediately after the nuclear accident. However, if ^{131}I has not been released from the reactor since the accident, because it has a half-life of 8 days and 2 months is more than seven times the half-life, 99% or more of ^{131}I in the environment would have disappeared. That is, the ecologically significant radionuclides in the case of the Fukushima nuclear accident are ^{90}Sr, ^{134}Cs, and ^{137}Cs.

The radioactive cesium released from the FDNPP is considered to have different species according to its physical and chemical properties. Gaseous radioactive cesium is deposited on the ground surface and exists as a cationic alkali earth metal. Radioactive cesium in the reactor is believed to have migrated to the radioactive plume as a low-boiling halide, but the mechanism is unknown. If deposited in the hydrosphere, it exists as dissolved species in water as cations. The ionized radioactive cesium deposited on the surface immediately reacts with clay minerals in the soil, penetrates and is fixed between the clay mineral layers. Ionized cesium is also strongly adsorbed to litter and plant surfaces. In general, radioactive cesium fixed to clay minerals plays an essential role in the transport process of radioactive cesium on the ground surface. However, radioactive cesium adsorbed to clay is thought to be excreted almost undigested even if organisms ingest it. When radioactive cesium in the ionized state enters the hydrosphere, it may continue to recycle in the aquatic ecosystem through the food chain. This recycling suggests that there is a high risk of long-term radiation contamination in aquatic ecosystems. The other form of cesium is ultrafine particles emitted from the reactor as a cesium solid-solution (References 40, 41, 44–46 in Chap. 3). Japanese researchers call such material a cesium ball. In a high-temperature reactor, radioactive cesium reacts with a heat-insulating material or the like to form a solid solution. Microfine particles may enter a living body and be engulfed by macrophages; such particles can move inside the body undigested to cause local high-dose exposure at a specific site, but this process is well understood.

This chapter shows the results of a 5-year monitoring survey of radioactive contamination of fishes and shellfishes in lightly radioactively contaminated aquatic ecosystems. One of these ecosystems is Tokyo Bay and the Edogawa water system in the Tokyo metropolitan area, and the other is a small closed pond called Kejonuma in Miyagi Prefecture, approximately 150 km north of the FDNPP. In these water systems, water-soluble radioactive cesium deposited in the hydrosphere immediately after the accident was taken into the aquatic ecosystem and is still recycled in the hydrosphere.

8.2 Samples and Methods

8.2.1 Radioactivity Measurements

The radioactivity of biological samples such as fishes was measured by gamma-ray spectrometry, as described in Sect. 3.2. In some marine organisms, ^{110m}Ag (658 keV, 252 days) was detected in addition to ^{131}I, ^{134}Cs, and ^{137}Cs. The detection limit of each radionuclide under appropriate conditions in Sect. 3.2 was also 0.6 Bq/kg in a fish sample. The radioactivities of fish and shellfish samples are indicated on a wet weight basis. Adult fishes were dissected and divided into muscle and internal organs, which were used as samples for measurement. The muscle was collected from the dorsal side around the spine. In frogs, leg muscles were measured. For fry and tadpoles approximately 10–20 g of their whole-body samples was measured. For shellfish, the shell was removed, and the entire inner tissue was measured.

The accuracy of the radioactivity measurement of fish samples was confirmed. The radioactivity ratios of ^{134}Cs and ^{137}Cs in fishes collected off the coast of Fukushima Prefecture in December 2011 were investigated with values decay-corrected to March 16, 2011. The results are shown in Table 8.1 and Fig. 8.1. The ^{134}Cs/^{137}Cs radioactivity ratio of all fish species measured was 1.001 ± 0.092, which was consistent within the counting error with the ^{134}Cs/^{137}Cs ratio released in the accident.

8.2.2 Sample Collection

Fish samples were collected in Tokyo Bay and the Edogawa River from August 2011 to July 2017 after the FDNPP accident. As shown in Chap. 6, the Edogawa River is the most crucial source of radioactive cesium contamination in Tokyo Bay from the FDNPP accident. Collaborator Ms. Yanagisawa collected many fish samples in Tokyo Bay and the Edogawa River. Fishes were captured by fishing or netting. A large fish was dissected, and its muscles and organs were measured as

Table 8.1 Concentrations of ^{110m}Ag, ^{134}Cs, and ^{137}Cs in the muscle of fishes caught in 10–20 km east from the FDNPP in December 2011

				Detected, Bq/kg wet				Corrected[a], Bq/kg wet				
Sampling date	Habitat	Species	Part	^{134}Cs	^{137}Cs	Total Cs	^{110m}Ag	^{134}Cs	^{137}Cs	Total Cs	^{110m}Ag	$^{134}Cs/^{137}Cs$
December 5–7, 2011	Migratory fish	Japanese sea bass	Muscle	388	513	901	nd	571	527	1100	–	1.083
			Muscle	41.4	58.3	99.7	nd	61.0	59.9	121	–	1.018
		Sea bream	Muscle	36.2	55.1	91.3	nd	53.3	56.5	110	–	0.942
		Mackerel	Muscle	22.0	29.0	51.0	nd	32.4	29.8	62.1	–	1.088
		Largehead hairtail	Muscle	19.2	25.2	44.4	nd	28.2	25.9	54.1	–	1.087
		Black scraper	Muscle	6.0	8.5	14.5	nd	8.8	8.7	17.5	–	1.010
		Amberjack	Muscle	4.4	7.8	12.2	nd	6.5	8.0	14.5	–	0.816
		Sardine	Whole	3.8	4.7	8.5	nd	5.6	4.8	10.4	–	1.166
	Bottom fish	Japanese rockfish	Muscle	753	1090	1840	nd	1110	1120	2230	–	0.991
		Fat greenling	Muscle	566	818	1380	nd	833	840	1670	–	0.992
			Muscle	478	622	1100	nd	704	638	1400	–	1.102
			Muscle	404	578	982	nd	595	593	1190	–	1.003
			Muscle	317	447	764	nd	467	459	926	–	1.016
			Muscle	282	406	688	nd	415	417	832	–	0.996
			Muscle	172	234	406	nd	253	240	493	–	1.054
			Muscle	53.5	79.0	133	nd	78.8	81.1	160	–	0.971
		Bastard halibut	Muscle	211	301	512	nd	311	309	619	–	1.006
			Muscle	134	184	317	nd	197	189	385	–	1.042
			Muscle	83.7	115	198	nd	123	118	241	–	1.047
			Muscle	64.3	97.0	161	nd	94.7	99.6	194	–	0.950
			Muscle	23.1	33.4	56.6	nd	34.0	34.3	68.4	–	0.991
		Batoidea	Muscle	133	220	378	nd	225	232	457		0.968
			Muscle	98.1	146	245	nd	144	150	295		0.960

		Gurnard		66.7	99.1	166	nd	98.2	102	200	–	0.965
			Muscle	27.0	45.7	72.7	nd	39.7	47.0	86.7	–	0.846
		L. microptera	Muscle	11.4	15.1	26.5	nd	16.8	15.5	32.3	–	1.085
		Stone flounder	Muscle	9.5	13.8	23.3	nd	14.0	14.1	28.2	–	0.993
			Muscle	8.4	13.1	21.5	nd	12.4	13.5	25.9	–	0.922
		Monkfish	Muscle	5.1	6.5	11.6	nd	7.5	6.7	14.2	–	1.123
	Ascidiacea	Ascidian	Muscle	7.7	10.2	17.9	18.7	11.3	10.5	21.8	32.4	1.072
								Average				1.001 ± 0.092

Radioactivity concentrations of the organs are shown in Table 8.8
[a]Decay-corrected value to March 16, 2011

Fig. 8.1 $^{134}Cs/^{137}Cs$ activity ratio of the muscles of fishes caught in 10–20 km east from the FDNPP in December 2011. Radioactivities of ^{134}Cs and ^{137}Cs were decay-corrected to March 16, 2011. ^{110m}Ag detected only in ascidian

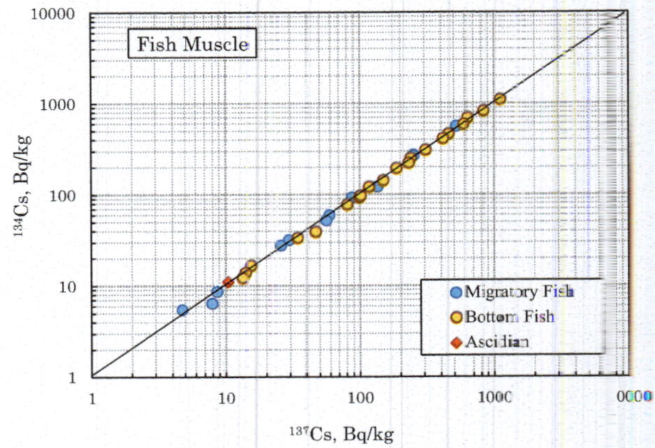

samples. For a small fish, the radioactivity was measured using the whole body. Shellfishes were caught from the coastal beaches around Tokyo Bay, and the fishers collected shellfishes inhabiting the deep sea. Several samples were purchased on the market. For radioactivity measurements, the shells of shellfish were removed, and the tissue was measured.

A monitoring survey was conducted on the dynamics of radioactive contamination of fishes in the radioactively contaminated lakes in Kejonuma, Osaki City, Miyagi Prefecture. No lakes or marshes could be found in Tokyo or the Tokyo metropolitan area to perform such a survey. Kejonuma pond is a small dammed lake with an area of 0.65 km^2 and a maximum depth of approximately 6 m, located approximately 120 km north of the FDNPP. The catchment area is approximately 10 km^2 and is mostly residential area surrounded by agricultural lands. Radioactive contamination from the FDNPP accident in this watershed was almost the same level as in central Tokyo. The fishes were caught between spring and autumn because the pond freezes in winter. Members of the NPO Ecopal-Kejonuma captured fishes, and the fish age was estimated by fish length for each fish species. The fishes were frozen immediately after being caught and transported to the author's laboratory. After thawing, the fish were dissected into muscle, skin, and internal organs, and their radioactivity concentrations were measured by gamma-ray spectrometry.

8.3 Summary of Radioactive Contamination in Offshore Fishes Caused by the FDNPP Disaster

Eight and a half years have passed since the FDNPP accident caused by the Great East Japan Earthquake that occurred on March 11, 2011. This manuscript is written the spring of 2020. A significant amount of radioactive material was released into the environment due to the melting and explosion of nuclear reactors that occurred between March 12 and 15 immediately after the earthquake. As a result, large amounts of radioactive material were deposited on land and sea around the FDNPP. A significant feature of environmental radioactive contamination due to the FDNPP accident is that radioactive materials leaked into the groundwater-rich subsurface of the nuclear power plant due to the meltdown of nuclear fuel and the destruction of containment vessels. The FDNPP is located on a gently sloping hilly coastal terrace. Therefore, the radioactive materials released in the accident not only polluted Japan's land and oceans via the atmosphere but also caused severe marine pollution due to the outflow of groundwater beneath the nuclear power plant. This marine radioactive contamination from groundwater continues in the present.

As shown in Fig. 1.3, the seawater at the FDNPP port is still polluted with radioactive cesium. This radioactive contamination of seawater is thought to result mainly from the inflow of contaminated groundwater rather than from the land. Since immediately after the accident, TEPCO has been conducting monitoring surveys on the radioactive contamination of seawater and marine organisms offshore the FDNPP and has published the results [5]. The JFA (Japanese Fisheries Agency) is also monitoring marine pollution offshore the FDNPP [6] Figure 8.1 shows the temporal changes in radioactive cesium concentrations in fishes published by TEPCO and the JFA. This figure shows the temporal variations in $^{134+137}$Cs radioactivity in the fishes that inhabit areas offshore Fukushima Prefecture. In addition, it shows $^{134+137}$Cs radioactivity of fishes collected in the FDNPP port. In these areas, greenling, flounders, halibut, rockfish, mullet, stingray, sea bream, cod, horse mackerel, sea bass, squid, octopus, *etc.* have been captured and measured. The number of fishes collected off the coast of Fukushima Prefecture by TEPCO and the Fisheries Agency is shown in the caption for Fig. 8.2.

One must be careful when evaluating the data on radioactive contamination in fishes as published by the TEPCO and the JFA and shown in Fig. 8.2 because this figure does not plot the data below the detection limit. Since TEPCO and the JFA measured many fish samples, they likely used a large number of measuring instruments. Therefore, the detection limits of individual measuring instruments are not shown. From the data, the detection limits of the measurement systems used for these measurements were estimated to be several Bq/kg for wet fish samples. In the original data, measurement results below the detection limit were indicated as "nd". Figure 8.2 does not include the "nd" data. The number of data points below the detection limit in a measured data population actively governs the nature of the data population. For example, the TEPCO data measured "within 20 km from the FDNPP" are shown for each year when fishes were captured.

In 2012, sampling was conducted from May to December, and 810 fishes were caught. Of these, 105 fishes were below the detection limit, and the contamination rate (CR) was 87.0%, assuming that the detection limit is the boundary between contamination and non-contamination. The data in Table 8.2 shown below were calculated similarly: 2013 (Jan–Dec, n = 1125, nd = 216, CR: 80.8%), 2014 (Jan–Dec, n = 1109, nd = 420, CR: 62.1%), 2015 (Jan–Dec, n = 1064, nd = 584, CR: 45.1%), 2016 (Jan–Dec, n = 1078, nd = 685, CR: 36.4%), 2017 (Jan–Oct, n = 538, nd = 370, CR: 31.2%). Despite the FDNPP facing the open ocean, within 20 km, the

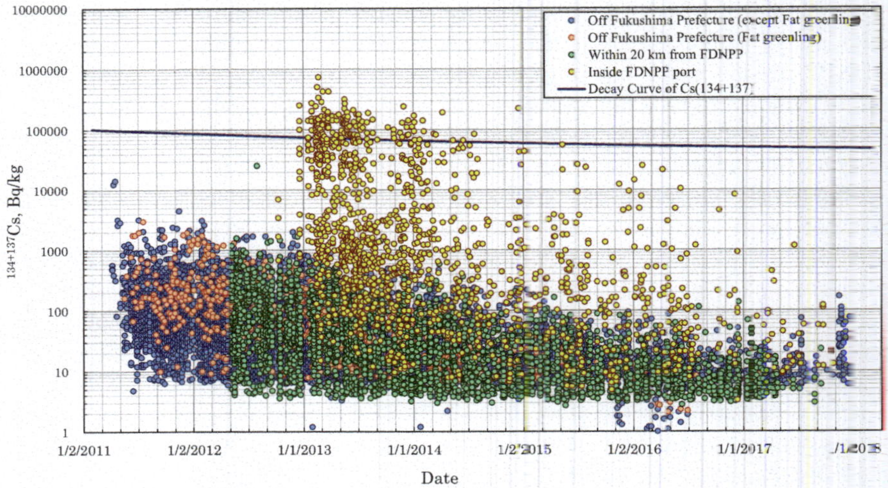

Fig. 8.2 Temporal changes in the $^{134+137}$Cs concentrations in fishes inhabiting the FDNPP coast. Data published by TEPCO [5] and JFA [6] plotted in this figure. The number of fish measured from after the FDNPP accident to the end of 2017 was approximately 5700 within 20 km of the FDNPP, 1400 within the FDNPP port, 47,800 for offshore Fukushima Prefecture except for fat greenlings, and 2070 for fat greenlings offshore Fukushima Prefecture

radioactive cesium contamination rate in fishes exceeded 30% 6 years after the FDNPP accident. Moreover, the fact that the fishes caught in open ocean areas were still found to have radioactive contamination implies that the FDNPP accident was severe.

The radioactive cesium concentration in seawater offshore the FDNPP is steadily decreasing while fish contamination is not smoothly decreasing; this issue is essential for management of public exposure, especially internal exposure. It can be inferred that new contamination of fish is still taking place off the coast of the FDNPP, considering the biological half-life of radioactive cesium in fish.

In the waters, offshore Fukushima Prefecture, the radioactive cesium concentration in fish reached 5000–10,000 Bq/kg immediately after the FDNPP accident. However, the apparent half-life of radioactive cesium contamination in fish was approximately 10 months based on the highest concentration in fish caught. While the concentration of radioactive cesium in seawater in this area has decreased rapidly, the radioactive cesium contamination of fish living there has continued for a long time. Table 8.1 shows the concentrations of radioactive cesium in the muscles of migratory and bottom fishes collected in December 2011 in the same area off the coast of Fukushima as those measured by TEPCO. The radioactive cesium concentrations were almost the same as those published by TEPCO and JFA. Although the radioactive cesium concentrations in the same species varied considerably among individuals, the bottom fish showed higher concentrations. This trend was also observed in the TEPCO and JFA data. This distribution of radioactive cesium

Table 8.2 Temporal changes in the contamination rate (CR) of fishes caught along the FDNPP coast

	TEPCO						JFA (Fisheries Agency, Japan)					
	Inside the FDNPP port			Within 20 km from the FDNPP[a]			Offshore Fukushima Prefecture (except fat greenling)[b]			Offshore Fukushima Prefecture (fat greenling)		
Caught year	Number captured	Number exceeding nd value[c]	CR[d], %	Number captured	Number exceeding nd value[c]	CR[d], %	Number captured	Number exceeding nd value[c]	CR[d], %	Number captured	Number exceeding nd value[c]	CR[d], %
2011	–	–	–	–	–	–	1838	1554	**84.6**	111	108	**97.3**
2012	17	17	**100**	810	705	**87.0**	5298	3058	**57.7**	283	262	**92.6**
2013	791	787	**99.5**	1125	912	**81.1**	7229	2193	**30.3**	329	233	**70.8**
2014	278	271	**97.5**	1108	686	**61.9**	8380	1328	**15.9**	333	144	**43.2**
2015	168	164	**97.6**	1064	513	**48.2**	8231	845	**10.3**	364	67	**18.4**
2016	118	113	**95.8**	1078	388	**36.0**	8336	464	**5.6**	364	25	**6.9**
2017	56	54	**96.4**	538	161	**29.9**	8428	212	**2.5**	283	4	**1.4**
Total	1428	1406	**98.5**	5723	3365	**58.8**	47,740	9654	**20.2**	2067	843	**40.8**

Figure 8.2 plots the data when the radioactive cesium concentrations exceed the detection limit (nd)

[a]In addition to fish, crab and squid included

[b]Includes squid, crab, octopus, shrimp, abalone, sea urchin, sea cucumber, and cultured fish

[c]The nd values of the measurements seem to vary by laboratories and types of equipment. The nd values were estimated to be 5–20 Bq/kg based on the published data

[d]The CR value calculated by counting the number of samples caught and the number exceeding the nd values from the measurement results published by TEPCO [5] and the JFA [6]

suggests that the food chain is circulating radioactive cesium even in open oceans. In this area, ^{110m}Ag was detected in sea squirts. As shown in Sect. 8.4, ^{110m}Ag was also detected in some shells in Tokyo Bay. Because sea squirts accumulate not only vanadium but also heavy metal elements such as copper and zinc, ^{110m}Ag, which is a copper group element, is also selectively concentrated in sea squirts.

Since TEPCO has set fishery rights for waters within 1.5 km from the quay of the FDNPP, no fish samples have been collected by organizations other than TEPCO. That is, the radioactive contamination in fish inside the FDNPP port has been measured only by TEPCO. Figure 8.2 plots the data from fall 2012, one and a half years after the accident, which can be found on the TEPCO website [5]. The highest concentration was observed with fat greenling (*Hexagrammos otakii*) with ^{134}Cs at 260,000 Bq/kg, ^{137}Cs at 480,000 Bq/kg, and $^{134+137}$Cs at 740,000 Bq/kg; this fish was caught on February 21, 2013. Fat greenling is a low-salt coastal fish that is carnivorous and eats polychaetes, crustaceans, mollusks, and small fish. It is not known whether this contaminated fish was exposed to radioactive cesium inside the port immediately after the accident or entered the harbor from the open sea after the accident and became contaminated. As shown in Fig. 1.3 (Reference 34 in Chap. 1), the concentration of radioactive cesium in seawater inside the port in 2013 was approximately 20–100 Bq/L, so it is estimated that this fish has the capacity to concentrate radioactive cesium. However, its apparent concentration factor is approximately 7400–37,000, which is enormous compared to the commonly accepted concentration factor of radioactive cesium for marine fish of approximately 100 [7, 8]. Compared to freshwater fishes, which drink very little water, marine fishes drink slightly more seawater, but the volume is only a few mL/kg h [9]. Thus, it is unlikely that contaminated fish directly took in radioactive cesium from seawater and concentrated it in the body. One hypothesis suggests that highly contaminated groundwater leached from the seafloor offshore the FDNPP may have contaminated the seafloor ecosystem, which may lead to high contaminations in fish via the food chain.

The variations in radioactive cesium concentrations in marine life have been reported in many papers before and after the FDNPP accident (ex. [10–15]). Interesting papers on the global diffusion of radionuclides by migratory fish have also been published [16–18]. However, this topic departs from the purpose of this book and is not discussed here.

8.4 Radioactive Contamination of Fishes Inhabiting Tokyo Bay and Its Inflowing Rivers

Radioactive nuclides released during the FDNPP accident were transported into the Tokyo metropolitan area on March 16 and 22, and they were washed out by rainfall and radioactively contaminated the ground surface [19–24]. The author analyzed the radioactive contamination in the soils of the Tokyo metropolitan area due to the

FDNPP accident [25]. According to the results, 5–15% of radioactive cesium deposited in the Tokyo metropolitan area during the 5 years after the accident may have flowed into Tokyo Bay (Chap. 5). On the other hand, spatiotemporal analysis of the radioactive cesium contamination of the Tokyo Bay sediments was also conducted [26] (Chap. 6). This analysis revealed that approximately 20% of radioactive cesium deposited in a highly contaminated area in eastern Tokyo and northern Chiba Prefecture flowed out of the Edogawa River and was deposited in the Old Edogawa River estuary in the inner part of Tokyo Bay. The estuary of the Edogawa River is the few remaining breeding areas for fishes in the Tokyo Bay ecosystem. It is also a source of fresh seafood called "Edo-Mae", which means fish caught along the margins of Tokyo Bay. Hence, there is a high possibility that the FDNPP accident is causing the radioactive contamination of fish inhabiting Tokyo Bay. Of course, radioactive cesium derived from the FDNPP continues to flow into Tokyo Bay [26]. In this section, the concentrations of radionuclides in fish collected from Tokyo Bay and the Old Edogawa, and Edogawa rivers since the FDNPP accident were measured, and the current status of and fluctuations in the radioactive contamination in fish were investigated.

8.4.1 Radioactivity of ^{110m}Ag, ^{134}Cs, and ^{137}Cs in Fish Inhabiting Tokyo Bay

Radioactive plumes spread over the Tokyo metropolitan area on March 16 and 22, 2011, immediately after the FDNPP accident, and Tokyo Bay and its catchment area suffered large-scale radioactive contamination. Radioactive contamination of soil in the Tokyo metropolitan area and sediments in Tokyo Bay has already been reported [19–26]. However, radioactive contamination of fish and shellfish inhabiting Tokyo Bay and its inflowing rivers has not yet been reported.

Tables 8.3 and 8.4 and Fig. 8.3 show the concentrations of radioactive cesium in fish and shellfish collected in Tokyo Bay from August 2011 to May 2013. Since the radioactivity ratio of ^{134}Cs and ^{137}Cs released in the FDNPP accident was almost 1.0 [27–29], the $^{134+137}$Cs radioactivity in 2013 should decrease almost 0.75 times from immediately after the accident due to the radioactive decay of ^{134}Cs. Throughout the study period, the radioactive cesium contamination in fish inhabiting Tokyo Bay showed low values. A large amount of radioactive cesium flowed into Tokyo Bay from the catchment area. However, since radioactive cesium is strongly adsorbed on sediments in the estuaries, to date, the ecosystem of Tokyo Bay is considered not to have received much contamination by this radionuclide. However, the radioactive cesium in carnivorous bottom fishes such as goby, mullet, stingray, and sea bass showed higher concentrations than those in surface migratory fishes. In particular, gobies, which mainly prey on small fish and shellfish, were highly contaminated, and the activity of $^{134+137}$Cs reached a maximum of approximately 34 Bq/kg.

Table 8.3 Concentrations of radioactive cesium in fishes and shellfishes inhabiting Tokyo Bay

Area	Species	Sampling date	Activity, Bq/kg wet		
			^{134}Cs	^{137}Cs	$^{134-137}Cs$
Tokyo Bay	Yellowfin goby (*Acanthogobius flavimanus*)	2011/10/21	6.5	6.3	13.3
		2011/12/7	4.4	5.3	9.7
		2011/12/18	nd	0.2	0.2
		2011/12/18	14.6	19.2	33.8
		2011/12/22	1.2	1.4	2.6
	Flathead mullet (*Mugil cephalus*)	2011/8/20	1.3	1.5	2.8
		2011/12/20	1.2	1.7	2.9
		2011/12/20	5.0	8.4	13.4
	Stingray (*Batoidea*)	2011/12/7	4.8	5.7	10.5
	Marbled sole (*Pseudopleuronectes yokohamae*)	2012/4/10	2.5	1.1	3.6
		2013/5/10	nd	nd	nd
	Gurnard (*Chelidonichthys spinosus*)	2012/4/10	1.3	2.0	3.3
	Surfperch (*Ditrema temmincki*)	2013/5/31	nd	3.7	3.7
	Japanese rockfish (*Sebastes inermis*)	2013/5/10	0.6	1.0	1.6
		2013/5/10	nd	nd	nd
	Japanese sea bass (*Lateolabrax japonicus*)	2011/12/22	2.1	3.3	5.4
		2011/12/22	1.5	3.7	5.2
		2012/11/25	1.2	1.7	2.9
		2012/11/25	nd	nd	nd
		2012/11/25	nd	2.3	2.3
		2013/5/10	0.7	1.3	2.0
		2013/5/12	nd	nd	nd
	Gafftopsail goby (*Myersina filifer*)	2012/6/27	nd	nd	nd
	Japanese whiting (*Sillago japonica*)	2012/6/28	nd	0.7	0.7
		2013/5/31	nd	nd	nd
	Big-eyed flathead (*Suggrundus meerdervoortii*)	2012/6/27	nd	nd	nd
		2013/5/31	nd	nd	nd
	Red barracuda (*Sphyraena pinguis*)	2011/10/24	2.3	3.2	5.5
		2013/5/10	nd	nd	nd
		2013/5/10	nd	0.3	0.3
	Chub mackerel (*Scomber japonicus*)	2011/10/23	0.2	0.5	0.7
		2013/5/10	nd	0.1	0.1
	Spotnape ponyfish (*Nuchequula nuchalis*)	2012/6/28	nd	nd	nd
	Japanese sardine (*Sardinops melanostictus*)	2012/4/10	nd	nd	nd
	Japanese jack mackerel (*Trachurus japonicus*)	2012/9/8	nd	nd	nd
	Japanese anchovy (*Engraulis japonicus*)	2012/9/8	nd	nd	nd
	Swimming crabs (*Portunidae*)	2012/4/2	2.8	2.9	5.7
	Manila clam (*Ruditapes philippinarum*)	2011/10/1	nd	nd	nd
		2012/4/10	nd	nd	nd

(continued)

Table 8.3 (continued)

Area	Species	Sampling date	Activity, Bq/kg wet ^{134}Cs	^{137}Cs	$^{134+137}$Cs
	Hard clam (*Mercenaria mercenaria*)	2011/8/20	nd	nd	nd
		2011/10/30	0.5	1.0	1.5
		2012/4/2	nd	nd	nd
	Mirugai[a] (*Tresus keenae*)	2011/10/1	nd	nd	nd
		2011/10/1	0.5	0.7	1.2
	Pen shell[a] (*Atrina pectinata*)	2011/10/1	nd	nd	nd

nd: not detected (<0.6 Bq/kg)
[a] ^{110m}Ag was detected with some shellfish organs (see Table 8.9)

8.4.2 Radioactivity of ^{134}Cs and ^{137}Cs in Fish Inhabiting the Edogawa River

The Edogawa catchment area located in eastern Tokyo, northern Chiba Prefecture, is the region that received the highest radioactive contamination in the Tokyo metropolitan area due to the FDNPP accident. Previous studies have revealed that radioactive cesium flows into the Edogawa River due to rainfall from this highly contaminated zone. Hence, there is a high possibility that ecosystems such as fish habitats in the Edogawa River are highly contaminated with radioactive cesium.

Ten eels (*Anguilla japonica*) were caught from March 2013 to October 2016 in the Edogawa River. The eating habits of eels change from benthic organisms such as earthworms and ragworms to shrimp and fish as they grow, and eels are at the top of the food chain system in the Edogawa River. Additionally, freshwater prawns (referred to here as "shrimp"), which are the predominant prey of eels, were collected. Radioactivities were measured on the edible muscles of eels and the whole body for shrimp. The results are shown in Table 8.4 and Fig. 8.3. Fishes captured in the Edogawa River by 2014 were contaminated with higher concentrations of radioactive cesium than those in Tokyo Bay. This result is consistent with the time when the Edogawa River water system was the most contaminated (Chap. 6). The cesium concentrations in eels are 3–5 times higher than those in shrimp prey, which indicates that the accumulation of radioactive cesium by the food chain is proceeding in the Edogawa River ecosystem.

8.4.3 Relationship Between Radioactive Cesium Contamination in Eels and Body Weight

As shown in Fig. 8.4, a clear correlation was found between the $^{134+137}$Cs activity in eels and the body weight. This correlation indicates that radioactive cesium

Table 8.4 Concentrations of radioactive cesium in fishes inhabiting the Edogawa River

Area	Species	Sampling date	Activity, Bq/kg wet		
			^{134}Cs	^{137}Cs	$^{134-137}$Cs
Edogawa River	Japanese eel (*Anguilla japonica*)	2013/3/9	54.5	93.0	***148***
		2013/4/28	33.2	69.3	***103***
		2013/4/28	29.1	63.3	97.4
		2013/4/29	44.7	84.9	***130***
		2013/5/4	35.6	65.9	***103***
		2014/5/4	17.0	47.4	64.4
		2014/8/19	4.5	11.7	16.2
		2015/5/5	nd	nd	nd
		2016/10/2	0.9	5.4	6.3
		2016/10/2	1.2	5.2	6.4
	Freshwater prawn (*Macrobrachium*)	2012/6/20	10.1	14.0	24.1
		2013/6/6	8.8	18.0	26.8
		2013/6/13	12.3	22.2	34.5
		2014/6/18	3.5	9.5	13.0
		2014/6/22	3.0	8.8	11.8
		2015/6/7	2.8	10.7	13.5
		2017/7/9	0.8	6.0	6.8
	Yellow goby (*Acanthogobius flavimanus*)	2013/6/6	12.8	22.9	35.7
		2013/6/13	14.0	25.3	39.3
		2015/5/5	nd	nd	nd
		2015/5/29	nd	4.8	4.8
		2015/9/22	nd	nd	nd
		2015/9/27	3.2	10.7	13.9
		2016/10/2	1.2	4.8	6.0
		2016/10/2	1.0	4.8	5.8
		2017/7/9	0.5	2.9	3.4
	Channel catfish (*Ictalurus punctatus*)	2014/5/2	7.8	25.1	32.9
	Japanese common catfish (*Silurus asotus*)	2014/8/19	6.9	14.8	21.7
	Japanese sea bass (*Lateolabrax japonicus*)	2015/8/28	1.7	7.9	9.6
		2016/10/2	1.9	11.2	13.1
		2016/10/2	1.7	9.9	11.6
	Japanese sardinella (*Sardinella zunasi*)	2015/8/8	nd	nd	nd
	Japanese barbel (*Hemibarbus barbus*)	2016/9/10	0.9	5.4	6.3

The new regulation values for radioactive contamination of Japanese food came into effect on April 1, 2012 [30]. Common foods containing fish are 100 Bq/kg, infant food and milk are 50 Bq/kg, drinking water is 10 Bq/kg. nd: not detected (<0.6 Bq/kg)

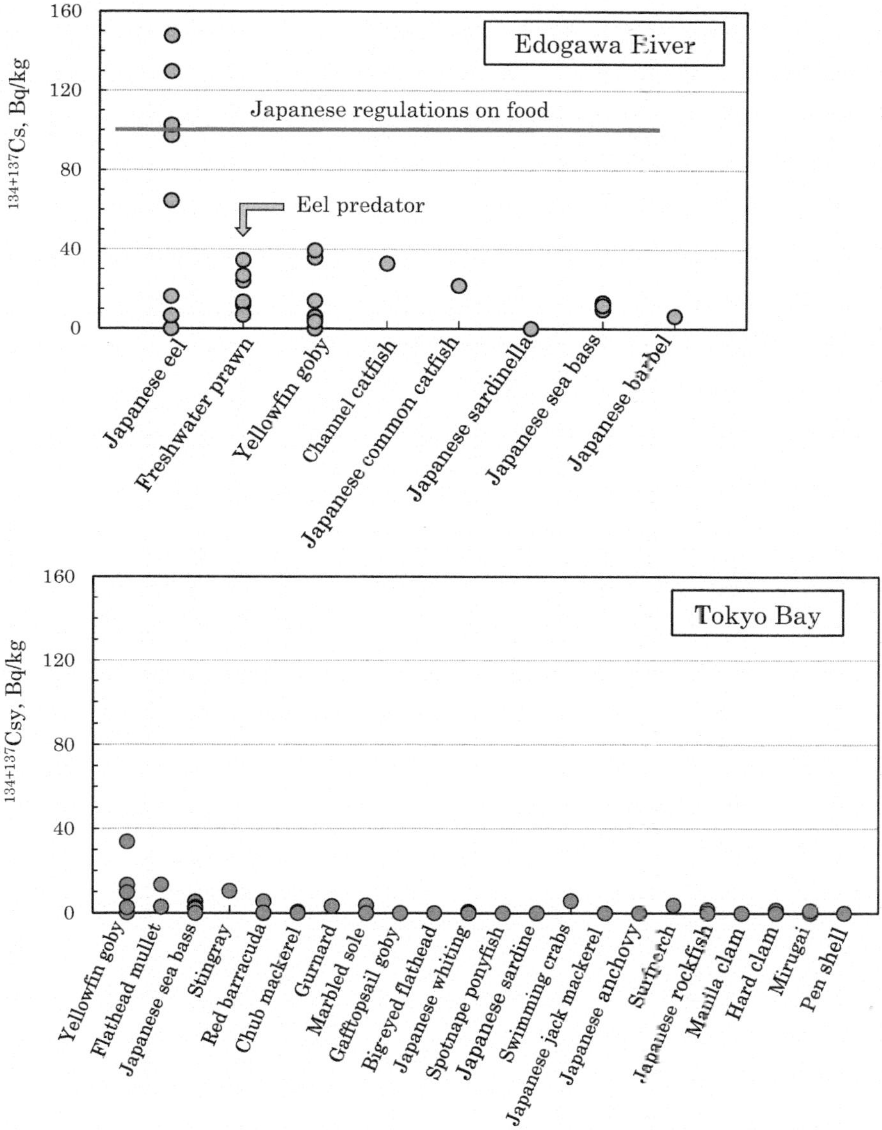

Fig. 8.3 $^{134+137}$Cs concentrations in the muscle of fishes and shellfish inhabiting Tokyo Bay and the Edogawa River. Fishes were captured from August 2011 to July 2017

accumulated in the muscles with the growth of eels. The activity of $^{134+137}$Cs in shrimp gradually decreased according to the radioactive decay of ^{134}Cs. However, in 2012–2017, the activity of $^{134+137}$Cs was 6.8–13.5 Bq/kg, which indicates that a radioactive cesium recycling process through the food chain had been established in

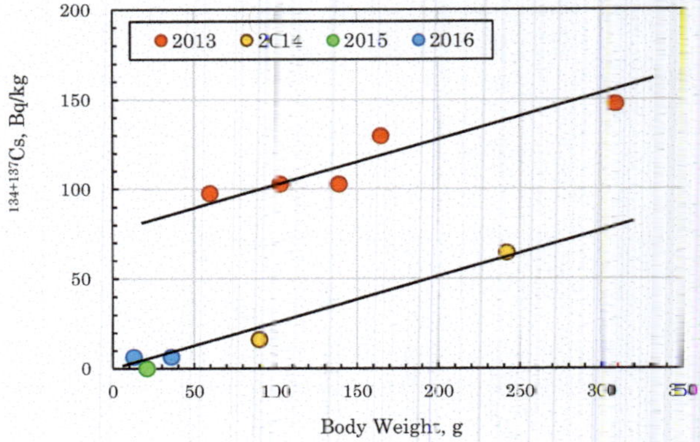

Fig. 8.4 Relationship between the $^{134+137}$Cs concentrations in the muscle of eels and their body weights. Eels were captured from March 2013 to October 2016 in the Edogawa River. In 2011 and 2012, eels could not be caught on the Edogawa River

the Edogawa River ecosystem. It also suggests that radioactive pollution in eels preying such contaminated food is higher than in shrimp and that radioactive cesium may accumulate in the higher species up the food chain. The possibility is discussed in Sect. 8.8.

In 2013, when radioactive cesium contamination was the highest in the Edogawa River [26], the $^{134+137}$Cs activities in four eels exceeded 100 Bq/kg, which is the control value for radioactivity contamination in foods set by the Japanese government [30]. The eels that exceeded the radioactivity regulation value were found in the Edogawa River water system in the Tokyo metropolitan area more than 200 km from the FDNPP, showing the seriousness of the FDNPP accident. The author reports that the radioactive cesium contamination in the sediments of the Old Edogawa River estuary was the worst in 2013 (Chap. 6). This contamination is consistent with the radioactive cesium contamination in eels from the Edogawa River, which showed the highest level in 2013. The concentrations of radioactive cesium in eels collected after 2014 decreased, but the contamination levels of grown eels did not decrease significantly. This change means that the recycling of radioactive cesium through fish in the food chain of the Edogawa ecosystem continues.

Additional Notes The author notified the JFA (Japanese Fisheries Agency) about radioactive contamination in eels inhabiting the Edogawa River according to the analytical results. The JFA also confirmed that the radioactive cesium concentrations in eels exceeded the regulation value for radiation contamination in Japanese foods [30]. As a result, the capture of eels from the Edogawa River ecosystem was prohibited by Tokyo and Chiba prefectures. This radioactive contamination of eels has been widely reported in various media in Japan.

8.5 Five-Year Changes in the Radioactive Contamination in Fishes from the Small Kejonuma Pond, Miyagi Prefecture

8.5.1 Kejonuma Pond and Fishes

Kejonuma is a closed dammed pond with an area of 0.65 km^2 and a depth of 6 m, located 120 km north of the FDNPP. The catchment area with scattered houses is approximately 10 km^2, most of which consists of fields. The radioactive cesium concentration in the soil is several hundred Bq/kg, which is almost the same pollution level as in central Tokyo. Kejonuma pond was designated a Ramsar Convention Wetland in 2008. Mallards and copper pheasants (see Table 9.1), as well as birds of prey such as the white-tailed eagle and sea eagle, also come here. Furthermore, Kejonuma pond is home to many species of native fish as well as exotic species such as black bass (*Micropterus*), bluegill (*Lepomis macrochirus*), and snakehead (*Channa argus*). These carnivorous alien species prey on native fish. To prevent the invasion of these alien species, the NPO Ecopal-Kejonuma has continued to control exotic species. The fishes analyzed in this section were caught during their program.

Between the spring of 2012 and the summer of 2017, 89 black basses, 151 bluegills, and 22 snakeheads were caught. Individual fishes were more than 1 year old, and muscles could be collected from them. Fry younger than 1 year of age also captured, but no muscle could be collected. The radioactivity of the fry was measured using the whole body. Of these fish samples, muscle and ovaries in the same individual could be analyzed in five black basses, 31 bluegills, and three snakeheads. In Kejonuma, 17 bullfrogs (*Lithobates catesbeiana*), tadpoles, legged tadpoles, and eggs were also collected.

Figure 8.5 shows a map of Kejonuma pond. The open circle is the pond water collection point, and the black circle is the sediment collection point. Surface water was collected in the summers of 2014 and 2015, and the radioactive cesium coprecipitated by the AMP method and its radioactive cesium concentration were measured. The results are shown in Table 8.5. The $^{134+137}$Cs concentrations were between approximately 10 and 20 mBq/L but were higher in 2015 than in 2014. The rainfall in the Kejonuma catchment 2 weeks before the collection of pond water was 13.0 mm in 2014 and 129 mm in 2015 [31]. Therefore, the difference in radioactive cesium concentrations in pond water was derived from the difference in inflows from the catchment area. Sediment samples collected in 2014 and their analytical results are shown in Table 8.6. The radioactive cesium concentrations tended to increase from Sites 1 and 2 to Sites 13, 14, and 15. Such fluctuations in the concentrations of radioactive cesium in the sediments indicate that the radioactive cesium flows from the catchment area into Kejonuma pond as suspended particles and then diffuses in the pond water and is deposited with the sediment.

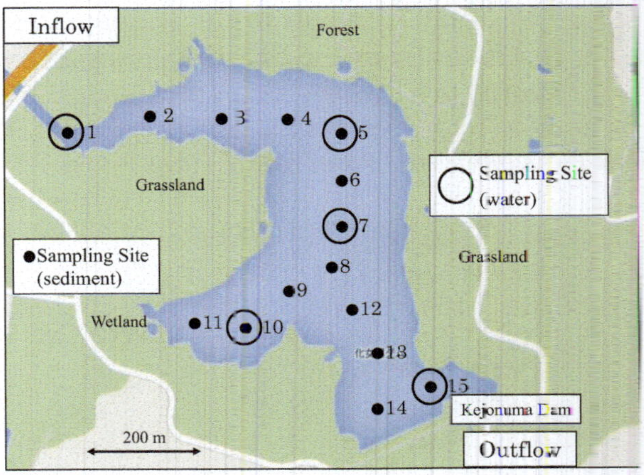

Fig. 8.5 Sampling sites in the Kejonuma pond, Miyagi Prefecture. The area is 0.65 km^2 and the catchment area is approximately 10 km^2

8.5.2 *Radioactive Cesium Concentrations in Exotic Fishes Inhabiting Kejonuma Pond*

From the spring of 2012 to the summer of 2017, the concentrations of $^{134+137}$Cs of exotic species from Kejonuma pond were 19.6–185 Bq/kg for snakehead (n = 22), nd–185 Bq/kg for black bass (n = 89), 1.7–89.8 Bq/kg for bluegill (n = 151), 9.5–133 Bq/kg for bullfrog (n = 17), and 3.4–60.8 Bq/kg for tadpoles (n = 24 sets). Radioactive cesium was not detected in bullfrog eggs (n = 3 sets).

Figure 8.6 shows the time series variations in the $^{134+137}$Cs concentrations in black bass and bluegill by fish age. The figure also shows the radioactive cesium concentrations in the ovaries obtained from these fishes. The changes in radioactive cesium concentrations over time in both black bass and bluegill differ depending on the age of each fish. Compared to fry and older fishes over the age of 4, 1- to 3-year-old fishes show slight temporary decreases in their radioactive cesium concentrations. This difference may reflect the life history of an individual fish. For example, if a 5-year-old black bass was caught in 2013, the fish hatched in 2008 and was exposed to radioactive cesium pollution from the FDNPP accident at the age of 3 years. For the next 2 years, this fish lived in lake water contaminated with radioactive cesium. Such an exposure history differs for each individual fish.

Therefore, the relationship between the birth year and the capture year for individual fish is illustrated in a 3D graph in Fig. 8.7. The dynamics of radioactive cesium contamination in the black bass and bluegill from Kejonuma pond depend on the exposure to the 2011 FDNPP accident. The fishes that were adults at this time quickly took up radioactive cesium precipitated into Kejonuma pond immediately after the accident. After this time, the radioactive cesium in their bodies was released slowly, and its concentration decreased over the years. On the other hand, fishes hatched after the accident in 2011 lived in pond water where the radioactive cesium concentrations remained steady, so their body concentrations did not increase.

Table 8.5 Concentrations of radioactive cesium in the surface water of the Kejonuma pond

Sampling date	Sampling site	Water depth, m	Water activity, mBq/L			$^{134}Cs/^{137}Cs$ ratio	Theoretical ratio[a]
			^{134}Cs	^{137}Cs	$^{134+137}Cs$		
2014/9/10	1	1.5	2.96	8.45	11.4	0.350	0.334
	7	3.5	2.09	6.23	8.32	0.335	
	15	6.0	1.82	6.18	8.00	0.294	
		Average	2.29 ± 0.60	6.95 ± 1.30	9.24 ± 1.88	0.327 ± 0.029	
2015/8/27	1	1.5	4.04	15.9	19.9	0.255	0.249
	5	3.0	3.54	14.5	18.0	0.245	
	7	3.5	3.49	14.7	18.2	0.237	
	10	3.0	3.18	13.6	16.8	0.234	
	15	6.0	3.30	13.6	16.9	0.243	
		Average	3.51 ± 0.33	14.4 ± 1.0	18.0 ± 1.3	0.243 ± 0.008	

Radioactive cesium in the pond water was collected by the AMP coprecipitation method [Reference 26 in Chap. 6]

[a]The radioactivity ratio of $^{134}Cs/^{137}Cs$ ratio immediately after the FNNPP accident is 1.0

Table 8.6 Concentrations of radioactive cesium in the sediments of Kejonuma pond

Sampling date	Sampling site	Water depth, m	Sediment activity, Bq/kg			^{134}Cs/^{137}Cs ratio	Theoretical ratio[a]
			^{134}Cs	^{137}Cs	$^{134+137}$Cs		
2014/9/10	1	1.5	76	222	298	0.343	0.334
	2	1.5	97	287	385	0.339	
	3	2.0	96	307	403	0.314	
	4	2.0	101	312	413	0.324	
	5	3.0	126	330	456	0.383	
	6	3.0	112	342	454	0.327	
	7	3.5	114	333	447	0.343	
	8	4.0	79	244	324	0.324	
	9	3.5	53	165	217	0.319	
	10	3.0	121	353	474	0.341	
	11	2.5	83	268	351	0.310	
	12	4.5	121	356	477	0.340	
	13	5.0	148	432	580	0.342	
	14	4.5	117	354	471	0.330	
	15	6.0	147	452	599	0.325	
		Average	106 ± 26	317 ± 74	423 ± 100	0.334 ± 0.018	

A Birge-Ekman grab sampler was used to collect the bottom sediment up to 5 cm above the surface

[a]The radioactivity ratio of ^{134}Cs/^{137}Cs immediately after the FNNPP accident is 1.0

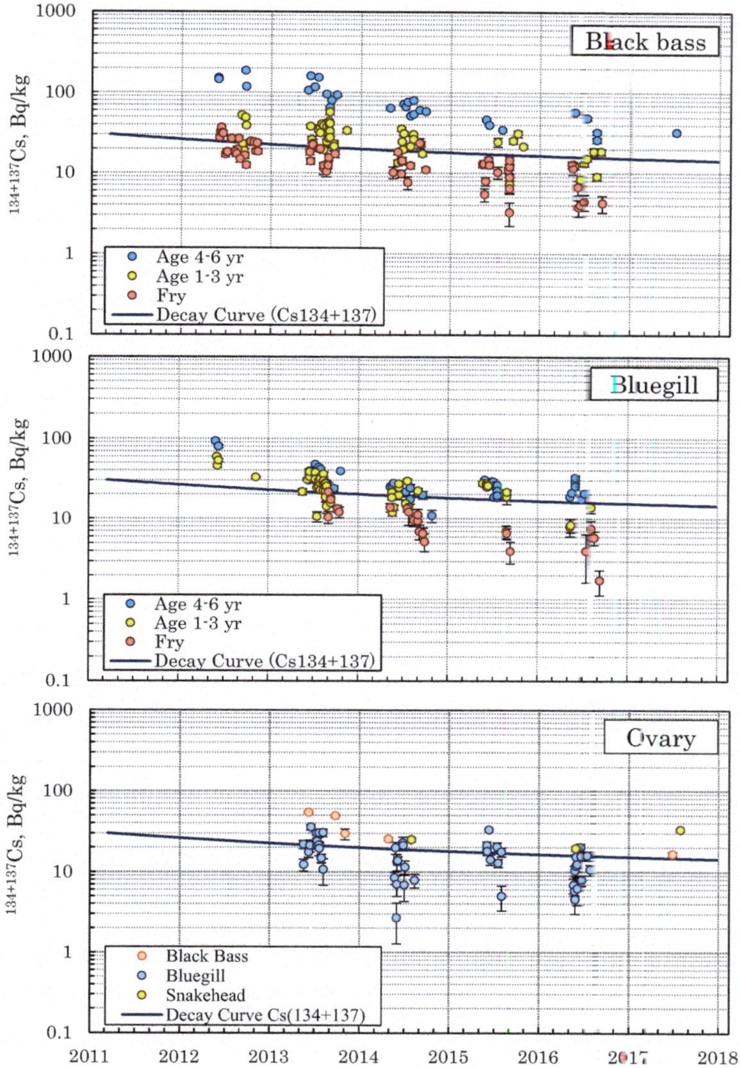

Fig. 8.6 Temporal changes in the $^{134+137}$Cs concentrations of fishes inhabiting the Kejonuma pond, Miyagi Prefecture. The fish ages were estimated from the fish lengths

Instead, the radioactive cesium concentration in the pond water gradually decreased with the decay of ^{134}Cs, so its concentration in the body also decreased with age.

Relatively high concentrations of radioactive cesium were found in adult fish living in Kejonuma pond. Its environment was exposed to large amounts of radioactive material from the FDNPP accident for only a short time. Since black bass and bluegill are carnivorous, the existence of a mechanism of enrichment in living organisms

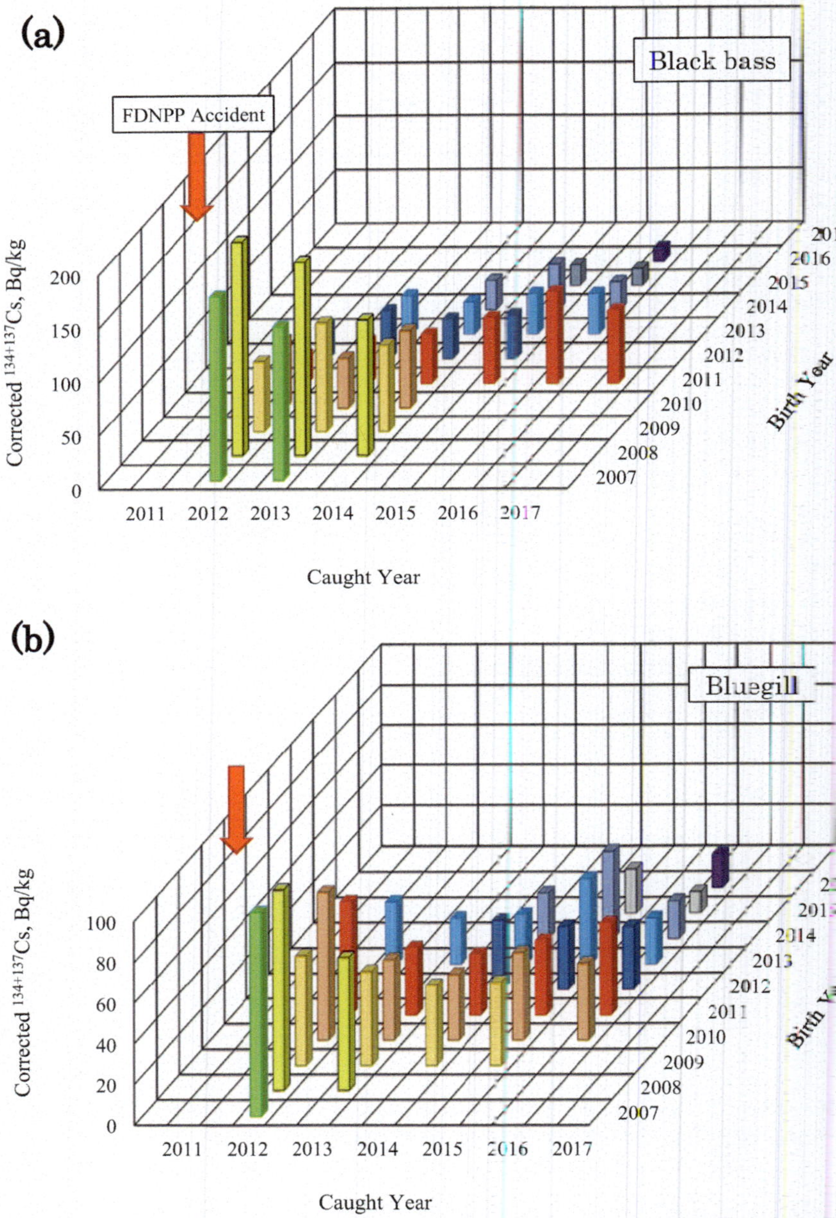

Fig. 8.7 (**a**, **b**) Temporal changes in the $^{134+137}$Cs contaminations of fishes in the Kejonuma pond and its relationship with fish ages. The birth year of fishes was estimated from their ages. The $^{134+137}$Cs activity was decay-corrected to March 16, 2011

through the food chain is expected. No analysis based on measured data was performed on the status of radioactive cesium contamination in Kejonuma pond immediately after the FDNPP accident. After the great earthquake and the accompanying nuclear reactor collapse accident, sufficient scientific observations could not be carried out in most areas of eastern Japan due to the disruption of infrastructure and information networks. However, as shown in Fig. 2.1, it was estimated that a high-concentration radioactive plume arrived over Miyagi Prefecture, where Kejonuma pond is located, on March 22, 2011. On this day, an average 7 m/s wind blew from the FDNPP toward Miyagi Prefecture, and the area around Kejonuma pond had a light rainfall of 2.5 mm/day [31]. Based on the soil concentrations, it was estimated that 10–30 kBq/m^2 of $^{134+137}$Cs precipitated onto the surface of Kejonuma pond at this time. This inventory is consistent with the results of the MEXT airborne monitoring [32]. The radioactive cesium deposited in the catchment area of Kejonuma pond was strongly adsorbed onto clay minerals and litter, so it is unlikely to have migrated and dissolved into the pond water immediately. In other words, fish that lived in Kejonuma pond may have taken up radioactive cesium precipitated directly onto the pond surface immediately after the FDNPP accident.

It is difficult to determine whether enrichment through the food chain or direct ingestion of radioactive cesium deposited on the pond surface was the dominant process of radioactive cesium contamination in black bass and bluegill that inhabited Kejonuma pond at the time of the FDNPP accident. However, as discussed in Sect. 8.8, the author believes that they are likely to have directly ingested radioactive cesium from pond water. In this case, since freshwater fish drink little water [9], soluble radioactive cesium is considered to be taken into the fish body through the gills. Of course, once radioactive cesium was incorporated into the Kejonuma ecosystem, a steady radioactive cesium supply system was established for these fish.

The distribution of radioactive cesium in the body was analyzed using the internal organs of the fish. The results are shown in Sect. 8.7. Moreover, the ovaries of these fish inhabiting Kejonuma pond could be collected. Figure 8.6 also shows temporal changes in the radioactive cesium concentration in the ovaries. Overall, the concentrations in ovaries were lower than those in muscles in adult fish. The details are discussed in Sect. 8.8.

8.6 Estimation of the Effective and Biological Half-Lives of Radioactive Cesium for Fishes in the Aquatic Ecosystem of Kejonuma Pond

Many papers have been published on the behavior of radioactive cesium in fish in rivers and lakes [33–47]. In addition, a biological half-life and a concentration factor have been summarized in various ways [8, 48–53]. The contamination routes from land to the ocean (ex. [54, 55]) and the exposure doses (ex. [55]) have also been evaluated. However, it is difficult to comprehensively compare the analytical results because the factors that affect evaluation of the environmental dynamics of biological samples differ greatly depending on the target environment.

This section presents the results of a simple simulation model of the temporal changes in radioactive cesium contamination of fish living in Kejonuma pond. Analyzing the radioactive contamination of living organisms in such a closed lake is essential for evaluating the dynamics of radioactive contamination in aquatic ecosystems. In general, the mechanisms of ingestion and excretion of radioactive materials in fish are examined in so-called aquarium experiments where fish are bred in aquariums to which radionuclides are added (ex. [57]). However, the environment of an aquarium is very different from the natural environment. Therefore, the results obtained in aquarium experiments do not necessarily reflect the actual situation of radioactive contamination of fish in the natural environment. Many factors affect the uptake and excretion of radioactive material in fish. However, experience and experimental results indicate that radioactive contamination in fish varies exponentially.

8.6.1 A Radionuclide Metabolism Model for Fish in an Enclosed Hydrosphere

Classically, the radioactivity concentration in fish (A) can be described by the half-life of the radionuclide (T_{phy}) and the biological half-life specific to the fish species (T_{bio}), as shown in the following equations. A_0 is the initial concentration of radioactivity in the fish body. The effective half-life (T_{eff}) can easily be described by the physical half-life (T_{phy}) and the biological half-life (T_{bio}).

$$\mathrm{A} = A_0 e^{\frac{-\ln 2 * t}{T_{phy}}} + A_0 e^{\frac{-\ln 2 * t}{T_{bio}}} = A_0 e^{\frac{-\ln 2 * t}{T_{eff}}}$$

$$\frac{1}{T_{eff}} = \frac{1}{T_{phy}} + \frac{1}{T_{bio}}$$

It can be assumed that the radioactivity concentrations in fish inhabiting Kejonuma pond (measured value) fluctuate according to several phases, as described below. The first phase (A_{in}) is a temporarily high concentration of contamination in the early stage of the FDNPP accident. In this phase, excretion can be ignored because excretion may be minimal compared to uptake. The next phase (A_{st}) begins when the initial accident situation has shown a transition to a steady state. At this time, both ingestion into the body and excretion outside the body occur. Over time, the radioactive contamination of the pond converges, and no new uptake of radionuclides into the fish occurs (A'_{st}).

$$Measured\ Value = \mathrm{A}_{in} + \mathrm{A}_{st} + \mathrm{A}'_{st}$$

Initial Phase In the early days of the FDNPP accident, large quantities of radionuclides were released into the environment. Radionuclides are estimated to have been loaded into the Kejonuma pond water by the radioactive plume on March 22 (see Fig. 2.1). In such an environment, the supply rate of radionuclides to fish is significantly higher than the excretion rate. Therefore, the effect of excretion on the estimation of radionuclides accumulated in the fish is ignored here. Such an approximation for the simplicity of the simulation is considered acceptable. In the author's simulations, it was assumed that A_0 was taken up by fish every week in the initial phase of the accident, and the cumulative amount of A_{in} that would accumulate in fish over time was calculated. Of course, the uptake for fish decreases according to the effective half-life (T_{eff1}) of the radionuclide.

$$Measured\ Value = A_{in}$$

$$A_{in} = A_0 e^{\frac{-\ln 2 * t}{T_{eff1}}}$$

Steady Phase 1 In this phase, radionuclides (B_0) that accumulated in the fish during the Initial Phase are excreted according to the effective half-life (T_{eff2}) of Steady Phase 1. On the other hand, fish would take in and accumulate radionuclides from a steady-state environment at a new uptake rate (S_0). This state continues until Steady Phase 2. The boundary accumulation amount between Steady Phase 1 and Steady Phase 2 is B'_0. Therefore, the amounts of radionuclides in fish observed during Steady Phase 1 is the sum of the amount taken up in the Initial Phase and the amount newly taken up in Steady Phase 1.

$$Measured\ Value = A_{in} + A_{st}$$

$$A_{in} = B_0 e^{\frac{-\ln 2 * t}{T_{eff2}}}$$

$$A_{st} = S_0 e^{\frac{-\ln 2 * t}{T_{phy}}} + B_0 e^{\frac{-\ln 2 * t}{T_{eff2}}}$$

Steady Phase 2 Kejonuma pond has not yet reached this phase due to continued radioactive contamination of the surrounding environment. If radioactive contamination converges, the uptake of radionuclides into fish will cease, and radionuclides in the body will be excreted. Therefore, the radioactive contamination in fish will be eliminated. Eventually, the radioactivity concentration in fish decreases as the amounts accumulated in the Initial Phase and Steady Phase 2 are excreted according to the effective half-life.

$$Measured\ Value = A_{in} + A'_{st}$$

$$A_{in} = B_0 e^{\frac{-\ln 2 * t}{T_{eff2}}}$$

$$A'_{st} = B'_0 e^{\frac{-\ln 2 * t}{T_{eff2}}}$$

Simulations of temporal changes in radionuclides in fish were performed, assuming effective half-life (T_{eff}) values of 100, 200, and 500 days, throughout the period. The physical half-life and the biological half-life depending on the fish species and the nuclide are included in the effective half-life. The length of the Initial Phase is 6 months, and 100 Bq/week is the uptake in fish in this phase. The uptake was counted weekly, but the decay according to the effective half-life was calculated daily, and the results were obtained as the cumulative amount of the radionuclide accumulated in the fish. Six months after the pollution had started, the total cumulative uptake amount would be B_0. The next 30 months mark Steady Phase 1. The radionuclide uptake at this time is assumed to have decreased to 10% of that in the Initial Phase, and an uptake of 10 Bq/week is set. During this period, excretion occurs along with uptake, and the amount of decrease is calculated daily. From 36 months onwards, the system is in Steady Phase 2; the uptake of radionuclides into fish has ceased, and excretion alone determines the amount of radionuclide accumulation in fish.

Figure 8.8 shows the results of a simulation using these parameters. Naturally, the longer the effective half-life is, the higher the accumulation of radionuclides in the fish, and the slower the reduction of contamination levels in the fish. There are various parameters in the natural environment other than those assumed here. For example, temporal fluctuations in the amount of intake during the study period cannot be simulated. The effectiveness of radionuclide recycling in a closed pond

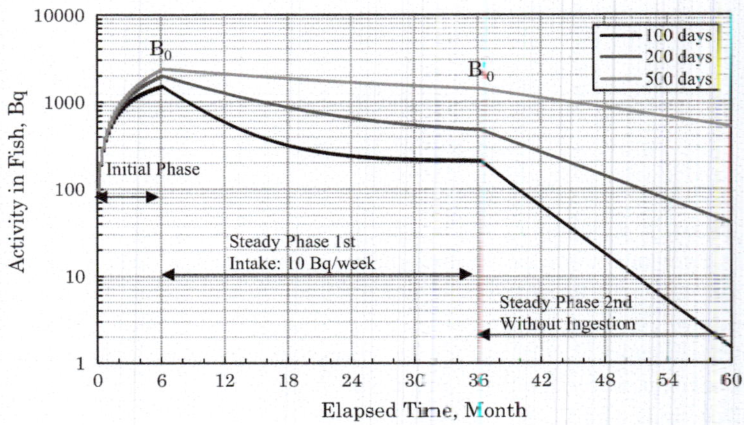

Fig. 8.8 A simple model for the accumulation and excretion of radioactive materials in vivo. The first 6 months of the accident are in the early stages, accumulating 100 Bq once a week. Reductions due to physical and biological decay were calculated daily. The next 30 months is the first steady phase, and it is assumed that the accumulated radionuclides will decrease to 10% of the initial phase. Fluctuations in intake during each phase were ignored in the calculation. The effective half-lives of the initial and steady phases are the same and are calculated at 100, 200, and 500 days. The second steady phase assumes no ingestion of radioactive materials

system is also difficult to quantify. However, by evaluating the results of simple simulations such as those shown here, it is possible to estimate the fluctuations in uptake and the proportion of radionuclides used for recycling. In the case of Kejonuma pond, the surrounding environment is still radioactively contaminated and has not reached Steady Phase 2.

8.6.2 Estimation of Effective Half-Life (T_{eff}) and Biological Half-Life (T_{bio}) of Radioactive Cesium for Fish Inhabiting Kejonuma Pond

Figures 8.9, 8.10, 8.11, 8.12, 8.13, and 8.14 show examples of fitting the above simulation model to the temporal changes in radioactive contamination by ^{134}Cs and ^{137}Cs for fishes inhabiting Kejonuma pond. The results are for fish aged more than 4 years, 1–3 years, 1-year or younger, and ovaries. Such analysis was performed on black bass, bluegills, and snakeheads in Kejonuma pond. In this simulation, A_0, B_0, S_0, T_{eff1}, T_{eff2}, and T_{bio} that best fit the measured values of radioactive cesium contamination in fish were determined, assuming the initial phase period to be 6 months, as in the model. In the case of black bass, the number of ovary samples for fitting was small, but the number of fish bodies was sufficient. Only three samples of the snakehead ovaries were collected in 6 years. Table 8.7 shows the parameters derived by obtaining the best fit for the fish in Kejonuma pond. The amount of uptake (A_0) in the Initial Phase was higher in aged carnivorous fish. This trend suggests that the intake of radioactive cesium by fish is dependent on food habits. In the case of snakeheads, the A_0 of ^{134}Cs and ^{137}Cs was estimated to be 8 Bq/kg week so that the daily intake would be approximately 1 Bq/kg. In the fry of black bass and bluegill, which are prey of snakeheads, the concentrations of ^{134}Cs and ^{137}Cs in the Initial Phase are approximately 10–40 Bq/kg (Fig. 8.6). Therefore, the value of A_0 obtained in this simulation is reasonable.

The intake (B_0) in Steady Phase 1 is 1/20 to 1/30 of A_0 in the Initial Phase. However, even with such a small input, the amount of cesium in the fish body rapidly increases during the first year of Steady Phase 1, and its concentration is maintained (see Figs. 8.9, 8.10, 8.11, 8.12, 8.13, and 8.14). Although the results of this simulation alone cannot be conclusive, they suggest that continued low-level radioactive contamination of the environment could result in high levels of contamination in living organisms.

The ^{137}Cs concentrations in Kejonuma pond water measured in 2014 were 6–9 mBq/L. The levels of ^{137}Cs in the muscles of black bass aged 4 years and older captured in 2014 were 13–65 Bq/kg. If the radioactive contamination of black bass originated from radioactive cesium in the Kejonuma pond water, its concentration factor would be approximately 1430–7020. The information [48, 58, 59] released by MAFF (Ministry of Agriculture, Forestry, and Fisheries of Japan) on its website indicates that the concentration factors of cesium are 400–3000 for

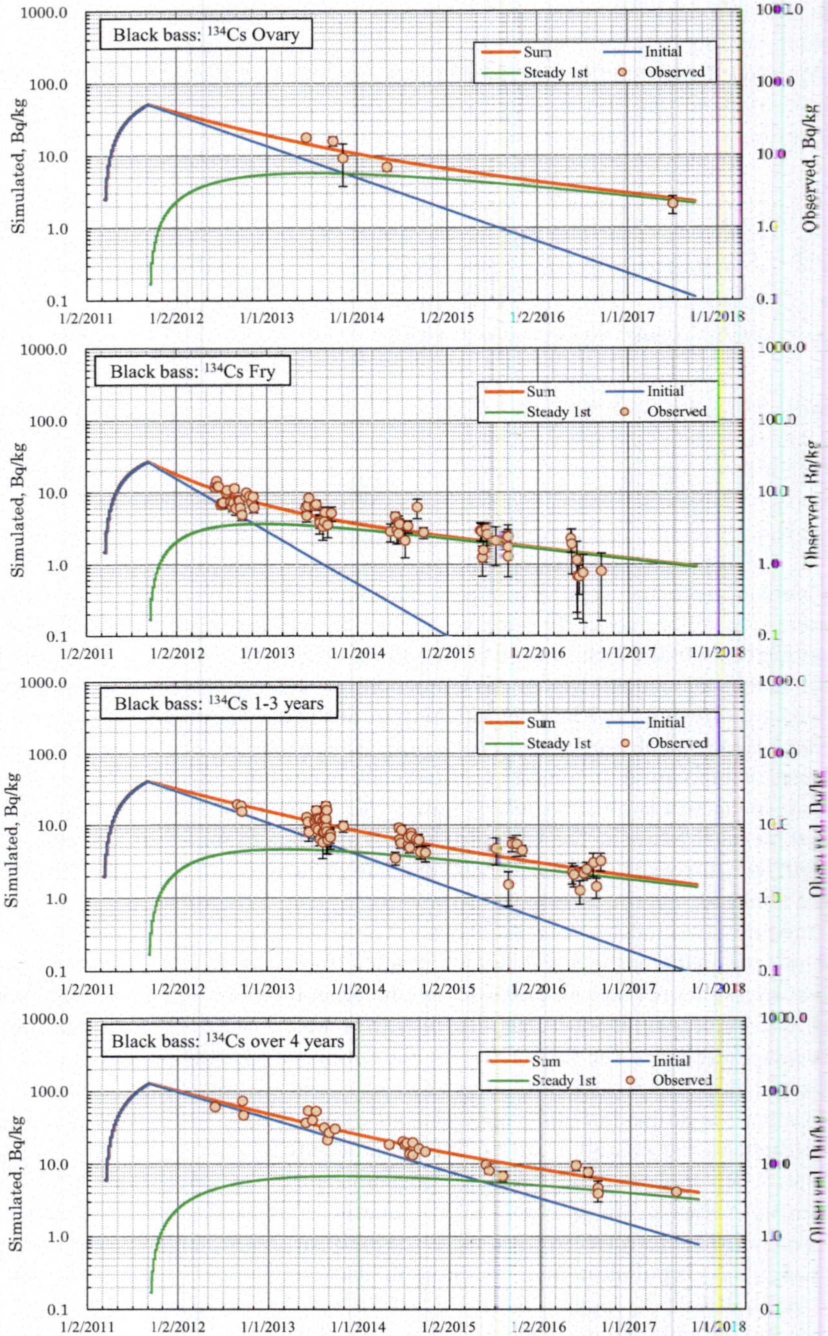

Fig. 8.9 Best fit curves for the concentration fluctuations of ^{134}Cs in the ovary, fry, 1–3 years old, and over 4 years old of black bass inhabiting the Kejonuma pond. The best-fitting parameters are shown in Table 8.7

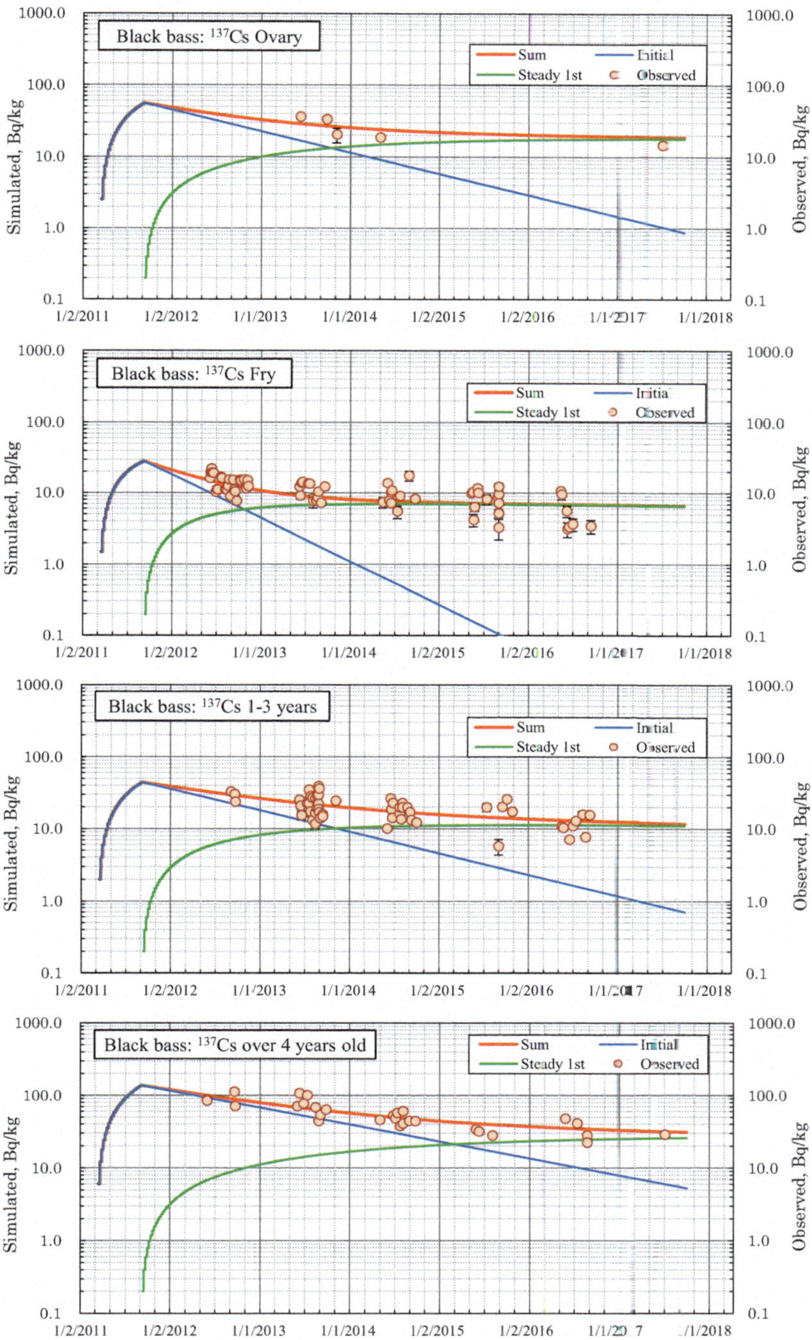

Fig. 8.10 Best fit curves for the concentration fluctuations of ^{137}Cs in the ovary, fry, 1–3 years old, and over 4 years old of black bass inhabiting the Kejonuma pond. The best-fitting parameters are shown in Table 8.7

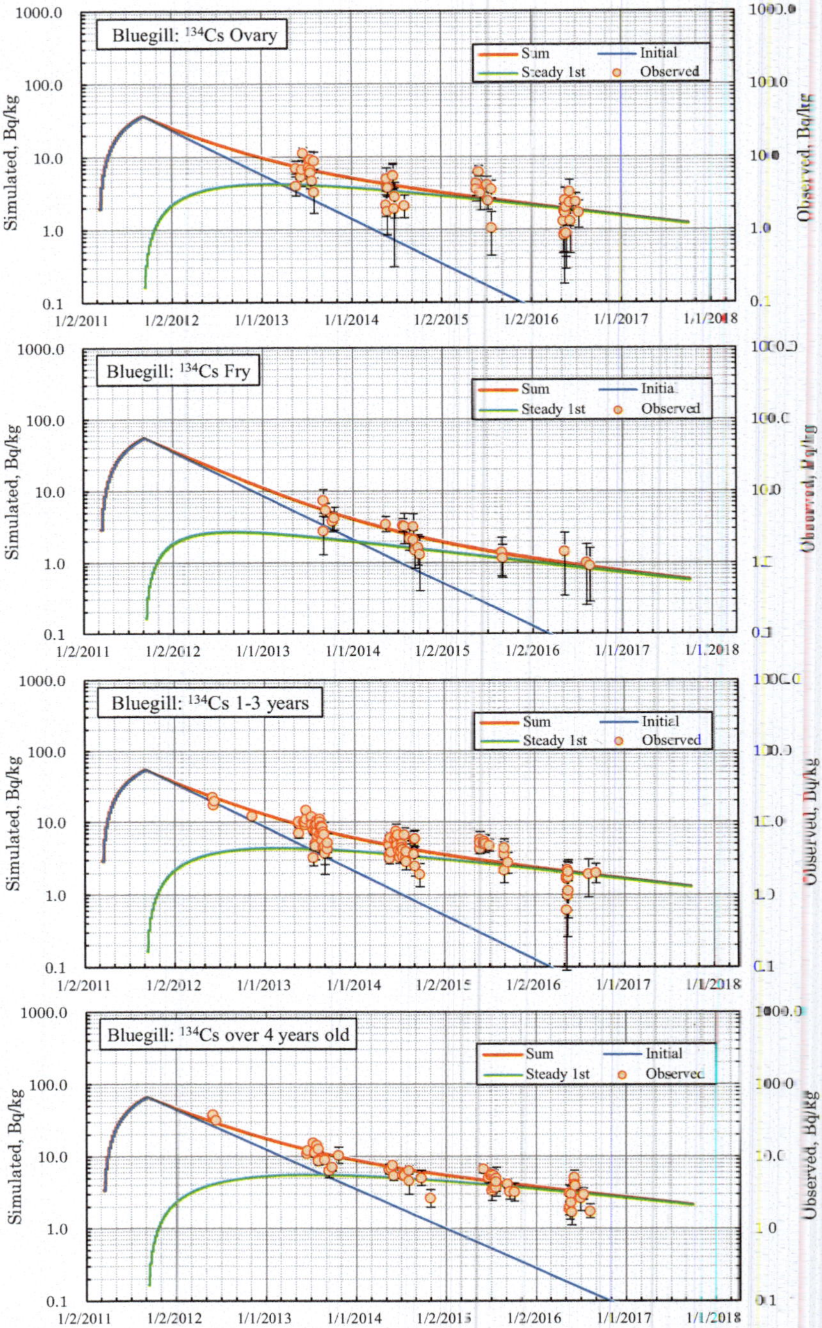

Fig. 8.11 Best fit curves for the concentration fluctuations of ^{134}Cs in the ovary, fry, 1–3 years old, and over 4 years old of bluegill inhabiting the Kejonuma pond. The best-fitting parameters are shown in Table 8.7

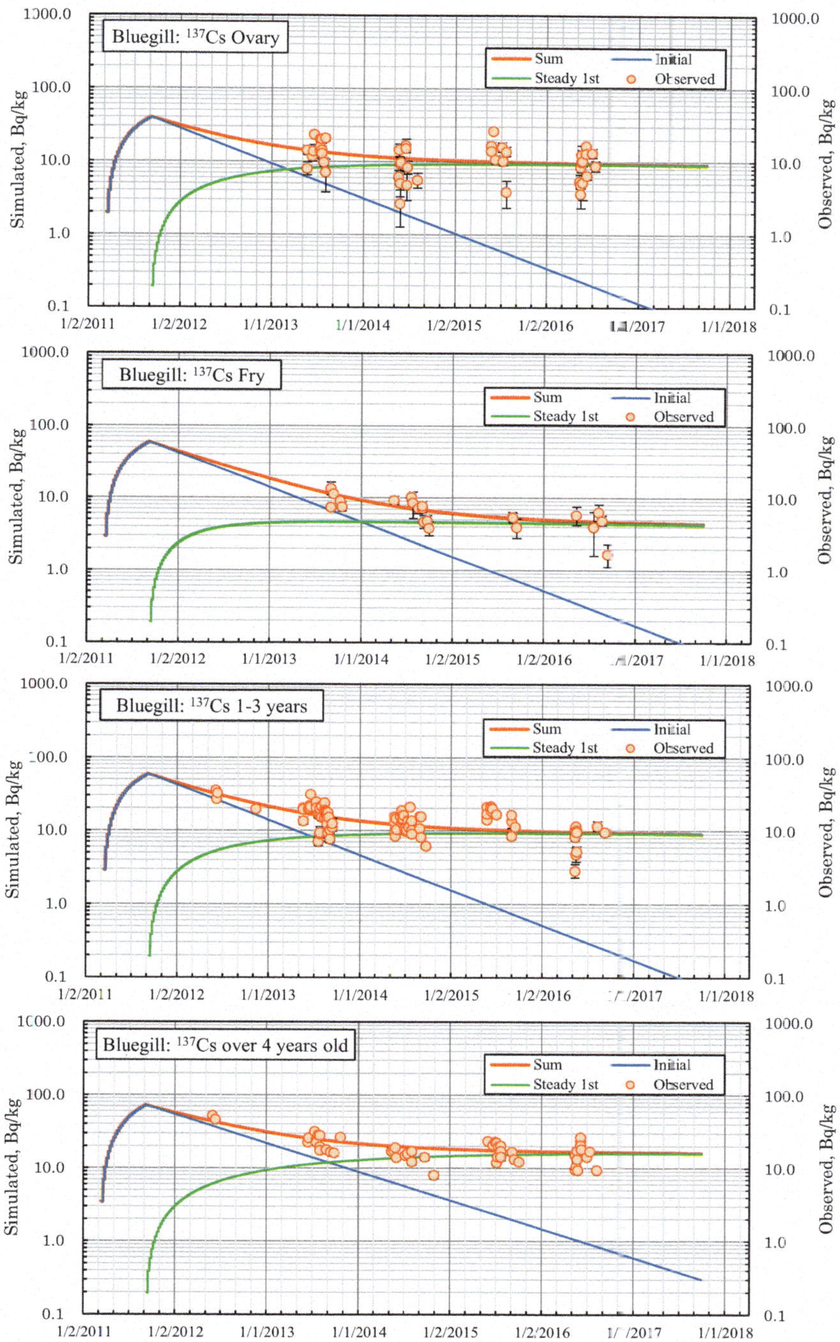

Fig. 8.12 Best fit curves for the concentration fluctuations of ^{137}Cs in the ovary, fry, 1–3 years old, and over 4 years old of bluegill inhabiting the Kejonuma pond. The best-fitting parameters are shown in Table 8.7

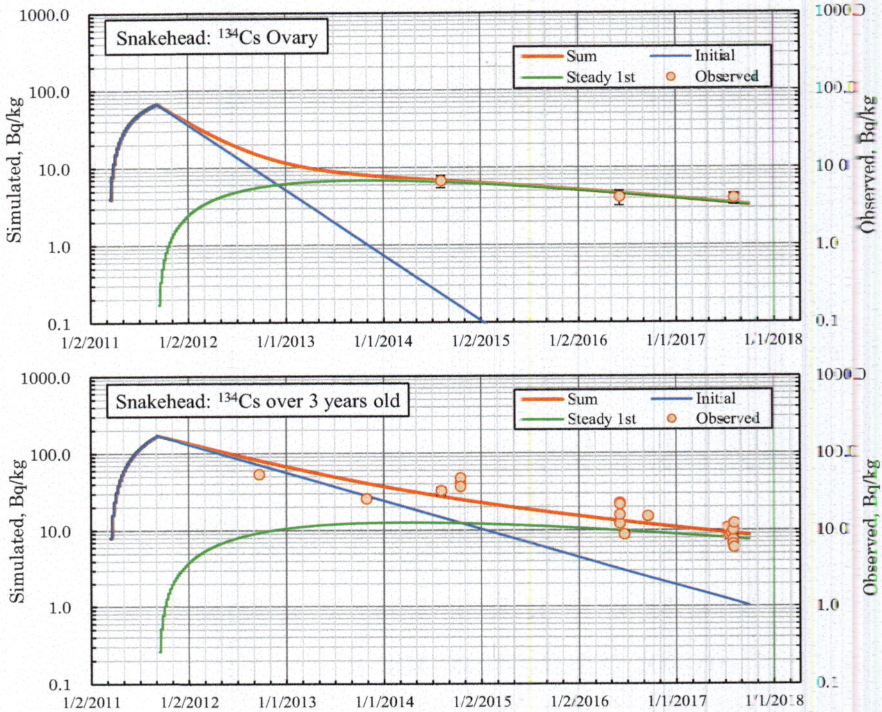

Fig. 8.13 Best fit curves for the concentration fluctuations of ^{134}Cs in the ovary and over 3 years old of snakehead inhabiting the Kejonuma pond. The best-fitting parameters are shown in Table 8.7

freshwater fish and 5–100 for seawater fish. In the same brief, the biological half-life of ^{137}Cs is given as 50–340 days for freshwater fish and 19–216 days for seawater fish. These values indicated by the MAFF were having obtained from an aquarium experiment in which radioactive cesium ions were added to water. The mechanism of radioactive contamination for fish in an actual radioactively contaminated environment is assumed to significantly differ from that in aquarium experiments. In Kejonuma pond, the biological half-life of ^{137}Cs for black bass aged 4 years and older obtained in this simulation is 863 days. According to JFA's brief [58], "^{137}Cs is excreted promptly in the natural world". Perhaps this statement expresses the official view of the Japanese government on radioactive cesium contamination of fish. However, the results of this simulation suggest that in a real aquatic environment, radioactive cesium contamination of fish will last much longer than the MAFF estimates.

The radionuclide concentration factors and biological half-life values for fish indicated by MAFF are values obtained from simple aquarium experiments. That is, the effects of the continuous uptake of radionuclides occurring in the natural environment and the effects of recycling between organisms along the food chain are not considered. Although extremely simplified, the parameter A_{in} introduced in

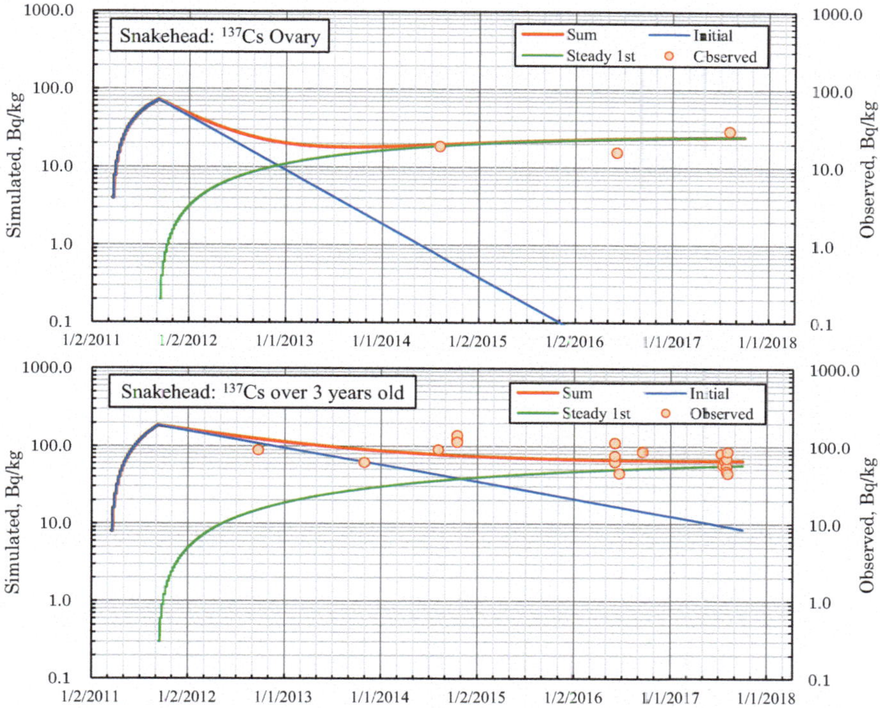

Fig. 8.14 Best fit curves for the concentration fluctuations of ^{137}Cs in the ovary and over 3 years old of snakehead inhabiting the Kejonuma pond. The best-fitting parameters are shown in Table 8.7

Steady Phase 1 of this simulation represents the effect of radionuclide enrichment in the fish body through sustained intake and recycling. Therefore, the biological half-life obtained in this simulation is considered to be close to the actual value in the natural environment. In other words, the above analytical results suggest that a significant source of radioactive cesium contamination in fish inhabiting this aquatic area contaminated by the FDNPP accident is ingestion through food. The possibility of the uptake of ionic radioactive cesium into fish through the gills, which has been noted in the past, is extremely low in the natural environment.

8.6.3 Concentration Factor and Biological Half-Life of Fish Inhabiting Kejonuma Pond

Many studies have been conducted on the incorporation of radionuclides into living bodies since the 1960s. In IAEA Safety Series No. 57 (1982) and Technical Report Series No. 472 (2010), which summarize these studies, the recommended values for

Table 8.7 Best fit parameters for temporal fluctuation of radioactive cesium concentrations of fishes and frogs inhabiting the Kejonuma pond, Miyagi Prefecture

			Initial phase (first 6 months)				Boundary point between initial phase and steady phase first			Steady phase first (next 78 months)											2014 Values	
			Intake (A_o), Bq/kg week		T_{eff1}, day		Conc. (B_o), Bq/kg		Ratio[a, b]	Intake (S_o)[b], Bq/kg week		T_{eff2}, day		**T_{bio}, day**			T_{bio}/T_{eff2} ratio		T_{bio}/T_{phy} ratio		Average, Bq/kg	**Concentration Factor[c]**
Species	Age	n	^{134}Cs	^{137}Cs	^{134}Cs	^{137}Cs	^{134}Cs	^{137}Cs	$^{134}Cs/^{137}Cs$	^{134}Cs	^{137}Cs	^{134}Cs	^{137}Cs	**^{134}Cs**	**^{137}Cs**	**$^{134}Cs/^{137}Cs$**	^{134}Cs	^{137}Cs	^{134}Cs	^{137}Cs	$^{134+137}Cs$	
Snakehead	>3	22	8	8	300	500	170	180	0.94	0.26	0.3	550	1600	**2050**	**1870**	1.10	3.73	1.17	2.73	0.170	154	**16,800**
(Carnivorous)	Ovary	3	4	4	130	160	66	72	0.92	0.17	0.2	400	750	**855**	**805**	1.06	2.14	1.07	1.14	0.073	25.0	**2720**
Black Bass	>4	29	6	6	300	470	130	140	0.93	0.17	0.2	400	800	**855**	**863**	0.99	2.14	1.08	1.14	0.078	64.6	**7020**
(Carnivorous)	1 < age < 3	60	2	2	250	370	41	44	0.93	0.17	0.2	220	300	**311**	**308**	1.01	1.41	1.03	0.41	0.028	23.6	**2570**
	<1 (Fry)	68[d]	1.5	1.5	150	180	26	28	0.94	0.17	0.2	160	180	**203**	**183**	1.11	1.27	1.02	0.27	0.017	13.2[e]	**1430**
	Ovary	5	2.5	2.5	250	370	51	55	0.93	0.17	0.2	300	500	**499**	**524**	0.95	1.66	1.05	0.66	0.048	25.4	**2760**
Bluegill	>4	54	3.5	3.5	200	280	67	73	0.92	0.17	0.2	300	450	**499**	**469**	1.06	1.66	1.04	0.66	0.043	21.1	**2290**
(Omnivorous)	1 < age < 3	97	3	3	180	230	56	60	0.93	0.17	0.2	210	250	**291**	**256**	1.14	1.39	1.02	0.39	0.023	18.0	**1960**
	<1 (Fry)	22[d]	3	3	180	230	56	60	0.93	0.17	0.2	110	120	**129**	**121**	1.07	1.17	1.01	0.17	0.011	9.7[e]	**1050**
	Ovary	46	2	2	180	230	37	40	0.93	0.17	0.2	200	250	**273**	**256**	1.07	1.37	1.02	0.36	0.023	11.7	**1270**
Bullfrog (Insectivorous)		17	10	10	190	250	190	200	0.95	0.17	0.2	150	160	**187**	**162**	1.15	1.25	1.01	0.25	0.015	25.7	**2790**
Tadpoles (with legs)[f]		13[d]	–	–	–	–	18	22	0.82	–	–	–	–	–	–	–	–	–	–	–	10.8[e, g]	**600[h]**
Tadpoles (without legs)[f]		11[d]	–	–	–	–	48	50	0.96	–	–	–	–	–	–	–	–	–	–	–	16.5[e, g]	**920[h]**

[a]Calculated assuming that the $^{134}Cs/^{137}Cs$ ratio at the immediately after the FDNPP accident is 1.0

[b]The theoretical $^{134}Cs/^{137}Cs$ ratio is 0.855. Physical half-lives (T_{phy}) is 752 days for ^{134}Cs and 10,990 days for ^{137}Cs

[c]The $^{134+137}Cs$ concentration in pond water in 2014 was 0.0092 ± 0.0019 Bq/L

[d]Each sample collected 10–20 g of the whole body

[e]Fry and tadpole are the whole bodies, other samples except for ovaries are muscles

[f]Tadpoles have a short lifespan and cannot suitable for this method

[g]Calculated from the 2015 values

[h]The $^{134+137}Cs$ concentration in 2015 was 0.0180 ± 0.0013 Bq/L

the concentration factors of radioactive cesium in fish are 100 for seawater fish (described in IAEA-TECDOC-211; a value of 50 is also shown) and 140–15,000 for freshwater fish muscle. In the RWMC (Radioactive Waste Management Center in Japan), EPM3 (Concentration Factors of Radionuclides in Freshwater Organisms: 1992) recommends 800 as the concentration factor of radioactive cesium for freshwater fish. Additionally, ORNL (Oak Ridge National Laboratory) ESDP No. 783 [49] suggests that the concentration factor of radioactive cesium in freshwater fish depends on the food habits of the fish, the concentration of potassium in the habitat, and the concentration of suspended substances. In an average freshwater environment with a potassium concentration of approximately 1 ppm, the concentration factor can be 1000–15,000. The concentration factors of radioactive cesium for fish in Kejonuma pond range from 1050 to 16,800, which is close to that from ORNL ESDP No. 783 [49]. The MAFF of Japan recommends a concentration factor of radioactive cesium of 400–3000 for freshwater fish [58]; this value is slightly underestimated.

As shown in Table 8.7, the concentration factors of radioactive cesium in black bass and bluegill, which were measured by dividing the fish by age into fry, 1–3 years, and 4 years or older, increase with the age of the fish. The increase in the concentration factor in the carnivorous fish black bass compared to the omnivorous fish bluegill may be due to differences in eating habits among fish species. The fry of black bass eat plankton, such as daphnia and chironomid, but after 1 year of age, they prey on small fish. It is known that the percentage of fish in the diet increases with growth. On the other hand, the bluegill is omnivorous and eats aquatic insects, crustaceans, small fish, fish eggs, and aquatic plants. In the food chain in a closed pond environment, radioactive cesium is highly likely to be sequentially concentrated via fish. Therefore, it is reasonable that cesium has a high concentration factor for fish located at a higher position in the food chain. Fish are a staple food of black bass and snakeheads.

The concentration factor of radioactive cesium for tadpoles is lower than that of the parent bullfrog. It is thought that this change occurs because the metamorphosis of a tadpole to a frog significantly changes eating habits. The radioactive cesium concentration for a frog is discussed in Sect. 8.8.

On the other hand, the biological half-life of radioactive cesium in freshwater fish is poorly known. The biological half-lives of cesium in fish from the aquarium experiments summarized in the RWMC EPM3 are 3.0–599 days for freshwater fish and 17.0–122 days for marine fish. Häsänen et al. suggested 55–350 days for freshwater fish [48]. Japan's MAFF recommends approximately 50 days for seawater fish and 50–340 days for freshwater fish. The estimated biological half-lives in Kejonuma pond differ significantly from the published values. Although there are data for radioactively contaminated fields following the FDNPP accident, they were often measured in transitional situations early in the accident. Under these conditions, it is challenging to estimate the effective half-life of radioactive cesium in a fish due to the large fluctuations in radioactive contamination of the aquatic environment. Naturally, the effective half-life includes the biological half-life, which depends on the metabolism of the organism, and the physical half-life, which depends on the decay of the radionuclide. However, if the radioactive contamination of the aquatic environment is in transition, the radionuclide concentration in the environment that the fish inhabits also fluctuates greatly, and the uptake of radionuclides into the fish changes over time. It is very chancy to introduce this new parameter into the effective half-life and biological half-life estimations based on the measured values in the natural environment. The reason is that the effective half-life and biological half-life obtained from these measurements may not reflect the dynamics of radioactive contamination related to the metabolic mechanism of fish. Therefore, the estimation of the effective and biological half-lives of a radionuclide in fish should be performed after environmental pollution has passed the transition to a steady state.

In the simulations performed here, A_0 and T_{eff1} of the Initial Phase and S_0 and T_{eff2} of the Steady Phase 1 were determined to best fit the respective temporal changes in the concentrations of ^{134}Cs and ^{137}Cs in the fish collected at Kejonuma pond. The T_{bio} of Steady Phase 1 was calculated from the value of T_{eff2}. Table 8.7 shows the obtained best parameters. The results for bullfrogs collected from the same pond are also shown for comparison with fish. The T_{bio} of ^{134}Cs and ^{137}Cs obtained for each fish species show almost the same value, confirming that there is no isotope effect of radioactive cesium associated with fish metabolism in such an ecosystem. The effective half-life T_{eff2} of ^{134}Cs of the Kejonuma pond fish in the Steady Phase 1 were 110–550. It was 120–1600 for ^{137}Cs. The values in ovaries were 200–400 and 250–750 for ^{134}Cs and ^{137}Cs, respectively. The biological half-lives of the radionuclides in fish were 129–2050 and 121–1870 for ^{134}Cs and ^{137}Cs, respectively. Those of the ovaries were 273–855 and 256–805. The biological half-life T_{bio} increased with age in black bass and bluegill. This fluctuation means that the metabolic rate of radioactive cesium decreases with growth, but it may also be affected by changes in eating habits and lifestyle (ecology) due to aging. The T_{bio}/T_{eff2} ratio of ^{137}Cs was almost the same and had a prolonged physical half-life, confirming that the effect of attenuation due to the physical half-life did not affect the effective half-life. Ovarian T_{bio} was slightly lower in all fish species than in adult fish, and the transfer of radioactive cesium from parents to the ovaries was suppressed. The distribution of radioactive cesium between fish parents and children is described in Sect. 8.8.

8.7 Radioactive Cesium Contents in Organs of Fish Inhabiting the Sea Offshore the FDNPP, Tokyo Bay, and Kejonuma Pond

The dynamics of radioactive contamination in fish can be evaluated primarily from the effective half-life of radionuclides and the biological half-life. However, these parameters are affected by fish habitats, physical and chemical species of radionuclides, and temporal changes in contamination levels. Therefore, to elucidate the mechanism of radionuclide uptake and excretion in fish, it is necessary to know the distribution and metabolic processes of radioactive substances in the body. Hence, the distributions of radioactive cesium in the internal organs of fishes collected from the sea offshore the FDNPP, Tokyo Bay, and Kejonuma pond were investigated. Tables 8.8, 8.9, and 8.10 show the analytical results.

Except for the viscera of flathead mullet caught in Tokyo Bay, the visceral concentration factors with respect to the muscle are less than 1.0. These values indicate that radioactive cesium is most concentrated in muscles in both seawater and freshwater fishes. Ions of cesium, which is an alkali element, are excreted outside the body together with potassium ions through gills in marine fish whose osmotic pressure is lower than that of seawater. On the other hand, in freshwater fish with higher osmotic pressure than the freshwater itself, cesium ions are ingested through the gills and accumulate in the body. Gills have less muscle but are rich in blood and blood vessels. However, both seawater and freshwater fishes showed enrichment factors less than 1 for viscera. This difference means that it is more reasonable to think that radioactive cesium intake in both seawater fish and freshwater fish is accomplished orally through food rather than by ionic cesium entering the body through the gills.

The results for Tokyo Bay shown in Table 8.9 are not very reliable, as discussed in Sect. 8.4, because fishes in Tokyo Bay had low levels of radioactive cesium contamination. Therefore, their internal organs also had low radioactive contamination that is close to the detection limit of measurement (<0.6 Bq/kg). The viscera of flathead mullet showed greater contamination than its muscle, but it is highly likely that the measurements detected sludge from Tokyo Bay sediment that had accumulated inside the viscera. The radioactive cesium concentration in the surface layer of the sediment in the sea area where this fish lived was approximately 100 Bq/kg. Except for this viscus, all fishes had the highest levels of radioactive cesium in their muscles. The characteristic radioactive contamination of Tokyo Bay organisms was that ^{110m}Ag detected in the muscles and internal organs of shellfish. The half-life of the radionuclide ^{110m}Ag is 150 days, but the fission yields of ^{235}U thermal and fast neutrons are not very high. However, the fission yield of ^{239}Pu produced in nuclear fuel is several tens of times higher than that of ^{235}U. There is no denying the possibility that ^{110m}Ag is produced by the (n, γ) reaction of ^{109}Ag in the FDNPP reactor. Although ^{110m}Ag was not detected in Tokyo Bay sediments, it was estimated that ^{110m}Ag from the FDNPP precipitated with radioactive cesium in Tokyo Bay, was metabolized by shellfish, and accumulated in their bodies. Many shellfish that bioconcentrate the copper and zinc group elements are well known. It is reasonable that ^{110m}Ag, a copper group element, was concentrated and detected in shellfish living in Tokyo Bay.

Table 8.8 shows the radioactive cesium concentrations in the internal organs of the fishes shown in Table 8.1. Fishes were collected 9 months after the FDNPP accident, and as illustrated in Fig. 8.2, the levels of contamination of these fish were average for marine fish offshore the FDNPP at this time. Although lower than the muscle concentration, the radioactive cesium concentration in the skin was higher than that of the internal organs. It is unknown whether radioactive cesium present in

Table 8.8 Concentrations of radioactive cesium in the organs of fishes caught in 10–20 km east from the FDNPP in December 2011

Sampling date	Habitat	Species	Organs	Detected, Bq/kg wet			Corrected[a], Bq/kg wet				
				^{134}Cs	^{137}Cs	$^{134}+^{137}Cs$	^{134}Cs	^{137}Cs	$^{134}+^{137}Cs$	$^{134}Cs/^{137}Cs$	Concentration factor[b]
December 5–7, 2011	Migratory fish	Japanese sea bass	Muscle	388	513	901	571	527	1100	1.083	1.000
			Skin	188	246	434	276	253	529	1.092	0.481
			Gills	84.6	133	218	125	137	262	0.909	0.238
			Liver	64.5	86.2	151	94.9	88.5	183	1.072	0.167
			Ovary	185	247	432	272	253	526	1.075	0.478
	Bottom fish	Fat greenling	Muscle	172	234	406	253	240	493	1.054	1.000
			Skin	113	165	278	166	169	335	0.984	0.680
			Gills	16.8	25.6	42.4	24.7	26.3	51.0	0.941	0.103
			Liver	50.5	70.1	121	74.3	72.0	146	1.033	0.297
			Ovary	103	140	243	151	144	295	1.051	0.598
		Bastard halibut	Muscle	23.1	33.4	56.6	34.0	34.3	68.4	0.991	1.000
			Skin	19.9	28.7	48.6	29.3	29.5	58.7	0.994	0.859
			Gills	7.2	16.4	23.6	10.6	16.9	27.4	0.625	0.401
			Liver	13.7	19.9	33.7	20.2	20.5	40.7	0.986	0.595
		Monkfish	Muscle	5.1	6.5	11.6	7.5	6.7	14.2	1.123	1.000
			Liver	nd	nd	nd	–	–	–	–	–
			Stomach	4.6	6.8	11.4	6.8	7.0	13.7	0.971	0.968
									Average	0.999 ± 0.116	

The organs were collected from the fishes in Table 8.1
[a]Decay corrected value to March 16, 2011
[b]Activity ratio to the muscle

Table 8.9 Concentrations of ^{110m}Ag, ^{134}Cs, and ^{137}Cs in the organs of fishes caught in Tokyo Bay

				Detected, Bq/kg wet				Corrected[a], Bq/kg wet				Concentration factor[b]	
Sample	Species	Sampling date	Organs	^{134}Cs	^{137}Cs	$^{134+137}Cs$	^{110m}Ag	^{134}Cs	^{137}Cs	$^{134+137}Cs$	^{110m}Ag	$^{134+137}Cs$	^{110m}Ag
Fish	Flathead mullet	2011/8/20	Muscle	1.3	1.5	2.8	nd	1.5	1.5	3.0	–	1.00	–
			Skin	nd	nd	nd	nd	–	–	–	–	–	–
			Gill	1.0	1.3	2.3	nd	1.2	1.3	2.5	–	0.83	–
			Viscera	12.9	14.6	27.5	nd	14.9	14.7	29.6	–	9.85	–
	Red barracuda	2011/10/24	Muscle	2.3	3.2	5.5	nd	2.8	3.2	6.1	–	1.00	–
			Skin	0.7	0.9	1.6	nd	0.9	0.9	1.8	–	0.29	–
			Gill	nd	nd	nd	nd	–	–	–	–	–	–
			Viscera	0.2	0.1	0.3	nd	0.2	0.1	0.3	–	0.06	–
			Bone	1.7	2.2	3.9	nd	2.1	2.2	4.3	–	0.71	–
	Stingray	2011/12/7	Muscle	4.8	5.7	10.5	nd	6.1	5.8	11.9	–	1.00	–
			Skin	4.0	4.7	8.7	nd	5.1	4.8	9.9	–	0.83	–
			Gill	1.2	1.3	2.5	nd	1.5	1.3	2.9	–	0.24	–
			Viscera	1.2	1.2	2.4	nd	1.5	1.2	2.8	–	0.23	–
			Liver	0.2	0.3	0.5	nd	0.3	0.3	0.6	–	0.05	–
	Flathead mullet	2011/12/20	Muscle	4.9	8.4	13.3	nd	6.3	8.5	14.8	–	1.00	–
			Viscera	2.9	3.4	6.3	nd	3.7	3.5	7.2	–	0.48	–
Shellfish	Mirugai	2011/10/1	Adductor Muscle	nd	nd	nd	nd	–	–	–	–	–	
			Incurrent Siphon	nd	nd	nd	1.0	–	–	–	1.7	–	–
	Mirugai	2011/10/1	Adductor Muscle	0.5	0.7	1.2	nd	0.6	0.7	1.3	–	1.00	
			Viscera	0.9	1.0	1.9	1.0	1.1	1.0	2.1	1.7	1.60	–
			Incurrent Siphon	nd	nd	nd	nd	–	–	–	–	–	–
	Pen shell	2011/10/1	Adductor Muscle	nd	nd	nd	0.4	–	–	–	0.7	–	1.00
			Viscera	nd	nd	nd	2.5	–	–	–	4.3	–	6.28

The organs were collected from the fish and shellfish in Fig. 8.3

[a]Corrected-value to March 16, 2011

[b]Activity ratio to the muscle

Table 8.10 Concentrations of radioactive cesium in the organs of snakehead caught in the Kejonuma pond

Species	Sampling date	Age	Part	Detected, Bq/kg wet			Corrected[a], Bq/kg wet			$^{134}Cs/^{137}Cs$	Concentration factor[b]
				^{134}Cs	^{137}Cs	$^{134+137}Cs$	^{134}Cs	^{137}Cs	$^{134+137}Cs$		
Snakehead	2017/8/1	6	Muscle	9.8	69.8	79.6	84.1	80.8	165	1.041	1.000
			Gills	4.5	32.3	36.7	38.2	37.4	75.6	1.022	0.458
			Liver	4.2	31.7	35.9	36.1	36.8	72.8	0.981	0.442
			Stomach	4.3	30.8	35.1	37.0	35.7	72.7	1.036	0.441
			Intestines	6.6	48.0	54.6	56.8	55.6	112	1.021	0.681
	2017/8/5	5.5	Muscle	12.1	85.2	97.3	104	98.7	203	1.055	1.000
			Gills	4.4	31.5	35.9	38.0	36.5	74.5	1.042	0.367
			Liver	7.2	53.1	60.3	61.5	61.5	123	1.000	0.607
			Stomach	4.2	32.6	36.8	35.8	37.8	73.6	0.948	0.363
			Intestines	6.2	47.7	53.9	53.4	55.2	109	0.967	0.535
			Ovary	4.0	29.3	33.2	33.9	33.9	67.9	1.000	0.335
								Average		1.010 ± 0.034	

The relationship between the radioactivity of stomach contents and muscle is shown in Table 8.12

[a]Decay-corrected value to March 16, 2011

[b]Activity ratio to the muscle

the skin of these fish migrated into the skin via the inner body or whether it adsorbed on the skin from seawater. It is also clear that radioactive cesium migrates more quickly to the ovaries than to the liver. The fact that the stomach concentrations of monkfish are almost equal to those of muscles suggests that the intake of radioactive cesium into fish depends on the diet.

In the freshwater Kejonuma pond, the internal organs of snakeheads at the top of the food chain were analyzed. In 2017, when snakeheads were collected, the radioactive cesium concentration in the pond water was estimated to be less than 10 mBq/L. The analytical results are shown in Table 8.10. The radioactive cesium concentration was lower in the internal organs than in the muscle, showing the same biodistribution as in the case of seawater fish. In terms of fish physiology, the behavior of radioactive cesium in the body has been considered to differ between freshwater fish and seawater fish, but such a difference was not found. This identity also suggests that radioactive cesium contamination of fish occurs via food.

8.8 Radioactive Cesium Recycling and Food Chain in the Aquatic Ecosystem

This section shows that radioactive cesium contamination of fishes in the hydrosphere occurs through feeding. The lowest species in the food chain in aquatic ecosystems are plankton. However, the concentration mechanism of radioactive cesium in plankton in the hydrosphere is not well understood [7, 8, 60]. In connection with the FDNPP accident, an extensive survey of zooplankton radioactive cesium contamination was conducted in the Pacific Ocean more than 500 km from the FDNPP approximately 10 months after the accident [61]. Naturally, the concentrations varied greatly depending on the place and timing of plankton collection, and the species of plankton. Their values were on the order of several to several tens of Bq/kg on a dry weight basis. Since the water content of plankton is over 90%, the concentration of the wet weight base is approximately one-tenth of the measured values. IAEA TRS No. 422 recommends concentration factors of 40 for zooplankton and 20 for phytoplankton on a dry basis.

Table 8.11 shows the radioactive cesium concentration of plankton collected offshore the FDNPP in September 2013, two and a half years after the accident. Sites A, B, and C are 2 km from the FDNPP, and Site D is 2.5 km. These samples contain phytoplankton and zooplankton because a fine plankton net was sampled by pulling it vertically. Interestingly, the radioactive cesium concentrations in seawater and sediments varied greatly, even in such a narrow open sea area. It can be understood that radioactive cesium is not uniformly mixed and diffused on the coast of the open ocean. Such heterogeneity is an essential factor when assessing radioactive contamination of fish. The concentration factor of plankton to seawater at Point A was approximately 2500 on a dry weight basis. It was 0.25 times the concentration of sediment. At Point D, the fish concentration factor for plankton

Table 8.11 Concentrations of radioactive cesium in the aquatic samples offshore the FDNPP

Point	Water depth, m	Sampling date	Sample	Layer	Detected, Bq/kg			Corrected[a], Bq/kg	
					^{134}Cs	^{137}Cs	$^{134+137}Cs$	$^{134+137}Cs$	$^{134}Cs/^{137}Cs$
A	11	2013/9/5	Seawater	Surface	0.0273	0.0599	0.0872	0.126	0.987
				Bottom	0.0197	0.0434	0.0631	0.0911	0.985
			Plankton[b]		60.0	130	190	275	1.004
			Sediment		237	527	764	1100	0.976
B	15		Seawater	Surface	0.0124	0.0249	0.0373	0.0547	1.080
				Bottom	0.0153	0.0308	0.0461	0.0677	1.077
			Sediment		47.4	114	161	229	0.902
C	17		Seawater	Surface	0.0112	0.0260	0.0372	0.0532	0.935
				Bottom	0.0137	0.0291	0.0428	0.0622	1.020
			Sediment		2400	5110	7510	10,900	1.015
D	24	2013/9/6	Plankton[b]		22.9	48.7	71.6	104	1.023
			Sediment		25.2	51.8	77	112	1.042
			Fish[c]		nd–135	nd–287	nd–422	nd–613	1.012[d]

Distance and direction from the FDNPP; A: 2 km south. B: 2 km southeast. C: 2 km east. D: 2.5 km south-southeast

[a]Decay-corrected value to March 16, 2011

[b]Plankton nets were moved from the seabed perpendicular to the water surface and collected. Radioactivity is on a dry weight basis

[c]Nine bottom and five migratory fishes were captured, and three fishes below the detection limit (nd, <0.6 Bq/kg). Radioactivity in fish muscle showed on a wet weight basis

[d]Average value of 11 fishes in which radioactivity detected

was less than 5.9. However, since most of the fishes analyzed were species that consumed small fish rather than plankton, this enrichment factor may not reflect radioactive cesium enrichment in the food chain. Since this area is open, fish species other than the settled bottom fishes are always migrating, making it difficult to quantitatively evaluate the concentration factor in such areas.

Kejonuma pond analyzed in Sect. 8.5 was estimated to have a very low density of plankton suspended in pond water, despite the abundant fish population in a mesotrophic lake. In all seasons, sample were unable to obtain the quantities of plankton needed to measure radioactive cesium concentrations. Therefore, the concentrations of radioactive cesium in the stomach contents of snakeheads at the top of the food chain were measured. Most of the stomach contents of snakeheads consisted of black bass and bluegill fry, but their digestion had progressed, and the content details are unknown. Table 8.12 shows the results obtained in 2017. The concentration factors of the muscle to stomach contents ranged from 1.6 to 4.6. These numbers revealed that radioactive cesium ingested from the food in snakeheads migrated into the muscles and was effectively concentrated. However, the average $^{134+137}$Cs concentration ratio in the snakehead muscles to the fry of black bass or bluegill in 2016 was approximately 16.1 (snakehead: 97.5 ± 29.3, n = 7; black bass: 6.78 ± 3.80, n = 8; bluegill: 5.31 ± 2.45, n = 5, Bq/kg). Such a significant difference in concentration indicates that diet has a significant effect on radioactive cesium levels in the body. Moreover, the radioactive cesium concentration in the stomach contents of snakehead is higher than that of the fry of black bass or bluegill. The radioactive cesium consumed may be concentrated in the stomach during the digestion process.

Since snakehead is known to prey not only on small fish but also on frogs, insects, and crustaceans, sampling of frogs that can easily collected at Kejonuma pond was conducted. Useful numbers of frogs were collected after 2014 and their tadpoles and eggs were also collected. Figure 8.15 shows the temporal change in radioactive cesium concentration. The apparent half-life of $^{134+137}$Cs contamination of bullfrog muscle is approximately 12 months. Section 8.6 discusses the effective half-lives and biological half-lives of ^{134}Cs and ^{137}Cs in the ecosystem of Kejonuma pond, divided into the initial phase and the steady phase. The apparent half-life obtained in Fig. 8.15 is not consistent with the simulation results shown in Table 8.7. This difference means that the apparent half-life from changes in the radioactivity concentration in a sample from an ecosystem where radioactive contamination is continuing is quite dangerous, and it is possible to misunderstand the actual state of radioactivity pollution. However, the total concentration of ^{134}Cs and ^{137}Cs in the bullfrog muscle at the boundary point is 390 Bq/kg in the simulation but can be read as 150–450 Bq/kg in Fig. 8.15. Since the data of individual bullfrogs vary widely, it is considered that the two values are in good agreement.

Tadpoles consume detritus and algae, which is very different from the eating habits of their parent frogs. Nevertheless, the temporal variations in the radioactive cesium concentration showed almost the same behavior as that for the parent bullfrog. However, a tadpole body changes shape when a tadpole transforms into a frog. At this time, the gills disappear, and the legs develop. Tadpoles that are

Table 8.12 Relationship between radioactive cesium concentrations in muscle and stomach contents of snakehead inhabiting the Kejonuma pond

Species	Sampling date	Age	Part	Detected, Bq/kg wet			Concentration factor[a]	Corrected[b], Bq/kg wet			$^{134}Cs/^{137}Cs$
				^{134}Cs	^{137}Cs	$^{134+137}Cs$		^{134}Cs	^{137}Cs	$^{134+137}Cs$	
Snakehead	2017/7/19	7	Muscle	8.4	56.4	64.8	4.63	71.3	65.2	137	1.09
			Stomach Contents	1.6	12.4	14.0		13.7	14.4	28.1	0.95
	2017/8/1	6.5	Muscle	7.4	54.7	62.2	1.56	63.5	63.4	127	1.00
			Stomach Contents	5.0	34.8	39.9		43.2	40.3	83.5	1.07
		6	Muscle	9.8	69.8	79.6	4.08	84.1	80.8	165	1.04
			Stomach Contents	2.3	17.2	19.5		19.5	20.0	39.5	0.98
	2017/8/5	6	Muscle	5.9	44.3	50.2	4.08	50.7	51.3	102	0.99
			Stomach Contents	1.4	10.9	12.3		12.4	12.6	25.0	0.98
					Average		3.59 ± 1.38				1.01 ± 0.05

[a] Activity ratio to muscle
[b] Decay-corrected value to March 16, 2011

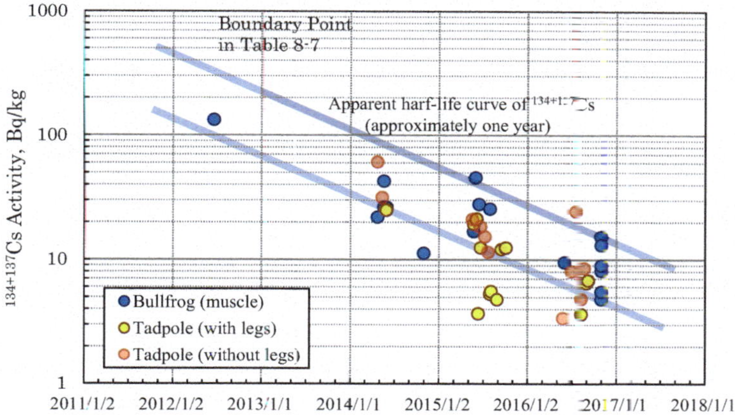

Fig. 8.15 Temporal changes in the concentrations of $^{134+137}$Cs in bullfrogs and their tadpoles inhabiting the Kejonuma pond

metamorphosing into frogs and have begun to grow legs stop eating temporarily because the mouth is also deformed.

Table 8.13 shows the relationship between the metamorphosis of bullfrogs and the concentration of radioactive cesium in Kejonuma pond. The $^{134+137}$Cs concentration is compared with the decay-corrected value on March 16, 2011. Differences in radioactive cesium levels between the muscles and legless tadpoles can be explained by differences in parent and child feeding habits. The $^{134+137}$Cs concentration in legged tadpoles was significantly lower than those of the parent bullfrog or legless tadpole. This distinct decrease suggests that the legged tadpoles were unable to ingest food during metamorphosis and that the balance of radioactive cesium in the body was being excreted. This fluctuation also suggests that radioactive cesium is incorporated into the food chain in aquatic ecosystems. The concentrations of radioactive cesium in the eggs of bullfrog were below the detection limit (<0.6 Bq/kg).

Of most significant concern with respect to the concentration of radioactive material in the food chain is the effect on the reproductive system. The accumulation of radioactive material in the reproductive tract increases the risk of radiation effects on offspring. Figure 8.16 shows the radioactive cesium concentration relationships between the ovaries and the maternal muscles of the black bass, bluegill, and snakehead inhabiting Kejonuma pond for each year of their capture. Except for two cases of bluegill, the concentration ratios of radioactive cesium in the ovaries relative to the maternal muscle were less than 1.0. Table 8.14 shows the enrichment factors of each fish. The more carnivorous fishes have lower enrichment factors. The decay-corrected $^{134+137}$Cs concentration to March 16, 2011, is shown to compare differences between fish species. The radioactive cesium concentrations in the ovaries were slightly higher in black bass than in other species but showed almost the same values within the deviation.

Table 8.13 Relationship between radioactive cesium contamination and frog transformation in the Kejonuma pond

Species	Sample	n	Corrected $^{134+137}Cs$[a], Bq/kg wet
Bullfrog	Muscle	17	41.1 ± 37.0
	Tadpole (with legs)	13[b]	18.7 ± 11.5
	Tadpole (without legs)	11[b]	32.9 ± 25.7
	Egg	3[b]	nd

nd: not detected (<0.6 Bq/kg)
[a]Decay-corrected value to March 16, 2011
[b]Collected approximately 20 g of a whole body

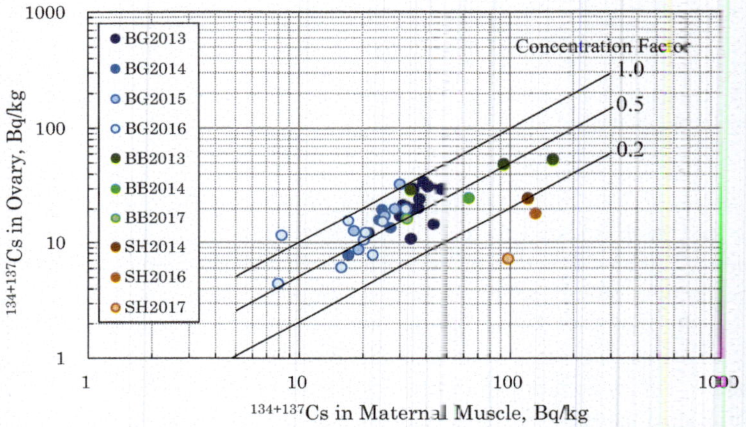

Fig. 8.16 Relationship between the $^{134+137}Cs$ concentrations in maternal muscle and ovary in the fishes inhabiting the Kejonuma pond. BB: Black bass, BG: Bluegill, SH: Snakehead

Table 8.14 Concentration factor of radioactive cesium in the ovary for the maternal muscle of fishes inhabiting the Kejonuma pond

			Corrected $^{134+137}Cs$[b], Bq/kg	
Species	n	Concentration factor[a]	Ovary	Maternal muscle
Snakehead	3	0.14 ± 0.06	30.4 ± 13.6	184 ± 78.4
Black bass	5	0.53 ± 0.19	52.4 ± 18.8	117 ± 70.0
Bluegill	31	0.65 ± 0.23	28.0 ± 11.4	43.6 ± 12.3

[a]Calculated by the detected value
[b]Decay corrected value to March 16, 2011

On the other hand, a clear significant difference was observed in the $^{134+137}Cs$ concentrations in the maternal muscles. The concentrations in snakehead reached approximately four times those in bluegill. Why radioactive cesium migrates to the muscles of carnivorous fish and is easily concentrated is unknown. Nevertheless, differences in the metabolic systems in the body can be assumed to contribute. In any case, from the viewpoint of the reproductive system, radioactive cesium is fortunately difficult to transfer to the ovaries. However, it must be noted that radioactive cesium is likely to accumulate in muscle through the food chain of ecosystems, including humans.

8.9 Conclusions

Radioactive cesium contamination of fish after the FDNPP accident was evaluated using data from TEPCO, JFA, and the author. The levels of pollution in saltwater fish were significantly decreased compared to those in the initial stage of the accident. However, although the FDNPP faces the open ocean, contaminated fish can still be found off the coast. Given the longevity of the fish, these fish were radioactively contaminated somewhere in the ocean. From the data, it can be inferred that the radioactive influx from the FDNPP to the ocean continues to be the source of radioactive contamination in fish. Perhaps the radioactive contamination in ocean fish is through another mechanism than food chain contamination. Unless the release of radioactive material from the FDNPP is completely stopped, the radioactive contamination of ocean fish will not end.

The effective half-life of radioactive contamination in terrestrial freshwater fish on is much longer than previously assumed because radioactive cesium is recycled in ecosystem. In addition, it is inferred that the diets of fishes largely control the radioactivity concentrations. In the future, radioactive cesium retained in the forest of the catchment area will flow out of the forest and into the terrestrial hydrosphere. Then, this radioactive cesium will inevitably recycle in the aquatic ecosystem. Most likely, the decontamination of forests cannot stop the release of radioactive cesium. It is necessary to continue to monitor the radioactive contamination of fish through more detailed research.

The FRA (Fishers Research Agency, Japan) has published comprehensive research results on radioactive contamination in fishes due to the FDNPP accident [62].

References

1. Zheng J, Tagami K, Uchida S (2013) Release pf plutonium isotopes into the environment from the Fukushima Daiichi Nuclear Power Plant accident. Environ Sci Technol 47:9584–9595. https://doi.org/10.1021/es402212v
2. Schneider S, Walther C, Bister S, Schauer V, Christl M et al (2013) Plutonium release from Fukushima Daiichi fosters the need for more detailed investigations. Sci Rep 3:2988. https://doi.org/10.1038/srep02988

3. Yang G, Tazoe H, Hayano K, Okayama K, Yamada M (2017) Isotopic compositions of ^{235}U, ^{239}Pu, and ^{240}Pu in soil contaminated by the Fukushima Daiichi Nuclear Power Plant accident. Sci Rep 7:13619. https://doi.org/10.1038/s41598-017-13998-6
4. Nitta W, Sanada T, Isogai K, Schlosser C (2014) Atmospheric ^{85}Kr and ^{133}Xe activity concentrations at locations across Japan following the Fukushima Dai-ichi Nuclear Power Plant accident. J Nucl Sci Technol 51:712–719. https://doi.org/10.1080/00223131.2014.888960
5. TEPCO (Tokyo Electric Power Company) (2020) Analytical results of radioactive materials around Fukushima Daiichi Nuclear Power Plant [in Japanese]. http://www.tepco.co.jp/decommission/data/analysis/, http://www.tepco.co.jp/decommission/data/analysis/archive/2020/202001-j.html
6. JFA (Fisheries Agency of Japan) (2020) About result of radioactive material investigation of marine products. https://www.jfa.maff.go.jp/e/index.html
7. Radioactive Waste Management Center (1996) Concentration factors in the marine organisms. Environmental Parameter Series 6. RWMC-96-P-18. https://www.rwmc.or.jp/library/other/file/Kankyou6-1.pdf
8. IAEA (International Atomic Energy Agency) (2004) Sediment distribution coefficients and concentration factors for biota in the marine environment. Technical Reports Series No. 422
9. Perrott MN, Grierson CE, Hazon N, Balment (1992) Drinking behaviour in sea water and fresh water teleosts, the role of the renin-angiotensin system. Fish Physiol Biochem 10:161–168
10. Brittain JE, Gjerseth JE (2010) Long-term trends and variation in ^{137}Cs activity concentrations in brown trout (*Salmo trutta*) from Øver Heimdalsvatn, a Norwegian subalpine lake. Hydrobiologia 642:107–113. https://doi.org/10.1007/s10750-010-0155-5
11. JCAC (Japan Chemical Analysis Center) (2011) Radioactivity survey data in Japan. Environmental and dietary materials. ISSN 0441-2516 https://www.kankyo-hoshano.go.jp/07/07.html
12. Buesseler KO (2012) Fishing for answers off Fukushima. Science 338:480–482. https://doi.org/10.1126/science.1228250
13. Oikawa S, Watabe T, Takata H, Suzuki C, Nakahara M et al (2013) Long term temporal changes of ^{90}Sr and ^{137}Cs in seawater, bottom sediment and marine organism samples - from the Chernobyl accident to immediately after the Fukushima accident. Bunseki Kagaku (Analytical Chemistry) 62:455–474. [in Japanese with English Abstract]
14. Shigenobu Y, Fujimoto K, Ambe D, Kaeriyama H, Ono T et al (2014) Radiocesium contamination of greenlings (*Hexagrammos otakii*) off the coast of Fukushima. Sci Rep 4:6851. https://doi.org/10.1038/srep06851
15. Horiguchi T, Yoshii H, Mizuno S, Shiraishi H (2016) Decline in intertidal biota after the 2011 Great East Japan earthquake and Tsunami and the Fukushima nuclear disaster: field observations. Sci Rep 6:20416. https://doi.org/10.1038/srep20416
16. Madigan DJ, Baumann Z, Fisher NS (2012) Pacific bluefin tuna transport Fukushima-derived radionuclides from Japan to California. Proc Natl Acad Sci 109:9483. https://doi.org/10.1073/pnas.1204859109
17. Madigan DJ, Baumann Z, Snodgrass OE, Ergül HA, Dewar H et al (2013) Radiocesium in Pacific bluefin tuna *Thunnus orientalis* in 2012 validates new tracer technique. Environ Sci Technol 47:2287–2294. https://doi.org/10.1021/es4002423
18. Madigan DJ, Baumann Z, Snodgrass OE, Dewar H, Berman-Kowalewski M et al (2017) Assessing Fukushima-derived radiocesium in migratory Pacific predators. Environ Sci Technol 51:8962–8971. https://doi.org/10.1021/acs.est7b00680
19. Amano H, Akiyama M, Chunlei B, Kawamura T, Kishimoto T et al (2012) Radiation measurements in the Chiba metropolitan area and radiological aspects of fallout from the Fukushima Dai-ichi Nuclear Power Plants accident. J Environ Radioact 111:42–52. https://doi.org/10.1016/j.jenvrad.2011.10.019
20. Furuta S, Sumiya S, Watanabe H, Nakano M, Imaizumi K et al (2011) Results of the environmental radiation monitoring following the accident at the Fukushima Daiichi Nuclear Power Plant - Interim report (Ambient radiation dose rate, radioactivity concentration in the air

and radioactivity concentration in the fallout). JAEA Rev 35:1–89. [in Japanese with English Abstract]. https://jopss.jaea.go.jp/pdfdata/JAEA-Review-2011-035.pdf

21. Nagakawa Y, Suzuki T, Kinjo Y, Miyazaki N, Sekiguchi M et al (2011) Measurement of the ambient radioactivity and estimation of human radiation exposure dose in Tokyo with regard to the radioactive substance leakage due to the Fukushima Daiichi Nuclear Power Plant accident. Radioisotopes 60:467–472. [in Japanese with English Abstract]. https://doi.org/10.3769/radioisotopes.60.467
22. Yoshida N, Kanda J (2012) Tracking the Fukushima radionuclides. Science 336:1115–1116. https://doi.org/10.1126/science.1219493
23. Tsuruta H, Nakajima T (2012) Radioactive materials in the atmosphere released by the accident of the Fukushima Daiichi Nuclear Power Plant. Chikyukagaku (Geochemistry) 46:99–111. [in Japanese with English Abstract]. https://doi.org/10.14934/chikyukagaku.46.99
24. SCJ (Science Council of Japan) (2014) Sectional Committee on Nuclear Accident. A review of the model comparison of transportation and deposition of radioactive materials released to the environment as a result of the Tokyo Electric Power Company's Fukushima Daiichi Nuclear Power Plant accident [in Japanese]. http://www.scj.go.jp/ja/info/kohyo/pdf/kohyo-22-h140902-e1.pdf
25. Ishida M, Yamazaki H (2017) Radioactive contamination in the Tokyo metropolitan area in the early stage of the Fukushima Daiichi Nuclear Power Plant (FDNPP) accident and its fluctuation over five years. PLoS One 12:e0187687. https://doi.org/10.1371/journal.pone.0187687
26. Yamazaki H, Ishida M, Hinokio R, Yamashiki YA, Azuma R (2018) Spatiotemporal distribution and fluctuation of radiocesium in Tokyo Bay in the five years following the Fukushima Daiichi Nuclear Power Plant (FDNPP) accident. PLoS One 13:e0193414. https://doi.org/10.1371/journal.pone.0193414
27. Chaisan K, Smith JT, Bossew P, Kirchner G, Loptev GV (2013) Worldwide isotope ratios of the Fukushima release and early-phase external dose reconstruction. Sci Rep 3:2520. https://doi.org/10.1038/srep02520
28. Komori M, Shozugawa K, Nogawa N, Matsuo M (2013) Evaluation of radioactive contamination caused by each plant of Fukushima Daiichi Nuclear Power Station using $^{134}Cs/^{137}Cs$ activity ratio as an index. Bunseki Kagaku (Analytical Chemistry) 62:475–483. [in Japanese with English Abstract]. https://doi.org/10.2116/bunsekikagaku.62.475
29. Nishizawa Y, Yoshida M, Sanada Y, Torii T (2016) Distribution of the $^{134}Cs/^{137}Cs$ ratio around the Fukushima Daiichi nuclear power plant using an unmanned helicopter radiation monitoring system. J Nucl Sci Technol 53:468–474. https://doi.org/10.1080/00223131.2015.1071721
30. MHLW (Ministry of Health, Labor and Welfare) (2012) New reference values of radioactive material in food [in Japanese]. https://www.mhlw.go.jp/shinsai_jouhou/dl/leaflet_120329.pdf
31. Japan Meteorological Agency (2020) Search past weather data [in Japanese]. https://www.data.jma.go.jp/obd/stats/etrn/select/prefecture00
32. MEXT (Ministry of Education, Culture, Sports, Science and Technology of Japan) (2011) MEXT's airborne monitoring results in Iwate, Shizuoka, Nagano, Yamanashi, Gifu, and Toyama prefectures and revisions to airborne monitoring that take into account the effects of natural nuclides [in Japanese]
33. Koulikov AO, Ryabov IN (1992) Specific cesium activity in freshwater fish and the size effect. Sci Total Environ 112:125–142
34. Smith JT, Kudelsky AV, Ryabov IN, Hadderingh RH (2000) Radiocesium concentration factors of Chernobyl-contaminated fish: a study of the influence of potassium, and "Blind" testing of a previously developed model. J Environ Radioact 48:359–369. S0265-931X(99) 00089-2 [pii]
35. Sundbom M, Meili M, Andersson E, Östlund M, Broberg A (2003) Long-term dynamics of Chernobyl ^{137}Cs in freshwater fish: quantifying the effect of body size and trophic level. J Appl Ecol 40:228–240
36. Rask M, Saxén R, Ruuhijärvi J, Arvola L, Järvinen M et al (2012) Short- and long-term patterns of ^{137}Cs in fish and other aquatic organisms of small forest lakes in southern Finland since the

Chernobyl accident. J Environ Radioact 103:41–47. https://doi.org/10.1016/j.jenvrad.2011.08.002

37. Shimada T, Ashizawa J, Hoshi M, Suzuki K, Shindo K et al (2012) Radioactive cesium concentration in sediment, mussels, fish and reeds in Lake Izunuma-Uchnuma and in its watershed. Izunuma-Uchinuma Wetland Res 6:13–19. [in Japanese with English Abstract]
38. Mizuno T, Kudo H (2013) Overview of active cesium contamination of freshwater fish in Fukushima and Eastern Japan. Sci Rep 3:1742. https://doi.org/10.1038/srep01742
39. Elliott JM, Hilton J, Rigg E, Tullett PA, Swift DJ et al (2013) Sources of variation in post-Chernobyl radiocaesium in fish from two Cumbrian lakes (North-West England). J Appl Ecol 29:108–119
40. Arai T (2014) Radioactive cesium accumulation in freshwater fishes after the Fukushima nuclear accident. Springer Plus 3:479. https://springerplus.springeropen.com/articles/10.1186/2193-1801-3-479
41. Yamamoto S, Yokoduka T, Fujimoto K, Takagi K, Ono T (2014) Radiocaesium concentrations in the muscle and eggs of salmonids from Lake Chuzenji, Japan, after the Fukushima fallout. J Fish Biol 84:1607–16013. https://doi.org/10.1111/jfb.12368
42. Yoshimura M, Yokoduka T (2014) Radioactive contamination of fishes in lake and streams impacted by the Fukushima nuclear power plant accident. Sci Total Environ 482-483:184–192. https://doi.org/10.1016/j.scitotenv.2014.02.118
43. Arita K, Ashizawa J, Fujimoto Y, Shimada T, Hayashi S et al (2015) Radioactive cesium accumulation during developmental stages of Largemouth bass, *Micropterus salmoides*. J JSCE G (Environ) 71(III):267–276. [in Japanese with English Abstract]
44. Wada T, Tomiya A, Enomoto M, Sato T, Morishita D et al (2015) Radiological impact of the nuclear power plant accident on freshwater fish in Fukushima: an overview of monitoring results. J Environ Radioact 151:144–155. https://doi.org/10.1016/j.jenvrad.2015.09.017
45. Haque ME, Gomi T, Sakai M, Negishi JN (2017) Seasonal variation in food web-based transfer factors of radiocesium on white-spotted char (*Salvelinus leucomaenis*) from headwater streams. Landsc Ecol Eng 14:45. https://doi.org/10.1007/s11355-016-0324-4
46. Mori M, Tsunoda K, Suzuki K, Yuasa Y, Kuge T et al (2017) Fractionation of radiocesium in soil, sediments, and aquatic organisms in Lake Onuma of Mt. Akagi, Gunma Prefecture using sequential extraction. Sci Total Environ 575:1247–1254. https://doi.org/10.1016/j.scitotenv.2016.09.205
47. Suzuki K, Watanabe S, Yuasa Y, Yamashita Y, Arai H et al (2018) Radiocesium dynamics in the aquatic ecosystem of Lake Onuma on Mt. Akagi following the Fukushima Dai-ichi Nuclear Power Plant accident. Sci Total Environ 622–623:1153–1164. https://doi.org/10.1016/j.scitotenv.2017.12.017
48. Häsänen E, Kolehmainen S, Miettinen JK (1968) Biological half-times of ^{137}Cs and ^{22}Na in different fish species and their temperature dependence. Proc First Int Cong Rad Protec 1:401–406. https://doi.org/10.1016/B978-1-4832-8312-8.50074-6
49. Oak Ridge National Laboratory (1975) ORNL-5002. Environmental Sciences Division Publication No. 783. Bioaccumulation factors for radionuclides in freshwater biota. https://nuclear.org/radiation/radioactivity/3445600598280.pdf
50. Radioactive Waste Management Center, Japan (1992) RWMC-92-P-14. Environmental Parameters Series 3. Concentration factors of Radionuclides in Freshwater Organisms [in Japanese]. https://www.rwmc.or.jp/library/other/file/kankyo3_1.pdf
51. Pröhl G, Ehiken S, Fiedler I, Kirchner G, Klemt E et al (2006) Ecological half-lives of ^{90}Sr and ^{137}Cs in terrestrial and aquatic ecosystems. J Environ Radioact 91:41–72. https://doi.org/10.1016/j.ienvrad.2006.08.004
52. Doi H, Takahara T, Tanaka K (2012) Trophic position and metabolic rate predict the long-term decay process of radioactive cesium in fish: a meta-analysis. PLoS One 7(1):e29295. https://journals.plos.org/plosone/

53. Tagami K, Uchida S (2016) Consideration on the long ecological half-life component of ^{137}Cs in demersal fish based on field observation results obtained after the Fukushima accident. Environ Sci Technol 50:1804–1811. https://doi.org/10.1021/acs.est.5b04952
54. JFA (2013) Urgent research for elucidation of pollution source and pollution route of fish contaminated with radioactive cesium in high concentration [in Japanese]. https://www.jfa.maff.go.jp/j/housyanou/pdf/kouhyou.pdf
55. Fiévet B, Bailly-du-Bois P, Laguionie P, Morillon M, Arnaud M et al (2017) A dual pathway transfer model to account for changes in the radioactive caesium level in demersal and pelagic fish after Fukushima Dai-ichi nuclear power plant accident. PLoS One 12:e0172442. https://doi.org/10.1371/journal.pone.0172442
56. Fisher NS, Beaugelin-Seiller K, Hinton TG, Baumann Z, Madigan DJ et al (2013) Evaluation of radiation doses and associated risk from the Fukushima nuclear accident to marine biota and human consumers of seafood. PNAS 110:10670–10675. https://www.pnas.org/content/110/26/10670
57. Kinura Y, Ogawa Y, Honda Y, Katsurayama K (1981) Studies on the concentration and elimination of some radionuclides during development and growth of fish eggs, advanced fry and fingerlings. Ann Rep At Energ Inst, Kindai University 18:15–35 [in Japanese with English Abstract]. https://kindai.repo.nii.ac.jp/?action=pages_view_main&active_action=repository_view_main_item_detail&item_id=6231&item_no=1&page_id=13&block_id=21
58. JFA (2012) Radioactive substances in marine products: Iodine and Cesium [in Japanese]. https://www.jfa.maff.go.jp/j/kakou/Q_A/pdf/qanda-j.pdf
59. Kasamatu F (1999) Marine organisms and radionuclides - with special reference to the factors affecting concentration of ^{137}Cs in marine fish. Radioisotopes 48:266–282. [in Japanese]
60. Kaeriyama H, Watabe T, Kusakabe M (2008) ^{137}Cs concentration in zooplankton and its relation to taxonomic composition in the western North Pacific Ocean. J Environ Radioact 99:1838–1845. https://doi.org/10.1016/j.jenvrad.2008.08.006
61. Kitamura M, Kumamoto Y, Kawakami H, Cruz EC, Fujikura K (2013) Horizontal distribution of Fukushima-derived radiocesium in zooplankton in the northwestern Pacific Ocean. Biogeosciences 10:5729–5738. https://doi.org/10.5194/bg-10-5729-2013
62. Nakata K, Sugisaki H (eds) (2015) Impacts of the Fukushima Nuclear Accident on fish and fishing grounds. Springer, Cham. https://doi.org/10.1007/978-4-431-55537-7. ISBN: 978-4-431-55537-7 (eBook)

Chapter 9
Topics

Abstract This chapter summarizes how radioactive contamination in the Tokyo metropolitan area will change in the future and how this contamination affects the health of residents. The items are as follows.

- High concentration ^{235}U was found in a house in Tokyo.
- Estimation of radiation dose in the garden of a kindergarten that was radioactively contaminated in Tokyo.
- The current situation of ^{137}Cs in the aquatic sediment caused by the Nagasaki atomic bomb.
- Radioactive cesium contamination of terrestrial ecosystems such as birds and earthworms.
- Reliability of the radioactive contamination map by the airborne monitoring.

Keywords ^{235}U in Tokyo · Cumulative dose · Fate of radionuclides · Contamination in non-fish organisms · Airborne monitoring

9.1 Introduction

Japan has suffered many nuclear disasters, including the atomic bombings of Hiroshima and Nagasaki. Atmospheric nuclear tests by the U.S. exposed many seafarers on tuna vessels in the South Pacific to radiation and killed one seafarer due to radiation damage. In atmospheric nuclear tests by the U.S. and the Soviet Union, the precipitation of ^{137}Cs reached 1.59 kBq/m^2 year in Chiba City approximately 30 km west of central Tokyo in 1963 [1]. Additionally, the accumulated precipitation amount of ^{137}Cs for 45 years from 1954 to 2011 reached 6.75 kBq/m^2. The ^{137}Cs precipitation in 1986, when the Chernobyl nuclear accident occurred, was 0.16 Bq/m^2 year. Of course, ^{90}Sr corresponding to ^{137}Cs also precipitated. Therefore, Japanese people at this time would have received significant external and internal exposures, but no strict epidemiological assessment is known to have been conducted on health effects due to radiation exposure.

H. Yamazaki, *Radioactive Contamination of the Tokyo Metropolitan Area*,
https://doi.org/10.1007/978-981-15-7368-2_9

In 1999, the first criticality accident occurred in Japan at JCO Co., Ltd., in Tokai village, Ibaragi Prefecture, approximately 110 km northwest of central Tokyo. In this region, there are many research institutes of the Japan Atomic Energy Agency and two nuclear power plants of Japan Atomic Power Co. (one reactor is being decommissioned), and the area is the center of Japan's nuclear power research and development. Although JCO was processing uranium fuel with an enrichment of 18.8% for a fast breeder reactor, the falsification of the operating procedure by an employee caused a criticality accident in which three persons were exposed. Two people exposed to neutron radiation estimated at 16,000–20,000 mSv and 6000–10,000 mSv died within 1 year of radiation exposure. One patient was exposed to 1000–4500 mSv but survived. In this accident, a total of 667 people, including JCO employees, paramedics, and nearby residents, were recognized by the Japan Nuclear Safety Commission as exposed persons. The highest dose was estimated at 120 mSv, and there were six exposed people with 50 mSv and above.

Because of the FDNPP (Fukushima Daiichi nuclear power plant) accident, the Japanese government is considering disposal methods for the contaminated soil widely distributed in Fukushima Prefecture and eastern Japan. The Japanese government defines radioactive waste with a concentration of over 8000 Bq/kg as designated waste that needs to be managed and stored. Moreover, they intend to dispose of radioactive waste under 8000 Bq/kg by using the contaminated waste for public works such as road foundations and embankments. Radioactive wastes other than these designated wastes are estimated to amount to 14 million cm^3 in Fukushima Prefecture alone, and there is a high possibility that these wastes will be spread as radioactively contaminated wastes from Fukushima Prefecture to all of Japan under the guise of public works. When radioactive cesium solidifies with concrete, it becomes soluble due to the alkalinity of cement and dissolves easily in rainwater. The use of radioactive waste with 8000 Bq/kg or less for public works means the spread of radioactive cesium contamination throughout Japan. If such a plan is implemented, it will increase the health risk from radiation exposure.

One hundred seventy-five children who were under 18 years old at the occurrence of the FDNPP accident by the end of March 2019 developed malignant thyroid cancer, according to the official announcement by Fukushima Prefecture [2]. An additional 56 children have suspected thyroid cancer but have not received a definitive diagnosis. At the time of the accident, the population of children in Fukushima Prefecture was approximately 340,000, so the incidence was approximately 1 in 2000. The prevalence of this childhood thyroid cancer in Russia and Belarus during the Chernobyl accident was also between 2000 and 2500 [3, 4], so the incidences are approximately the same. Moreover, the incidence of thyroid cancer continues to increase since both the Fukushima and Chernobyl accidents. According to data from the National Cancer Center, Japan, the incidence of childhood thyroid cancer in Japan before the Fukushima nuclear accident was approximately 0.1–0.5 per 100,000 annually [5, 6]. To date, the Japanese government has not recognized the causal relationship between the occurrence of thyroid cancer and the FDNPP accident. In addition, more than 9 years after the accident, the results of various epidemiological surveys of Japanese health have been published [7, 8]. These results show that some diseases such as heart disease, have increased since the Fukushima nuclear accident.

This chapter summarizes how radioactive contamination in the Tokyo metropolitan area will change in the future and how this contamination affects the health of residents.

9.2 ^{235}U Detected in the Atmospheric Dust of Tokyo

In the FDNPP accident, three reactors melted down, but massive amounts of volatile radionuclides, such as radioactive noble gas, radioactive iodine, and radioactive cesium, were released from the broken reactors. As shown in Fig. 2.13, the author detected ^{95}Zr-^{95}Nb in atmospheric dust in Higashi-Osaka City, approximately 600 km southwest of the FDNPP. Since this nuclide is a high-boiling fission product, the presence of such a high-boiling nuclide in the atmosphere far from the FDNPP is due to damage to the reactor pressure vessel and containment vessel, which is evidence that nuclear fuel escaped into the atmosphere. Of course, uranium and plutonium isotopes derived from nuclear fuel have been found in soil in Fukushima Prefecture near the FDNPP [References 42, 43, and 47 in Chap. 3]. Uranium and plutonium have radiotoxicity as well as high metal toxicity. If these fission nuclides were transported to the densely populated Tokyo metropolitan area due to the FDNPP accident, it would be a severe problem in terms of the health effects on residents.

The lower part of Fig. 9.1 shows the gamma-ray spectrum of atmospheric dust collected in Higashi-Osaka City, approximately 600 km from the FDNPP. When the radioactive plume arrived, 390 m^3 of air was collected by a glass fiber filter with a retained particle size of 0.5 μm. In addition to ^{131}I, ^{134}Cs, and ^{137}Cs, radionuclide peaks belonging to natural radioactive decay series such as ^{210}Pb, ^{212}Pb, ^{208}Tl, and ^{212}Bi were observed. Further, ^{7}Be is produced by the ^{14}N (p, 2α) reaction between solar radiation protons and atmospheric nitrogen. From this spectrum, it can be seen that fine rock fragments floated as atmospheric dust in the air sample collected in Higashi-Osaka City. Most likely, radioactive iodine and radioactive cesium were adsorbed onto the suspended dust and transported to Higashi-Osaka City. The concentration of uranium in typical Japanese rocks and soils is approximately 1–3 μg/g. Since the abundance ratio of ^{235}U is 0.72%, the concentration of ^{235}U is 0.07–0.2 μg/g. Therefore, the ratio of the decay series nuclides in the atmospheric dust of Higashi-Osaka to the gamma-ray spectrum intensity of ^{235}U is typical.

On the other hand, the upper part in Fig. 9.1 is the spectrum of an air conditioner filter. The air conditioner was installed in a concrete apartment house in Tokyo, where residents and their families lived. The filter was a mesh filter made of polyethylene and had an area of approximately 600 cm^2. It was used for approximately 2 years starting in October 2010 and was replaced in August 2012. This air conditioner was used for 1 year and 5 months after the FDNPP accident. The filter

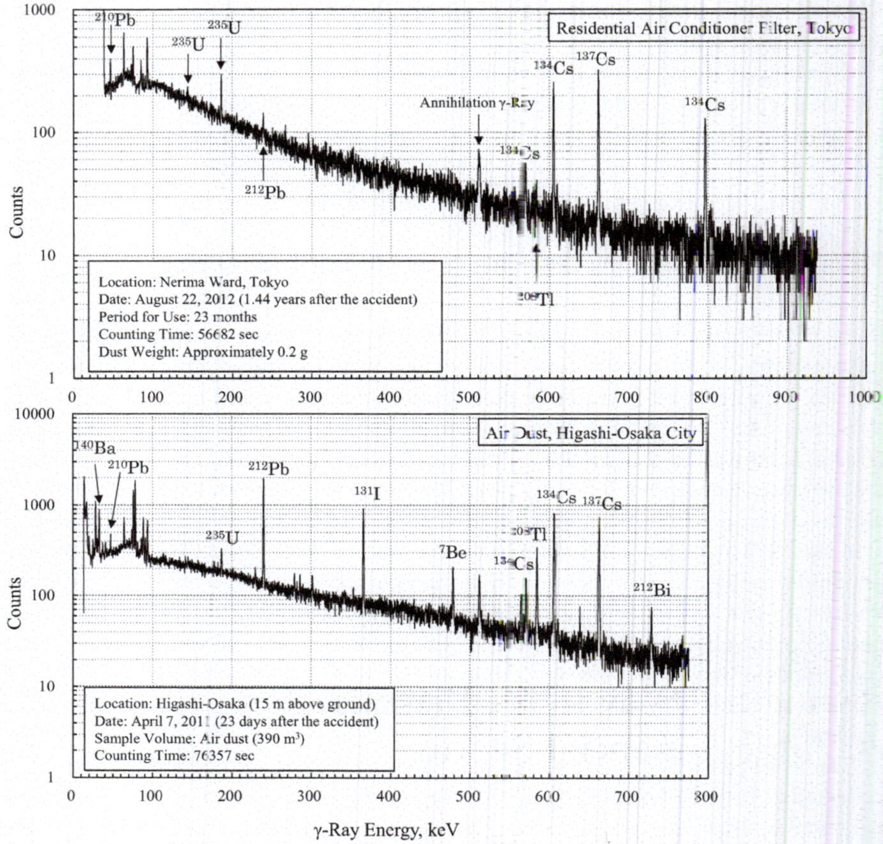

Fig. 9.1 Gamma-ray spectrum of an air-conditioning air filter used in central Tokyo. This air conditioner has been used for approximately 2 years since October 2010 in the living room of a concrete apartment where the family lives. The lower spectrum is atmospheric dust collected in Higashi-Osaka City

retention diameter for dust is not known, but it seems to be quite large. Fiber dust had adhered to the filter.

Almost no natural decay series nuclide peaks were detected from the air conditioner filter used in Tokyo. This filter did not retain the atmospheric rock and soil dust. Nevertheless, an apparent ^{235}U peak appeared. The presence of uranium from sources other than rock and soil was suggested by this spectrum. Calculated from the peak intensity, approximately 2 μg of ^{235}U was retained on this filter. Since the abundance ratio of ^{235}U in natural uranium is 0.72%, this implies 280 μg of uranium. If this ^{235}U originated from the FDNPP accident, the uranium enrichment of the FDNPP Unit 1 and 2 reactors is 3.7%, which gives approximately 55 μg for $^{235+238}$U. The Unit 3 reactor used plutonium MOX fuel, so the enrichment of ^{235}U is 1.2%, in which case the amount of $^{235+238}$U is 170 μg. In other words, a considerable amount of uranium spread to Tokyo via the air due to the FDNPP accident. Fine particles derived from the FDNPP accident containing high concentrations of uranium have been found in various parts of eastern Japan [References 42, 43, and 47 in Chap. 3]. Uranium trapped in the air conditioner filter in Tokyo can also be considered to have spread from the FDNPP reactor as such fine nuclear fuel particles.

If this assumption is valid, then Tokyo residents would have inhaled not only radioactive iodine and radioactive cesium but also meltdown nuclear fuel materials. If this ^{235}U on the filter was released from the Unit 3 reactor, massive amounts of plutonium would also have arrived over Tokyo. Unfortunately, these data do not tell when the uranium reached Tokyo.

9.3 Estimation for Air Dose in a Kindergarten Schoolyard in Tokyo

Exposure doses that would be received if radioactive materials were not decontaminated from the garden soil at a kindergarten in Tokyo were estimated. Similar to other areas of Tokyo, this kindergarten was presumed to have been severely contaminated by radioactivity on March 16 and 22, 2011. The soil sample was collected on April 15, 2011. The radioactivity concentration and inventory converted to March 16, 2011, were 22,900 Bq/kg and 595 kBq/m^2 for ^{131}I, 2220 Bq/kg and 57.7 kBq/m^2 for ^{134}Cs, and 2210 Bq/kg and 57.5 kBq/m^2 for ^{137}Cs, respectively. The gamma-ray dose rate at a height of 1 m from the surface of the garden was estimated. The calculation was performed for each nuclide by the IAEA-TECDOC-1162 method, based on the inventories of these nuclides. Figure 9.2 shows the calculation results.

The degree of radioactive contamination in this kindergarten is an average contamination level in eastern Tokyo caused by the FDNPP accident. In this kindergarten, it seems that decontamination treatment was performed after the measurement, such as replacing contaminated soil. Therefore, it is unlikely that children and staff will be exposed to radiation, as shown in Fig. 9.2. However, radioactive decontamination was not performed in many parts of Tokyo. Many residents and workers live in radioactively contaminated towns

Until the authors conducted a soil contamination survey, the kindergarten staff did not know that the garden was radioactively contaminated. Such deficiencies in information mean that children and staff were almost unprotected from ^{131}I exposure immediately after the accident. In the first month after the accident, the cumulative dose of ^{131}I, ^{134}Cs, and ^{137}Cs reached 0.5 mSv. Since children who play in the garden and their instructors are also affected by internal exposure from inhaling suspended soil dust, this radiation dose is not small for health effects. Incredibly, the Japanese government has little done toward the quantitative assessment of internal exposure due to the FDNPP accident. Not only the highly contaminated areas in Fukushima but also the dust raised from the ground in the Tokyo metropolitan area,

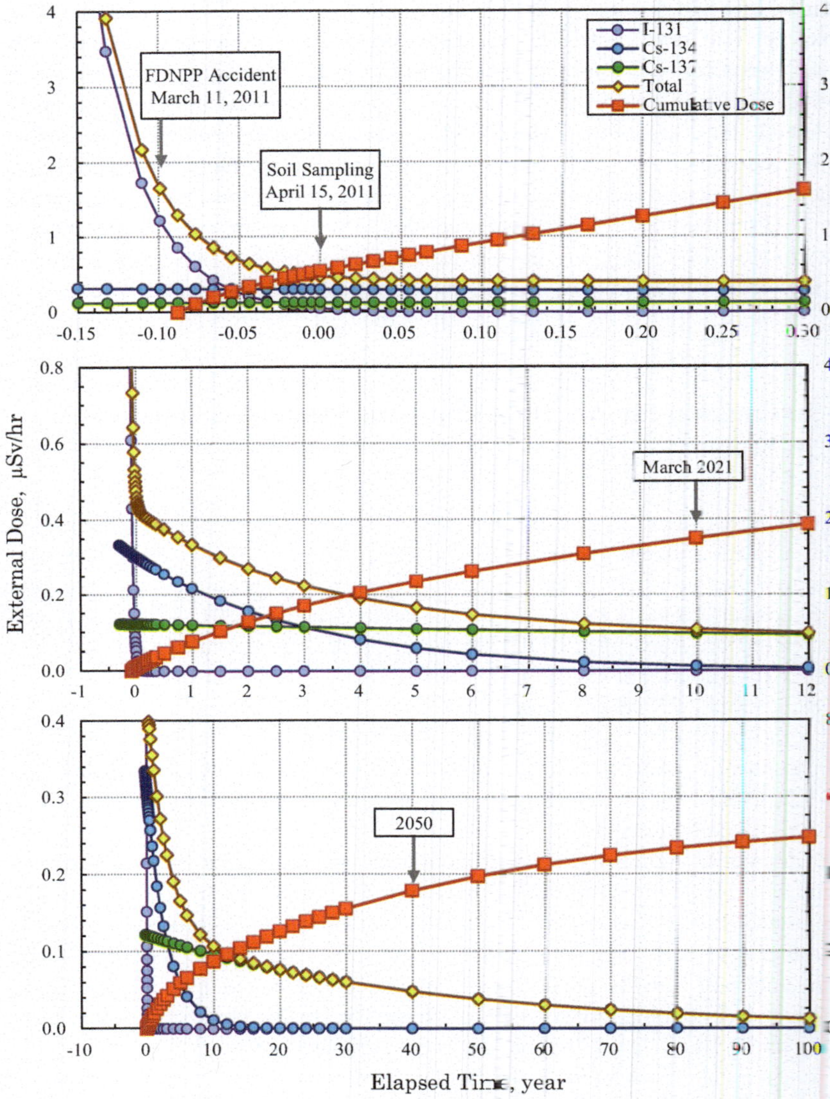

Fig. 9.2 External dose estimated from the radioactive contamination of garden soil at a kindergarten in Tokyo. The external dose was estimated by the IAEA-TECDOC-1162 method. The soil sample was collected on April 15, 2011, 2 cm deep from the surface in the center of the garden. The activity and inventory of radionuclides that corrected the radioactive decay to March 16, 2011, were 22,900 Bq/kg and 595 kBq/m^2 for ^{131}I, 2220 Bq/kg and 57.7 kBq/m^2 for ^{134}Cs, and 2210 Bq/kg and 57.5 kBq/m^2 for ^{137}Cs

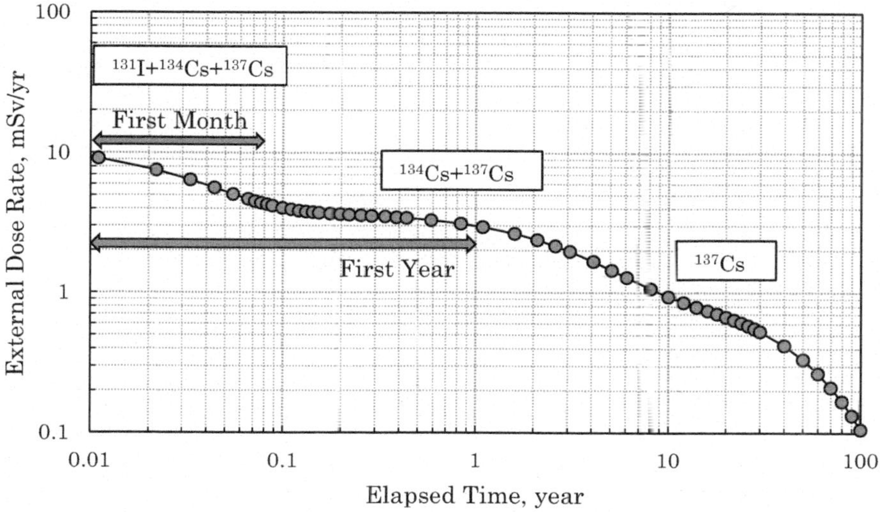

Fig. 9.3 Temporal change in the estimated dose rates since March 15, 2011, at a kindergarten garden in Tokyo

can become sources of internal exposure. Of course, this includes internal exposure from nuclides that emit alpha and beta rays.

In areas with the same pollution level as this kindergarten, the cumulative exposure dose will be approximately 40 mSv in 2050, 40 years after the FDNPP accident. As shown in Fig. 9.3, the dose rate decreases sharply over time. In the case of the FDNPP accident, the air dose rate decreases sharply in the first month while ^{131}I (half-life: 8.04 days) remains, but long-term exposure at a low dose rate due to ^{134}Cs (half-life: 2.06 years) and ^{137}Cs (half-life: 30.1 years) cannot be ignored.

The MEXT estimated that external exposure dose to the students in the kindergarten, elementary school, and junior high school in Fukushima Prefecture for 200 days after the FDNPP accident was several mSv, and internal exposure from the schoolyard soils was 0.5–4.0% of the external exposure [9]. This estimate does not include the internal exposure associated with the passage of the radioactive plume. From June 2011, after the FDNPP accident, the Koto Ward government in Tokyo started the monitoring of radioactive contamination in elementary and junior high schoolyards in the Koto Ward, and the results have been published on the website [10]. Continuation of such surveys is essential to determine the risks of radiation exposure and ensure the safety of students, staff, and residents.

The ICRP 2007 Recommendation (ICRP Publication 103) and the IAEA Safety Standards (General Safety Guide No. GSG-8. 2018) sets dose limits for the public to 1 mSv/year or less, and dose limits for professionals to 100 mSv/5 year and 50 mSv/year or less are well known. There are various ways of thinking about the risk assessment of radiation exposure to the public during a nuclear disaster, and internationally unified dose limits are not recognized. There are various debates about the health risks of low-dose exposure [11–25]. If these arguments are valid, the dose limits recommended in the ICRP and IAEA should be tightened. Compared to the TDI (Tolerable Daily Intake) and ADI (Acceptable Daily Intake) are shown for many hazardous chemicals, these radiation dose limits are not acceptable for public safety.

9.4 The Fate of Radioactive Nuclides Released from the FDNPP Accident

Radionuclides released to the environment during a nuclear disaster are eventually transported to the hydrosphere, where they diffuse through the water and become diluted or are deposited with sediments. A typical nuclide of the former group is tritium, and the latter type includes radioactive cesium, radioactive strontium, uranium, and plutonium. Tritium also diffuses into the atmosphere in hydrogen and water molecules, but large amounts of radioactive noble gases released from nuclear facilities also diffuse into the atmosphere. Although the half-lives of ^{41}Ar (1.83 h) and ^{133}Xe (5.25 days) are short, ^{85}Kr (10.7 years) is as long as tritium (12.3 years), so it cannot be ignored from the viewpoint of health effects. Tritium and ^{85}Kr emitted from nuclear power plants and nuclear fuel reprocessing facilities around the world continue to accumulate in the atmosphere. In the FDNPP accident, a total of 500 PBq of noble gases was released from the Unit 1, 2, and 3 reactors [26]. In the FDNPP accident, 3.4 PBq of tritium existed in the reactors [27], and 0.1–0.5 PBq was estimated to have been released to the environment immediately after the accident. The average annual release of tritium from Japanese nuclear power plants before the FDNPP accident was estimated to be 0.36 PBq/year [28], but a similar amount was released into seawater from a nuclear fuel reprocessing facility in Aomori Prefecture [29, 30]. This book does not discuss radioactive noble gases and tritium because author have not measured tritium and radioactive noble gas nuclides from this accident, and they have often been uncertain about the behavior of these nuclides the environment.

Figures 2.4 (Numazawa Lake, Fukushima Prefecture) and 6.9 (Central Tokyo Bay) show the time-series distributions of radioactive cesium in sediment cores. Both cores recorded changes in ^{137}Cs global fallout due to atmospheric nuclear weapon tests after World War II. These distributions indicate that radioactive cesium deposited on the surface migrated to the aquatic sediment quite early. Moreover, the migration process continued for a long time. Radioactive cesium released to the environment during the Chernobyl and FDNPP accidents will, of course, follow a similar process. However, to understand its behavior strictly, it is necessary to analyze the state of radioactive cesium.

The fate of radioactive nuclides discharged into the environment by nuclear events is complex. The first global radioactive contamination we know was from a Trinity nuclear test in New Mexico on July 16, 1945. However, the first to be used

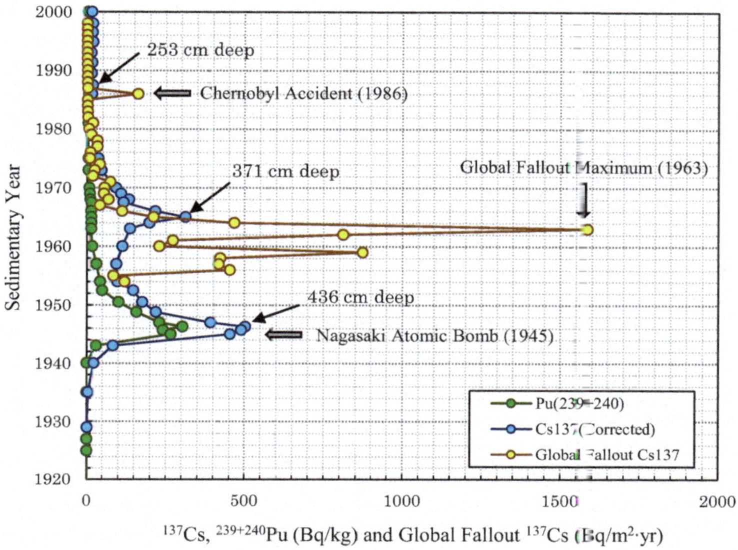

Fig. 9.4 Vertical distributions of ^{137}Cs and $^{239+240}$Pu in the core sediment of the Nishiyama reservoir, Nagasaki Prefecture [32, 33]. The ^{137}Cs activity was decay-corrected according to the deposited time of the sediment. Global fallout ^{137}Cs was measured by the NIRS in Chiba City [1]

against humans was the atomic bombs dropped on Hiroshima and Nagasaki on August 6 and 9, 1945. After more than 75 years, these cities still hold traces of the atomic bombs. Hiroshima and Nagasaki, Japan, are the oldest places in the world where we can learn about the fluctuations of radioactive nuclides in the human environment. Nishiyama Dam is the second oldest concrete dam in Japan, located 2.6 km west of Nagasaki City. It has a small reservoir completed in 1903 with an area of 0.14 km^2, a catchment area of 4.5 km^2, and a storage capacity of 1.47 million ton. The reservoir is still used as a water source in Nagasaki City. The so-called “Black Rain” containing high concentrations of radioactive material is known to have fallen immediately after the atomic bomb was dropped on the watershed northwest of Nishiyama Dam. The authors collected sediment cores from this reservoir and analyzed them for radionuclides [31–33].

Figure 9.4 shows the age distribution of ^{137}Cs and $^{239+240}$Pu in the sediment cores. The “Fat Man” bomb dropped in Nagasaki was a plutonium bomb, so the “Black Rain” contained a high concentration of plutonium. This area is the most plutonium-polluted place in Japan. Although the lake is dammed, the vertical distributions of these nuclide indicates that there was little physical vertical mixing of sediments. The sedimentary layers for the 1940s include ^{137}Cs and $^{239+240}$Pu from the Nagasaki atomic bomb, but in the early 1950s, the vertical distribution of ^{137}Cs in sediments began to be affected by global fallout from atmospheric nuclear tests. Nuclear testing at the Nevada Test Site in the United States began in 1951. The distribution of plutonium in sediments gradually decreased from the early 1950s to approximately 1970, and this plutonium came from the catchment contaminated by the Nagasaki atomic bomb but was hardly affected by global fallout. The disparity in the two temporal distributions indicates differences in the transport and diffusion processes of cesium and plutonium in the atmosphere on a global scale. The peak of the ^{137}Cs global fallout occurred in 1963, but the peak in the sediment appeared 2 years later in the 1965 layer. This difference represents the time required for ^{137}Cs to travel from the catchment to the sediment. In Sect. 6.7, radioactive cesium in the Edogawa catchment took approximately 2 years to migrate to the Tokyo Bay sediments. This time lag is an essential parameter for managing radioactive cesium contamination on the ground surface.

The most crucial point from the distributions of ^{137}Cs and $^{239+240}$Pu in the Nishiyama cores is that these nuclides that migrated to the sediment had moved little within the sediment after 75 years. Recent studies have shown that organic materials play an essential role in the long-term control of plutonium in these cores [34]. These nuclides are completely retained in sediments. This phenomenon means that unless sediment is subjected to waves, floods, and artificial physical mixing, these nuclides will remain in place indefinitely and eventually be buried underground. It is attractive to be able to store and manage radioactive materials without spending large amounts of money and building individual facilities.

9.5 Radioactive Cesium Contamination in Non-fish Organisms

Game meat is a popular ingredient in Japan. Wild boar, deer, bear, rabbit, mallard, pheasant, *etc*., are eaten. However, as discussed in Chap. 7, radioactive cesium deposited in forests is strongly adsorbed and retained not only on soil but also on trees and litter. Many birds and beasts are living in radioactively contaminated forests in eastern Japan contaminated with radioactivity. These contaminated animals travel far. Some birds cross the Eurasian continent. In other words, wild birds and animals not only cause radioactive contamination of humans through the food chain but also act as carriers for the spread of radioactive contamination. According to the "Decontamination Information Website" of the MOE (Ministry of the Environment, Japan) [35], the average spatial radiation dose in forests in Fukushima Prefecture was 0.91 μSv/h in August 2011 ($n = 362$), but it decreased to 0.45 μSv/h ($n = 1230$) in March 2016. This change in radiation dose is the decrease associated with radioactive cesium decay, indicating that in the last 5 years, radioactive cesium has not migrated from forests.

Regarding the removal of radioactive contamination in the environment related to the FDNPP accident, guidelines for radioactive decontamination of forests were outlined in the MOE's "Decontamination Guidelines (May 2013, supplemented March 2018)" [36]. The method involves removing decaying leaves and litter in an area up to 20 m from the edge of the forest when decontaminating a forest near

Table 9.1 Relationship between the species of birds and radioactive contamination in the Kejonuma pond (see Chap. 8)

Species	Feeding habits	Sampling date	Part	Detected activity, Bq/kg wet		
				^{134}Cs	^{137}Cs	^{134}+^{137}Cs
Copper pheasant	Vegetable Omnivorous (Earthworm, Insects, *etc.*)	2012/12/23	Muscle	3.6	4.9	8.5
				2.2	3.1	5.3
				3.2	4.3	7.5
Mallard	Aquatic and vegetable Omnivorous (River Snail, Water Plant, *etc.*)	2012/12/23	Muscle	119	219	338
		2013/1/13		95.2	164	259
				125	221	346

houses. In other words, the Japanese government has abandoned the decontamination of most areas of radioactively contaminated forests.

Radioactive contamination of freshwater fish is discussed in detail in Chap. 8. Although many wild beasts move across large areas of the forest, it is difficult to assess the process of radioactive contamination affecting them. In this section, birds with relatively limited habitats are analyzed. Table 9.1 shows the concentrations of radioactive cesium in the muscles of copper pheasant and mallard collected from around Kejonuma pond in Miyagi Prefecture. Copper pheasant (*Syrmaticus soemmerringii*) is endemic to Japan and is omnivorous, eating mostly vegetation but also preying on earthworms, insects, and crustaceans. Mallards (*Anas platyrhynchos*), on the other hand, are migratory birds that fly from continents to Japan in winter, overwinter in Japan and return to the continent in summer. Mallards are aquatic omnivore, and they eat aquatic plants and shellfish, such as river snails. They are also said to prey on fish.

Despite living in the same area, the concentrations of radioactive cesium in copper pheasants and mallards varied greatly. While the two birds are ecologically similar, the significant differences in levels of radioactive contamination may reflect differences in food intake. Chapter 8 discusses aquatic organisms such as fish contaminated with radioactive cesium. High contamination by radioactive cesium in freshwater organisms was found even in areas with low levels of contamination. Mallards are omnivorous but prefer eating aquatic organisms, so it is reasonable for them to have higher concentrations of radioactive cesium than copper pheasants. However, copper pheasants also prey on earthworms and other organisms that are believed to be contaminated with radioactive cesium.

Therefore, radioactive cesium contamination of earthworms living in contaminated soil was investigated. Earthworms (*Oligochaeta*) were collected in Date City, Fukushima Prefecture, together with the soil that they inhabited, and were bred in the laboratory for 10 days in the soil. Then, the earthworms were removal from the soil, and the mud attached to the body surfaces was washed away with distilled water. Water on the body surfaces was wiped with paper, and the earthworms were placed in a plastic measuring container and measured for radioactivity by gamma-ray

Table 9.2 Radioactive cesium in earthworm inhabiting the soil contaminated by the FDNPP accident

Sampling date	Sample	Sample weight, g	Activity, Bq/kg			Ratio to soil, %
			^{134}Cs	^{137}Cs	$^{134+137}$Cs	
2013/3/19	Earthworm Inhabiting Soil		49,800	91,600	141,000	–
	Living Earthworms[a]	3.45	3750 ± 99	6560 ± 137	10,400 ± 169	7.4
	Fasted Earthworms[a]	3.08	86 ± 17	106 ± 16	192 ± 23	0.14
	Earthworm Feces[a]	0.35	24,800 ± 527	47,500 ± 746	72 300 ± 914	51.1

[a]Wet samples

spectrometry. The earthworms were still alive after the measurements were finished. The earthworms were wrapped in wet tissue paper, and food was withheld for 36 h. The concentrations of radioactive cesium in the fecal pellets and the earthworm bodies were also measured. The results are shown in Table 9.2.

Since the numbers of fecal pellets that could be measured was small, the calibration accuracy of the geometric efficiency when measuring the gamma-ray spectrum with a Ge semiconductor detector was not very good. However, the results are significant. Although the earthworms lived in soil with extremely high concentrations of contamination, it has been suggested that little of the radioactive cesium was absorbed into their bodies. Earthworms consume soil and digest and absorb proteins, lipids, polysaccharides, and cellulose. Nevertheless, radioactive cesium has hardly migrated into earthworms. At least, radioactive cesium in the soil of Date City, where the earthworms lived, does not form complexes or aggregates with these organic substances. Radioactive cesium is strongly adsorbed to clay minerals that are not digested in the earthworm's digestive organs, suggesting that it passes through the earthworm's body. Earthworms are at the bottom of the food chain of many wild birds and beasts. This finding is significant for assessing the dynamics of radioactive cesium in radioactively contaminated forest ecosystems.

9.6 Reliability of the Radioactive Contamination Map by the MEXT Airborne Monitoring

The MEXT (Ministry of Education, Culture, Sports, Science, and Technology, Japan) was the first to comprehensively and systematically evaluate environmental radioactive pollution from the FDNPP accident. At the time, based on a press release released by the MEXT, the air dose rate in the atmosphere was measured on March 26, 2011, 2 weeks after the accident, on a straight flight north and south approximately 50 km west of the FDNPP [37]. The aircraft flew through snow at an altitude of 1000–2000 m and a speed of 260 km/h. The obtained air dose was 0.02–0.03 μSv/h (n = 4). Since the air dose rate measured over Fukushima Prefecture during the nuclear emergency drill held in October 2008 before the accident was 0.01–0.03 μSv/h, the air dose obtained from the first airborne monitoring after the accident did not differ from that before the accident. The next published monitoring took place on April 2 [38]. At this time, the helicopters used flew at an altitude of 300 m and a speed of 120 km/h approximately 40 km from the FDNPP. The air dose was 0.03–0.60 μSv/h (n = 10). The highest dose was observed over the Iitate-mura Village, Fukushima Prefecture, in this monitoring. Because the highly concentrated radioactive plumes passed the area around the Iitate-mura Village, the radioactive contamination in this area is extensive.

Meanwhile, airborne monitoring was conducted from April 6 to 29 to create a pollution map. From this monitoring, the air dose at 1 m above the ground surface, and the amounts of ^{134}Cs and ^{137}Cs precipitated on the ground surface were estimated. The first pollution map was published on May 6, 2011 [39]. According to this pollution map, a region 5–15 km wide and 50 km long with an area of 350 km^2 northeast of the FDNPP had $^{134+137}$Cs contamination of 3000–30,000 kBq/m^2 (air dose of 1 m above ground level: 9.5–91 μSv/h). The $^{134+137}$Cs precipitation in the surrounding area of 900 km^2 was 1000–3000 kBq/m^2 (3.8–9.5 μSv/h). Such airborne monitoring gradually expanded its measurement range, and on November 11, 2011, a radioactive contamination map for all of eastern Japan was completed [40]. The creation of a pollution map by airborne monitoring was initially carried out in cooperation with the United States [39, 41]. Afterward, the airborne monitoring performed by the MEXT was carried out using helicopters.

In airborne monitoring, the flight altitude was 150–300 m above the ground surface, and the average of a circular area 300–600 m in diameter below the aircraft was measured. The width of the wake was approximately 3 km. A NaI scintillation detector was used for the measurement, and the counts of ^{134}Cs (605 and 796 keV) and ^{137}Cs (662 keV) were estimated from the difference in respective cumulative counts between the gamma-ray energy over 450 keV and over 900 keV. One test line was set up for each monitored prefecture to convert the counts to air dose rates and ^{134}Cs and ^{137}Cs precipitation. A conversion factor for the air dose was estimated from the ratio of the air dose rate measured at the ground surface of the site and the count values obtained by the detector on the helicopter in the same site.

The press release states that the radioactive cesium precipitation calculated using the net air dose rate obtained by subtracting the average value of the air dose rate of natural radionuclides in eastern Japan from the air dose rate was obtained by the above method. However, this method is unreliable for estimating the correct precipitation of radioactive cesium. Because the geological structure of Japan is very complex, the concentrations of potassium, thorium, and uranium, which cause variations in the natural background radiation intensity, are locally unevenly distributed [42]. Therefore, it is incorrect to use the average background intensity throughout eastern Japan to determine the net air dose. There was a complaint against the

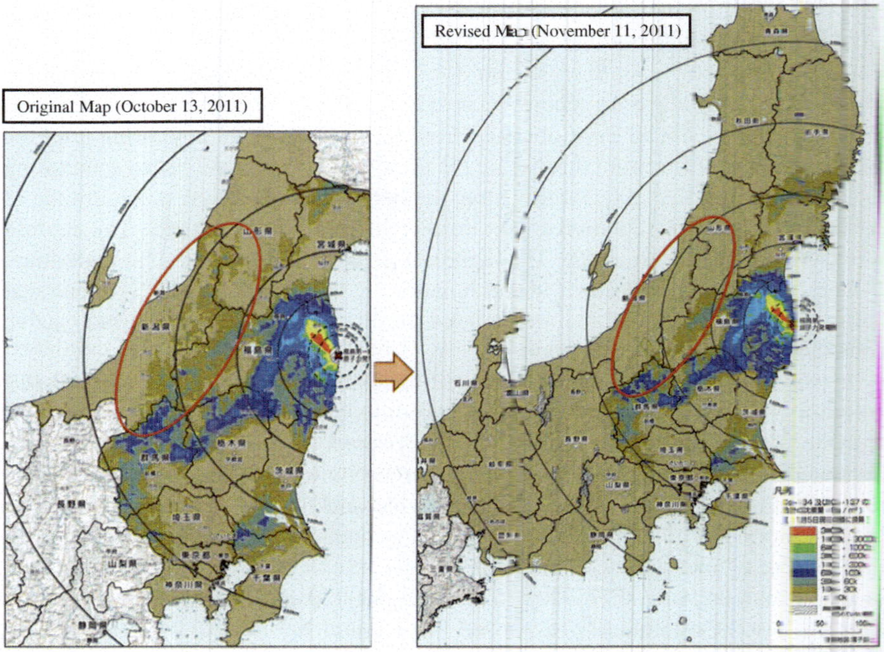

Fig. 9.5 Transition of the aircraft monitoring map of MEXT. As indicated by the red circles, the revised map [40] compared to the original map [43] has areas where the precipitation of radioactive cesium significantly reduced

pollution map published on October 12, 2011 [43], which forced a press release on November 11, 2011 [40], to revise the results of past airborne monitoring.

A new assessment of radioactive cesium precipitation performed by the MEXT was described in a press release on November 11, 2011, but did not provide any details other than a summary of the revision. In gamma-ray spectrometry, the background intensity on the low-energy side increases due to the scattering of high-energy gamma rays such as those from ^{40}K (1461 keV), ^{214}Bi (1765 keV), and ^{208}Tl (2615 keV). Therefore, when the radioactivity of radioactive cesium is low, and the intensity of gamma rays is low, the accuracy of the measurement decreases. When ^{134}Cs and ^{137}Cs were not clearly observed in the gamma-ray spectrum measured on the helicopter and the background of natural radiation was strong, the MEXT estimated the amount of radioactive cesium precipitation using a method different from the conventional method. The precipitation amount was estimated to be lower with this revised method than with the conventional method. Figure 9.5 shows the original contamination map obtained by the conventional method and the revised contamination map. For example, in the red area, the revised $^{134+137}$Cs precipitation is lower. The areas where the radioactive cesium precipitation estimated in the new and old pollution maps fluctuated considerably do not always agree with the measured concentration distributions of uranium and thorium, which govern the natural background radiation intensity. Their concentration distributions are shown in the Geochemical Map of Sea and Land of Japan published by the GSJ (Geological Survey of Japan) [42]. This inconsistency suggests that the method for estimating the precipitated radioactive cesium by the MEXT is incomplete. In other words, the revision by the MEXT is inappropriate for evaluating the background intensity in the energy region where the peaks of ^{134}Cs and ^{137}Cs appear, and there is a risk that the amount of radioactive cesium precipitated is underestimated.

This revision refers to the effect of scattered radiation intensity on the low-energy side due to high-energy gamma rays emitted from decay series radionuclides. Nevertheless, the revised method does not mention the effects of ^{208}Tl (583 keV), ^{212}Bi (727 keV), annihilation gamma rays (511 keV), *etc.*, whose spectral peaks overlap with those of the radioactive cesium used for estimating the precipitation amount.

The release of this pollution map without a clear basis for the revision significantly reduces the reliability of the MEXT radioactivity pollution map. Airborne monitoring continued after 2012 in northeastern Japan. However, the monitoring in Tokyo, in Saitama, Kanagawa, and Yamanashi prefectures and in central and southern Chiba Prefecture only took place only in 2011. Figure 3.2 shows the revised $^{134+137}$Cs pollution map, and Fig. 3.3 shows the air dose pollution maps in 2011 and 2018.

There is no publicly available pollution map showing the time course of environmental radiation pollution in the metropolitan area and surrounding areas where the population is large, and the pollution level is not low. This lack means that the Japanese government's stance on environmental pollution from the FDNPP accident is to abandon its fundamental role in protecting its citizens from radiation exposure. Currently, the website of the NRA (Nuclear Regulatory Authority, Japan) publishes the results of airborne monitoring [44].

References

1. JCAC (Japan Chemical Analysis Center). Radioactivity survey data in Japan. Environmental and dietary materials. http://www.kankyo-hoshano.go.jp/en/07/07.html
2. Fukushima Prefecture Government (2019) Status of thyroid examination [in Japanese]. https://www.pref.fukushima.lg.jp/uploaded/attachment/365363.pdf, https://www.pref.fukushima.lg.jp/site/portal/kenkocyosa-kentoiinkai-b14.html
3. IAEA (2006) Environmental consequences of the Chernobyl accident and their remediation: twenty years of experience. Radiological assessment reports series. https://www-pub.iaea.org/MTCD/Publications/PDF/Pub1239_web.pdf
4. IAEA (2006) Chernobyl's legacy: health, environmental and socio-economic impacts and recommendations to the governments of Belarus, the Russian Federation and Ukraine. The Chernobyl Forum: 2003–2005. https://www.iaea.org/sites/default/files/chernobyl.pdf
5. Shirahige Y, Ito M, Ashizawa K, Motomura T, Yokoyama N et al (1998) Childhood thyroid cancer: comparison of Japan and Belarus. Endocr J 45:203–209

6. National Cancer Center Japan (2020) Latest cancer statistics. https://ganjoho.jp/reg_stat/statistics/stat/summary.html
7. Tsuda T, Tokinobu A, Yamamoto E, Suzuki E (2015) Thyroid cancer detection by ultrasound among residents ages 18 years and younger in Fukushima, Japan: 2011 to 2014. Epidemiology 27:316–322. https://doi.org/10.1097/EDE.0000000000000385
8. Murase K, Murase J, Mishima A (2019) Nationwide increase in complex congenital heart diseases after the Fukushima nuclear accident. J Am Heart Assoc 8:e009486. https://doi.org/10.1161/JAHA.118.009486
9. MEXT (Ministry of Education, Culture, Sports, Science and Technology, Japan) (2011) Calculation results and the basis for internal exposure examined in compiling the "provisional idea" [in Japanese]. https://radioactivity.nsr.go.jp/ja/contents/4000/3912/24/1305995_0512_1.pdf
10. Koto Ward Government, Tokyo (2020) Measurement results of radioactivity in soil. https://www.city.koto.lg.jp/380303/bosai/daishinsai/hoshano/sokute/61429.html
11. Wing S, Richardson D, Armstrong D, Crawford-Brown D (1997) A reevaluation of cancer incidence near the Three Mile Island nuclear plant: the collision of evidence and assumptions. Environ Health Prespect 105:52–57
12. American Academy of Pediatrics (1998) Risk of ionizing radiation exposure to children: a subject review. Pediatrics 101:717–719
13. European Communities (2001) Low dose ionizing radiation and cancer risk. Radiation protection 125. https://ec.europa.eu/energy/sites/ener/files/documents/125.pdf
14. Muirhead CR (2001) Cancer after nuclear incidents. Occup Environ Med 58:482–488. https://doi.org/10.1136/oem.58.7.482
15. Rothkamm K, Löbrich (2003) Evidence for a lack of DNA double-strand break repair in human cells exposed to very low X-ray doses. PNAS 100:5057–5062. https://www.pnas.org/content/100/9/5057
16. Bonner WM (2003) Low-dose radiation: thresholds, bystander effects, and adaptive responses. PNAS 100:4973–4975. https://www.pnas.org/content/100/9/4973
17. Tondel M, Hjalmarsson P, Hardell L, Carlsson G, Axelson O (2004) Increase of regional total cancer incidence in north Sweden due to the Chernobyl accident? J Epidemiol Community Health 58:1011–1016. https://doi.org/10.1136/jech.2003.017988
18. EPA (U.S. Environmental Protection Agency) (2004) Risk assessment guidance for superfund. Volume I: Human health evaluation manual (Part E, Supplemental guidance for dermal risk assessment). EPA/540/R/99/005
19. Tondel MT, Lindgren P, Hjalmarsson P, Hardell L, Persson B (2006) Increased incidence of malignancies in Sweden after the Chernobyl accident – a promoting effect? Am J Indust Med 49:159–168. https://doi.org/10.1002/ajim.20271
20. Laurier D, Jacob S, Bernier MO, Leuraud K, Samson E et al (2008) Epidemiological studies of leukaemia in children and young adults around nuclear facilities: a critical review. Radiat Prot Dosim 132:182–190. https://doi.org/10.1093/rpd/ncn262
21. ECRR (2010) The health effects of exposure to low doses of ionizing radiation. 2010 Recommendations of the European Committee on Radiation Risk. ISBN: 978-1-897761-16-8
22. EPA (2011) EPA radiogenic cancer risk models and projections for the U.S. population. EPA 402-R-11-001
23. Wing S, Richardson DB, Hoffmann W (2011) Cancer risks near nuclear facilities: the importance of research design and explicit study hypotheses. Environ Health Perspect 119:417–421. https://doi.org/10.1289/ehp.1002853
24. Busby C (2011) The health outcome of the Fukushima catastrophe. Initial analysis from risk model of the European Committee on Radiation Risk ECRR. Green Audit. Occasional Paper 2011/7. http://www.greenaudit.org/wp-content/uploads/2012/11/24.pdf
25. Leuraud K, Richardson DB, Cardis E, Daniels RD, Gillies M et al (2015) Ionizing radiation and risk of death from leukaemia and lymphoma in radiation-monitored workers (INWORKS): an

international cohort study. Lancet Haematol 2:e276. https://doi.org/10.1016/52352-3026(15)00094-0
26. TEPCO (2012) Estimation of the amount of radioactive material released into the atmosphere at the Fukushima Daiichi Nuclear Power Plant accident [in Japanese]. http://www.tepco.co.jp/cc/press/betu12_j/images/120524j0105.pdf
27. Nishihara K, Yamagishi I, Yasuda K, Ishimori K, Tanaka K et al (2012) Radionuclide release to stagnant water in Fukushima-1 Nuclear Power Plant. Trans At Energ Soc Jpn 11:13–19. [in Japanese with English Abstract]. https://doi.org/10.3327/taesj.J11 040
28. METI (Ministry of Economy, Trade and Industry, Japan) (2013) Properties of tritium [in Japanese]. https://www.meti.go.jp/earthquake/nuclear/osensuitaisaku/committee/takakusyu/pdf/008_02_02.pdf
29. METI (2020) Report of subcommittee on handling of treated water such as multi-nuclide removal equipment [in Japanese]. https://www.meti.go.jp/earthquake/nulear/osensuitaisaku/committee/katakusyu/report.html
30. AEC (Japan Atomic Energy Commission) (2017) White paper on nuclear energy. https://www.aec.go.jp/jicst/NC/about/gaiyo_1_e.pdf [English summary], https://www.aec.go.jp/jicst/NC/about/hakusho/ha/cusho2018/index_html.html
31. Saito-Kokubu Y, Yasuda K, Magara M, Miyamoto Y, Sakurai S et al (2007) Geographical distribution of plutonium derived from the atomic bomb in the eastern area of Nagasaki. J Radioanal Nucl Chem 273:183–186. https://doi.org/10.1007/s10967-007-0733-9
32. Saito-Kokubu Y, Esaka F, Yasuda K, Magara M, Miyamoto Y et al (2007) Plutonium isotopes derived from Nagasaki atomic bomb in the sediment of Nishiyama reservoir at Nagasaki, Japan. Appl Radiat Isot 65:465–468. https://doi.org/10.1016/j.apradiso.2006.10.010
33. Saito-Kokubu Y, Yasuda K, Magara M, Miyamoto Y, Sakurai S et al (2008) Depositional records of plutonium and ^{137}Cs released from Nagasaki atomic bomb in sediment of Nishiyama reservoir at Nagasaki. J Environ Radioact 99:211–217. https://doi.org/10.1016/j.jenvrad.2007.11.010
34. Lin P, Xu C, Kaplan DI, Chen H, Yeager CM et al (2019) Nagasaki sediments reveal that long-term fate of plutonium is controlled by select organic matter moieties. Sci Total Environ 678:409–418. https://doi.org/10.1016/j.scitotenv.2019.04-375
35. MOE (Ministry of Environment, Japan) (2020) Decontamination Information Website. About decontamination of forest [in Japanese]. http://josen.env.go.jp/about/efforts/forest.html
36. MOE (2020) Decontamination guidelines. Supplemented March 2018 [in Japanese]. http://josen.enx.go.jp/material/pdf/josen-g100_ver2_supplement_1803.pdf
37. MEXT (2011) Airborne monitoring results around Fukushima Daiichi Nuclear Power Plant [in Japanese]. https://radioactivity.nsr.go.jp/ja/contents/4000/3783/24/1304280_2610.pdf
38. MEXT. About measurement result of air dose rate by helicopter more than 30km from Fukushima Daiichi Nuclear Power Plant [in Japanese]. https://radioactivity.nsr.go.jp/ja/contents/4000/3778/24/1304280_040213h.pdf
39. MEXT. Measurement results of airborne monitoring by MEXT and US Department of Energy [in Japanese]. https://radioactivity.nsr.go.jp/ja/contents/4000/3710/24/1305820_20110506.pdf
40. MEXT. About airborne monitoring results in Iwate, Shizuoka, Nagano, Yamanashi, Gifu, and Toyama prefectures by MEXT and revisions to airborne monitoring results that take into account the effects of natural nuclides [in Japanese]. https://radioactivity.nsr.go.jp/ja/contents/5000/4899/24/1910_111112.pdf
41. MEXT. Measurement results of airborne monitoring by MEXT and US Department of Energy [in Japanese]. https://radioactivity.nsr.go.jp/ja/contents/9000/8969/24/110616_2nd.pdf
42. AIST (Advanced Industrial Science and Technology, Japan) (2020) Geological survey of Japan. Geochemical map of sea and land of Japan [in Japanese]. https://gbank.gsj.jp/geochemmap/
43. MEXT. About airborne monitoring results in Niigata and Akita prefectures by MEXT [in Japanese]. https://radioactivity.nsr.go.jp/ja/contents/5000/4898/24/1910_101012.pdf
44. NRA (Nuclear Regulation Authority, Japan) (2020) Monitoring information of environmental radioactivity level. https://radioactivity.nsr.go.jp/en/

Correction to: Spatiotemporal Distribution of Radionuclides in Soil in the Tokyo Metropolitan Area

Correction to:
H. Yamazaki, ***Radioactive Contamination of the Tokyo Metropolitan Area*****,**
https://doi.org/10.1007/978-981-15-7368-2_3

The book was inadvertently published with errors in Table 3.4 in Chap. 3. It appears correctly in the eBook version, and only the print version has been updated.

The updated online version of the chapter can be found at
https://doi.org/10.1007/978-981-15-7368-2_3

H. Yamazaki, *Radioactive Contamination of the Tokyo Metropolitan Area*,
https://doi.org/10.1007/978-981-15-7368-2_10

The manufacturer's authorised representative in the EU is Springer Nature Customer Service Centre GmbH, Europaplatz 3, 69115 Heidelberg, Germany. If you have any concerns regarding our products, please contact ProductSafety@springernature.com

Printed and bound by CPI Group (UK) Ltd, Croydon, CR0 4YY
27/07/2026
02177388-0001